Physics Unit 2 for CAPE®

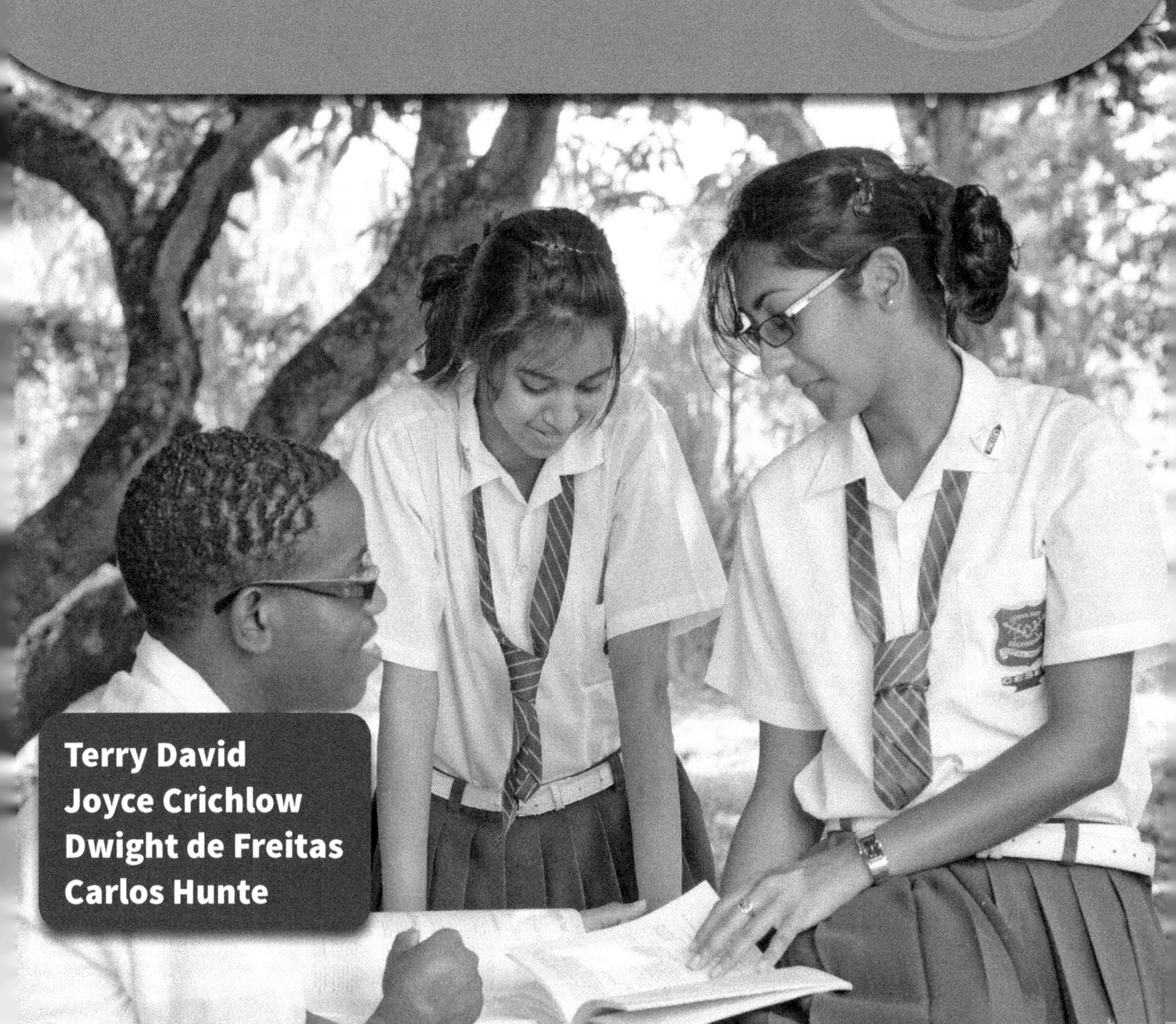

Terry David
Joyce Crichlow
Dwight de Freitas
Carlos Hunte

Great Clarendon Street, Oxford, OX2 6DP, United Kingdom

Oxford University Press is a department of the University of Oxford. It furthers the University's objective of excellence in research, scholarship, and education by publishing worldwide. Oxford is a registered trade mark of Oxford University Press in the UK and in certain other countries

First published by Nelson Thornes Ltd in 2013
This edition published by Oxford University Press in 2014

British Library Cataloguing in Publication Data
Data available

978-1-4085-1764-2

17

Printed and bound by CPI Group (UK) Ltd, Croydon, CR0 4YY

Acknowledgements

Cover photograph: Mark Lyndersay, Lyndersay Digital, Trinidad www.lyndersaydigital.com
Illustrations: GreenGate Publishing Services, Tonbridge, Kent
Page make-up: GreenGate Publishing Services, Tonbridge, Kent

Although we have made every effort to trace and contact all copyright holders before publication this has not been possible in all cases. If notified, the publisher will rectify any errors or omissions at the earliest opportunity.

Contents

Chapter 10 Digital electronics

Module 3 Atomic and nuclear physics

Chapter 11 The particulate nature of electromagnetic radiation

Chapter 12 Atomic structure and radioactivity

Chapter 13 Analysis and interpretation

Introduction

This Study Guide has been developed exclusively with the Caribbean Examinations Council (CXC®) to be used as an additional resource by candidates, both in and out of school, following the Caribbean Advanced Proficiency Examination (CAPE®) programme.

It has been prepared by a team with expertise in the CAPE® syllabus, teaching and examination. The contents are designed to support learning by providing tools to help you achieve your best in CAPE® Physics and the features included make it easier for you to master the key concepts and requirements of the syllabus. *Do remember to refer to your syllabus for full guidance on the course requirements and examination format!*

Inside this Study Guide is an interactive CD which includes electronic activities to assist you in developing good examination techniques:

- **On Your Marks** activities provide sample examination-style short answer and essay type questions, with example candidate answers and feedback from an examiner to show where answers could be improved. These activities will build your understanding, skill level and confidence in answering examination questions.
- **Test Yourself** activities are specifically designed to provide experience of multiple-choice examination questions and helpful feedback will refer you to sections inside the study guide so that you can revise problem areas.
- **Answers** are included on the CD for multiple-choice questions and questions that require calculations, so that you can check your own work as you proceed.

This unique combination of focused syllabus content and interactive examination practice will provide you with invaluable support to help you reach your full potential in CAPE® Physics.

1 Electrostatics

1.1 Electrostatics

Learning outcomes

On completion of this section, you should be able to:

- state that there are two types of charges
- describe and explain charging by friction and induction
- show that like charges repel and unlike charges attract
- distinguish between conductors and insulators.

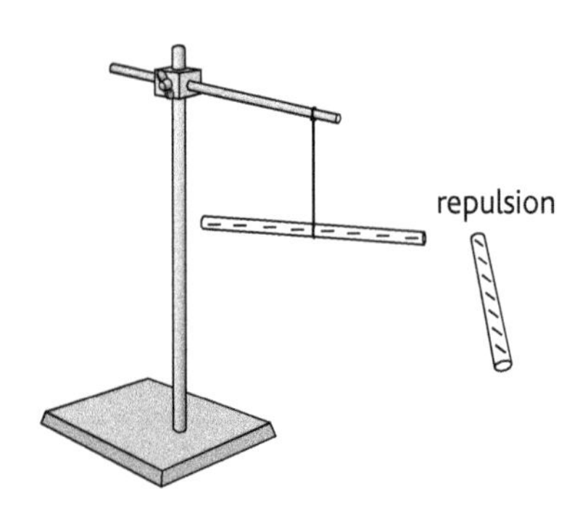

Figure 1.1.1 *Repulsion of two charged polythene rods*

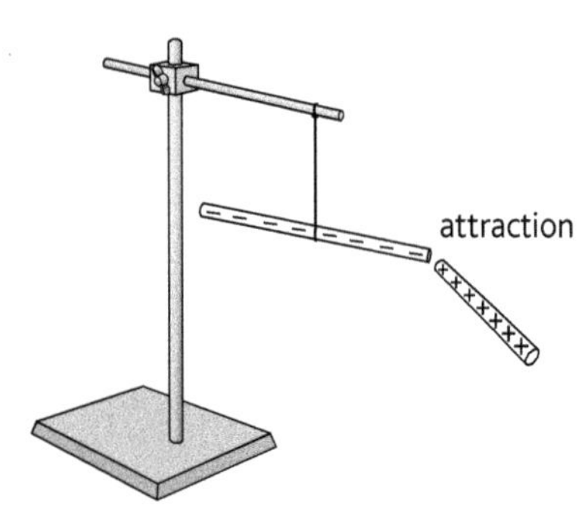

Figure 1.1.2 *Attraction between a polythene and a perspex rod*

Charging objects

Lightning is an example of an effect of static electricity. Clouds become charged as they move through the atmosphere. This charge is able to move to other clouds or to objects on the surface of the Earth such as trees or tall buildings. In order to explain this phenomenon, you must first understand how objects become charged.

Charging by friction

A polythene rod is rubbed with a piece of wool and suspended from an insulating thread so that it is free to move. Another polythene rod is rubbed with a piece of wool and placed close to the suspended rod. The suspended rod is repelled.

When a perspex rod is rubbed with a piece of wool it attracts the suspended polythene rod.

Experiments of this nature have shown that:

- there are two types of charge
- similarly charged objects repel each other
- objects that have unlike charges attract each other.

When electrostatic experiments were first conducted, it was arbitrarily decided that glass acquires a positive charge when rubbed with silk and ebonite (a type of hard rubber) acquires a negative charge when rubbed with fur. This convention was made up long before the existence of electrons became known.

It is important to note that charge is not created by the rubbing action. When the polythene rod was rubbed with the piece of wool, electrons were transferred from the surface of the wool to the surface of the polythene rod. The polythene rod acquired a **negative charge**. At the same time the piece of wool acquired an equal **positive charge**, because it lost some electrons. In this process, charge is conserved.

To summarise, polythene and ebonite acquire a negative charge when rubbed with wool. Perspex and glass acquire a positive charge when rubbed with wool.

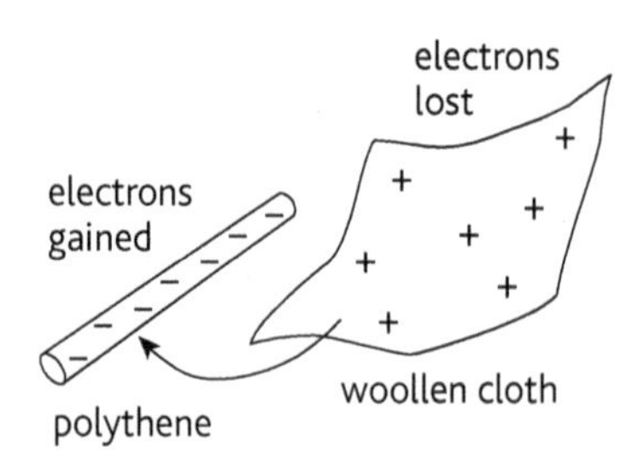

Figure 1.1.3 *Explaining charging by friction*

Charging by induction

An uncharged object can be charged by a charged object by a process called induction. A negatively charged rod is brought close to an uncharged metal sphere. The negatively charged rod causes the electrons in the uncharged metal sphere to move to one side. The metal sphere is then earthed by briefly touching it and the rod is taken away. The metal sphere has then become positively charged.

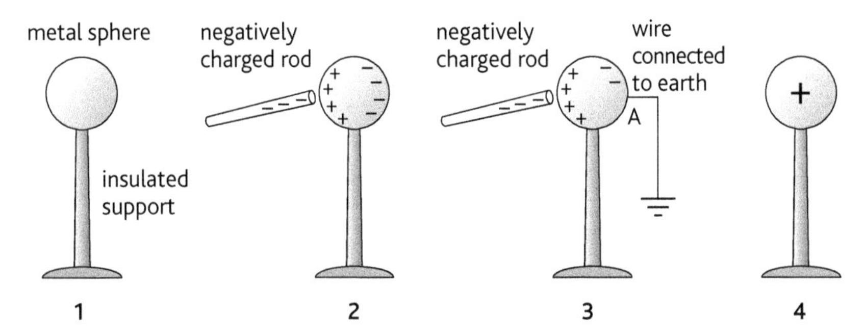

Figure 1.1.4 *Charging by induction*

Conductors and insulators

Metals are very good **conductors** of electricity. In a metal atom, the electrons that orbit the nucleus are situated in 'shells'. The electrons in the innermost shells are held tightly by the nucleus. In the outermost shells, the electrons are held less tightly by the nucleus. These electrons are able to escape the electrostatic force of attraction of the nucleus. These electrons are free to move throughout the metal structure. Since these electrons are free to move, they are said to be delocalised. Each electron has a tiny charge on it. The movement of charge constitutes an electric current. It is because of a high concentration of free, mobile electrons that metals are excellent conductors of electricity. Copper, silver and iron are good conductors of electricity. Graphite is an example of a non-metal that is a conductor of electricity.

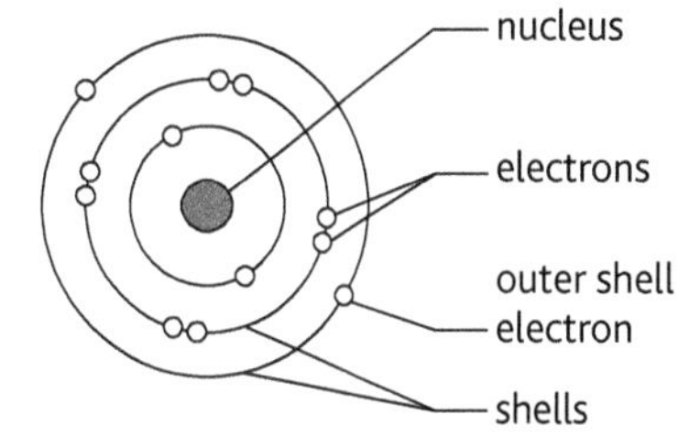

Figure 1.1.5 *Simple structure of an atom*

Some materials are poor conductors of electricity. These materials are called **insulators**. In these types of materials the type of bonding found in the structure is different from that of a metal. All the electrons are held tightly in the bonds throughout the structure. This means that there are no free mobile electrons to allow for electrical conduction to take place. Plastic and rubber are good insulators.

Materials that are neither good conductors nor good insulators are called **semiconductors**.

Key points

- There are two types of charge: positive and negative charge.
- Objects can be charged by friction or by induction.
- Charge is always conserved when an object is charged.
- Like charges repel.
- Unlike charges attract.
- Materials that conduct electricity are called conductors.
- Materials that are poor conductors of electricity are called insulators.

1.2 Applications of electrostatics

Learning outcomes

On completion of this section, you should be able to:

- describe practical applications of electrostatic phenomena
- appreciate the potential hazards associated with charging by friction.

Practical applications of electrostatics

Agricultural spraying

There are many problems associated with cultivating agricultural land. One such problem is pest control. Pests can be very harmful to development of the plants. They feed on the leaves of plants, usually living on the underside of the leaves. This makes spraying with pesticides a tricky process. A technique has been developed where the nozzle of the spray is connected to a high voltage supply. This high voltage supply has two effects. It charges the spray droplets positively and induces an opposite charge on the ground and the plants. This causes the spray droplets to be attracted to both sides of the leaves. This technique also prevents wastage during the spraying process.

Dust extraction

Dust particles may be extracted from the flue gases released by industrial chimneys, using the electric field that exists between a wire and a metal cylinder.

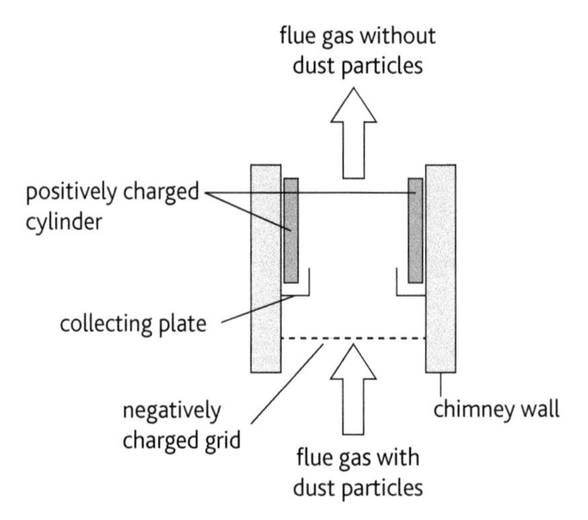

Figure 1.2.1 *Electrostatic dust extraction*

The inner wall of the chimney is fitted with a cylindrical piece of metal. A series of wires at high negative voltages are mounted inside the cylinder. A large electric field exists between the wires and the cylinder. The electric field removes electrons from some of the air molecules, thus forming ions. Free electrons in the air close to the wire are accelerated away. Electrons and negative ions in the air become attached to the dust particles. The charged dust particles are then attracted to the inner walls of the cylinder. Hammers are used to strike the cylinder to dislodge dust particles, which then fall into the traps at the bottom of the chimney.

Photocopiers and laser printers

Photocopiers use the principle of electrostatics to copy documents. A light-sensitive cylindrical drum is charged positively by a charged grid. An image of the document being copied is projected on to the drum. The areas on the drum exposed to light lose their positive charge.

A negatively charged powder, called the toner, is dusted over the drum. The toner is attracted to the positively charged image. A sheet of paper then receives a positive charge as it passes over the grid. The positively charged paper attracts toner from the drum and an image is formed on it. The image is made permanent by warming the final product.

A laser printer operates using a similar principle. When the printer receives a digital version of the document to be printed, a laser uses a series of mirrors and lenses to focus it on the light-sensitive cylindrical drum. The image to be printed is written on the drum by the laser.

Electrostatic paint spraying

Most car manufacturers use electrostatics to paint their vehicles. As the paint leaves the nozzle of the sprayer, the droplets are given a charge. Since all the droplets have the same charge, they repel each other so that the paint spreads out to form a large even cloud. The body of the car is charged with an equal and opposite charge. The result is that the paint sticks to the surface of the car tightly and less paint is wasted in the process.

Hazards of static electricity

Static electricity can be very hazardous. There is always a danger of electric shock from charged objects. For example, our bodies may become charged by friction. Then touching a metal door handle causes the charge to flow to the handle. The flow of charge results in an electric current and you experience an electric shock. It is possible to see and hear a spark as you touch the door handle. Sometimes, it is possible to be charged up to many kilovolts.

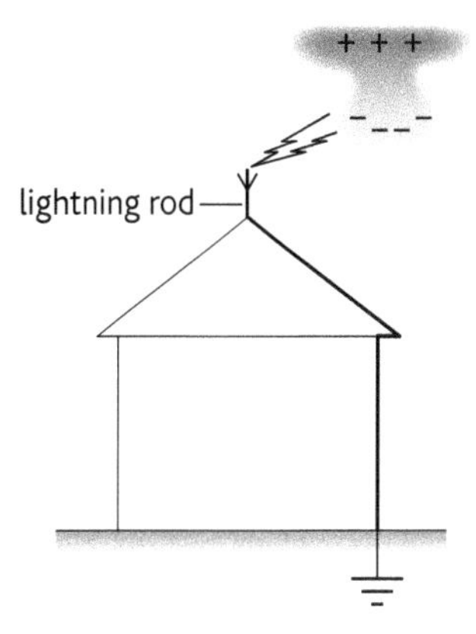

Figure 1.2.2 *Lightning rod*

Since there is always a danger of a spark being produced by electrostatics, great care must be taken when refuelling aircraft. A spark can ignite the fuel causing a dangerous explosion. A conducting cable is connected between the aircraft and the fuel tanker. This ensures that the aircraft and the fuel tanker are at the same electric potential. This ensures that no spark arises.

In thunderstorms, clouds are charged as they move through the atmosphere. These clouds have huge amounts of charge stored in them. When charge flows between clouds or towards tall buildings and trees, lightning is seen. Very large currents flow during this process. Lightning has been known to start fires, damage buildings and even kill people by electrocution.

A commonly used technique to protect buildings from lightning is to use lightning rods. A lightning rod takes the form of a thick piece of copper strip fixed to an outside wall reaching above the highest part of the building. The part of the rod above the highest part of the building consists of several sharp spikes. The other end of the copper strip is buried in the earth below. When lightning strikes, it usually strikes the highest point of a building and the current travels through the path of least resistance towards the earth. Since the lightning conductor is fixed to the ground it is effectively earthed. When a charged cloud passes by the building, a large electric field is set up between the cloud and the spikes of the lightning rod. This large electric field causes the air molecules to ionise. Electrons are stripped from the air molecules and ions are produced. The air breaks down and begins to conduct electricity. The electric current flows harmlessly through the lightning conductor towards the earth, without damaging the building.

Key points

- Electrostatics is used in electrostatic crop spraying, electrostatic paint spraying, dust extraction, photocopiers and laser printers.
- Charging by friction can create problems, but there are ways to reduce the effects.

2 Electrical quantities

2.1 Electric current and potential difference

Learning outcomes

On completion of this section, you should be able to:

- understand that an electric current is the flow of charge
- define charge and the coulomb
- recall and use $Q = It$
- define potential difference and the volt
- recall and use $V = \frac{W}{Q}$.

Definition

An electric current is the rate of flow of charge.

Equation

$Q = It$

Q – charge/C
I – current/A
t – time/s

Definition

1 coulomb is the quantity of charge that passes through any section of a conductor in 1 second when a current of 1 ampere is flowing.

1C = 1As

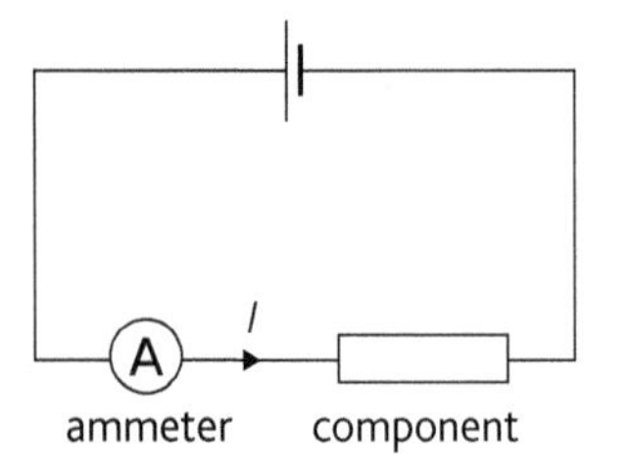

Figure 2.1.2 *Measuring an electric current*

Electric current and charge

Metals are good conductors of electricity. In a metal there are many atoms present. The outer shell electrons of the metal atoms are not tightly bound and are free to move throughout the lattice. Each electron carries a tiny amount of charge. When these charged particles (electrons) move in a particular direction, an **electric current** is produced. An electric current is the rate of flow of charged particles. The SI unit of electric current is the ampere (A).

When a salt such as sodium chloride is dissolved in water, sodium and chloride **ions** are produced. The ions present in the solution are charged. Therefore, a solution of sodium chloride will allow an electric current to flow through it.

Charged particles can have either a negative or a positive charge. The faster the charged particles move the greater the electric current.

The SI unit of charge is the **coulomb** (C). An electron has a charge of 1.6×10^{-19}C.

In order for the electrons in a metal to flow, energy must be supplied. In a simple circuit, an electric cell can be used to provide the energy needed to move the free electrons in the metal. The cell produces energy because of the chemical reactions taking place inside it.

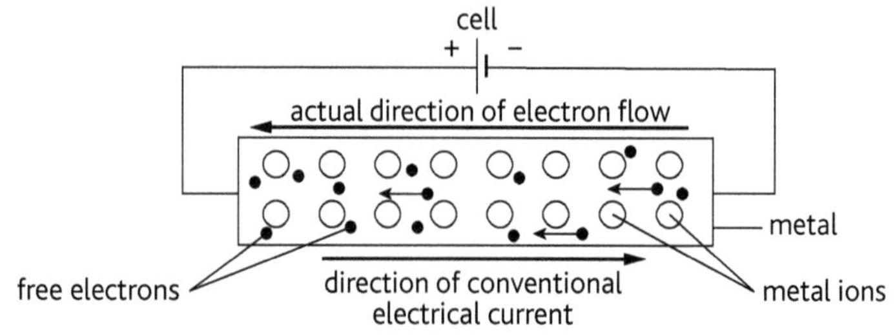

Figure 2.1.1 *Diagram showing the direction of electron flow in a metal*

In Figure 2.1.1 the electrons are flowing through the metal from right to left. The conventional electric current, however, is defined as flowing in the opposite direction. In an electrical circuit we think of the electric current as flowing from the positive terminal of the cell to the negative terminal. It is important to remember that if a stream of electrons is flowing in one direction, it can be thought of as the conventional electric current flowing in the opposite direction. If a stream of positive ions is flowing in one direction, the electric current will be flowing in that same direction.

Electrical currents are measured using an **ammeter**. In order to measure the electric current flowing through a component in a circuit, an ammeter is connected as shown in Figure 2.1.2. The ammeter is connected in series with the component. An ideal ammeter has zero resistance.

Example

In a cathode ray tube, there is a current of 160μA in the vacuum between the anode and cathode. Calculate:

a the time taken for a charge of 2.5C to be transferred between the cathode and anode

b the number of electrons emitted per second from the cathode.
(Charge on one electron $e = -1.6 \times 10^{-19}$ C)

a $t = \frac{Q}{I} = \frac{2.5}{160 \times 10^{-6}} = 1.56 \times 10^{4}$ s

b Charge on one electron $= e = -1.6 \times 10^{-19}$ C
Charge flowing per second $= 160\,\mu$C
Number of electrons emitted per second $= \frac{160 \times 10^{-6}}{1.6 \times 10^{-19}} = 1 \times 10^{15}$

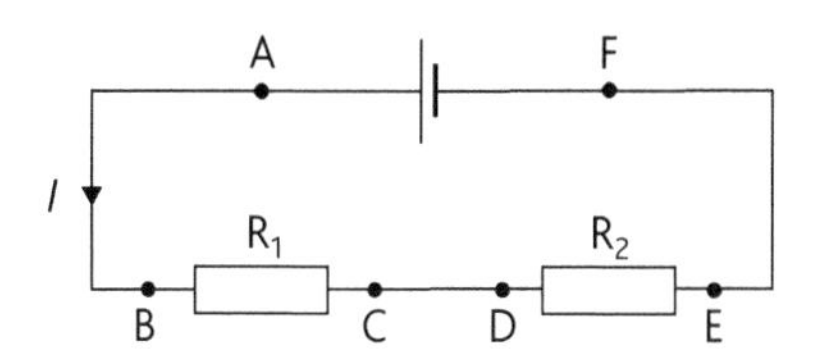

Figure 2.1.3

Definition

The potential at a point in an electric field is defined as the work done in moving unit positive charge from infinity to that point.

Potential difference

Whenever the term *unit positive charge* is used, it refers to +1 C of charge. Energy is required to move charge around an electrical circuit. A cell is usually required to power an electrical circuit. The cell converts chemical energy into electrical energy. When charge passes through components in a circuit, energy is converted from electrical to other forms of energy. In a filament lamp, electrical energy is converted into light and thermal energy. In the case of a resistor, electrical energy is converted into thermal energy.

An electric current flows from one point to another because of a difference in **electric potential** between the two points. The potential at a point in an **electric field** is defined as the work done in moving unit positive charge from infinity to that point. The SI unit of potential is the **volt** (V).

In the circuit in Figure 2.1.3 a cell is being used to provide a current I through resistors R_1 and R_2. Figure 2.1.4 shows the variation of electric potential around the circuit relative to point F.

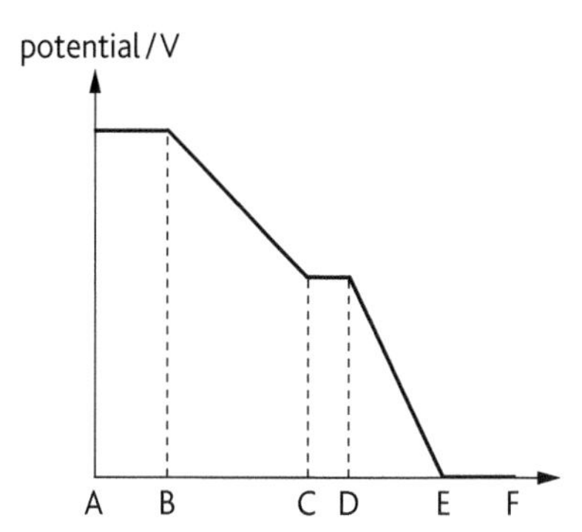

Figure 2.1.4 *Variation of electric potential around the circuit*

Electric potential cannot be measured directly. The difference in electric potential can, however, be measured. A **voltmeter** is used to measure **potential difference**. A voltmeter is connected in parallel with the component whose potential difference is being measured. An ideal voltmeter has an infinite resistance.

Definition

The potential difference V between two points in a circuit is the work done (energy converted from electrical energy to other forms of energy) in moving unit positive charge from one point to the other.

As charge passes between the two points, energy is converted from electrical to some other form. In the case of a resistor, electrical energy is converted into thermal energy. The resistor gets warm as charge flows through it. In the case of a filament lamp, electrical energy is converted into light and thermal energy.

The unit of potential difference is the volt (V).

1 volt is defined as the potential difference between two points in a circuit when 1 joule of energy is converted when 1 coulomb of charge flows between the two points.

$1\,\text{V} = 1\,\text{J}\,\text{C}^{-1}$

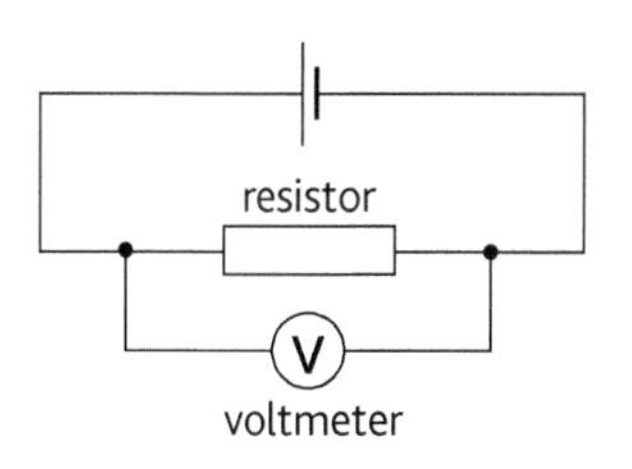

Figure 2.1.5 *Measuring potential difference*

Key points

- An electric current is the rate of flow of charge.
- 1 coulomb is the quantity of charge that passes through any section of a conductor in 1 second when a current of 1 ampere is flowing.
- The potential at a point is defined as the work done in moving unit positive charge from infinity to that point.
- The potential difference V between two points in a circuit is the work done (energy converted from electrical energy to other forms of energy) in moving unit positive charge from one point to the other.
- 1 volt is defined as the potential difference between two points in a circuit when 1 joule of energy is converted when 1 coulomb of charge flows between the two points.

Equation

$$V = \frac{W}{Q}$$

V – potential difference/V
W – work done/J
Q – charge/C

2.2 Drift velocity and power

Learning outcomes

On completion of this section, you should be able to:

- explain the term *drift velocity*
- derive an expression for drift velocity
- recall and use $P = IV$, $P = I^2R$ and $P = V^2/R$
- recall the symbols for commonly used circuit components.

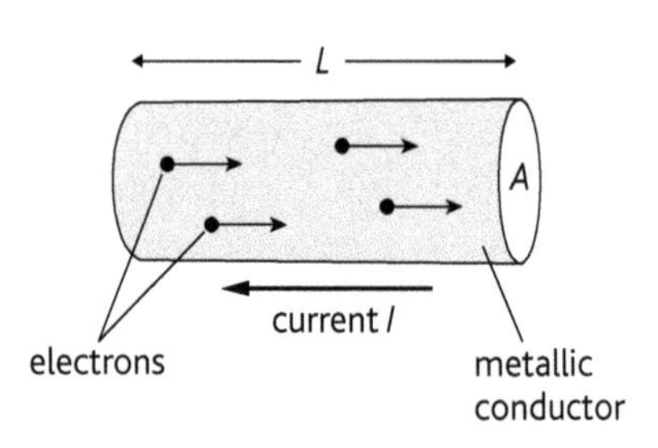

Figure 2.2.1 *Deriving an expression for drift velocity*

Exam tip

Notice that drift velocity in a metal is small, yet a light bulb turns on instantly when a light switch is turned on.

Free electrons are present in the conducting wires. As soon as the circuit is closed electrons start moving. A movement of charge is a current. Therefore the light bulb turns on instantly.

Drift velocity

When there is no current flowing in a metal, the electrons are moving about rapidly with a range of speeds, in random directions. When a potential difference is applied across a metal an electric field is set up. The free electrons begin moving under the influence of the electric field. As the free electrons accelerate they collide continuously with metal ions. This movement of the electrons is superimposed on the random motion. The **drift velocity** is the mean value of the velocity of the electrons in a conductor when an electric field is applied.

Consider a section of a metallic conductor of length L and cross-sectional area A. Let:

I = current flowing in the conductor/A

n = the number of free electrons per unit volume/m^3

e = charge on each electron/C

v = the mean drift velocity of the electrons/$m\,s^{-1}$

Volume of section $= AL$

Number of electrons in the section $= nAL$

Total amount of charge flowing

Time taken for electrons to travel from one end of the section to the next $= \frac{L}{v}$

$$\text{Electric current } I = \frac{Q}{t} = \frac{nALe}{L/v}$$

$$I = nevA$$

Example

A potential difference is applied across a piece of copper wire with cross-sectional area of $1.3 \times 10^{-6}\,m^2$. The current flowing through it is 1.2 mA. The concentration of free electrons in copper is $8.7 \times 10^{28}\,m^{-3}$. Calculate the drift velocity of the free electrons in the wire.

$$\text{Drift velocity} = v = \frac{I}{nAe} = \frac{1.2 \times 10^{-3}}{8.7 \times 10^{28} \times 1.3 \times 10^{-6} \times 1.6 \times 10^{-19}}$$

$$= 6.6 \times 10^{-8}\,m\,s^{-1}$$

In a metal, conduction is due to free electrons. In an electrolyte (e.g. a solution of sodium chloride) the mobile charge carriers are positive and negative ions (Na^+ and Cl^- in this case). In a semiconductor, the mobile charge carriers are electrons and 'holes'.

Example

Suppose a uniform glass tube of cross-sectional area A contains a salt solution (electrolyte). A current I flows through the solution. The current is carried equally by positive and negative ions. The charges on the positive and negative ions are $+2e$ and $-2e$ respectively. The number of each ion species per unit volume is n. Write down an expression for the current I flowing through the solution in terms of the drift velocity of charge carriers in the salt solution.

The positive and negative ions flow in opposite directions. Therefore, if the positive ions flow with a drift velocity v, the negative ions will flow with a drift velocity $-v$.

$$\text{Current } I = n(+2e)(v)(A) + n(-2e)(-v)(A) = 4nevA$$

Energy and power

Consider a steady current I, flowing through a resistor R for a duration of time t. As current flows through the resistor it dissipates energy. The energy dissipated is equal to the potential energy lost by the charge as it moves through the potential difference between terminals of the resistor.

From the definition of potential difference

$$V = \frac{W}{Q}$$

where

W is the energy dissipated in a time t

Q is the charge that flowed during a time t

V is the potential difference across the resistor

The **charge** that flows during time t is $Q = It$

$$\therefore \quad W = ItV$$

Power is defined as the rate at which energy is converted.

$$P = \frac{W}{t}$$

$$\therefore \quad P = \frac{ItV}{t}$$

$$P = IV$$

The SI unit of power is the **watt** (W).

From the definition of resistance (see 2.3)

$$V = IR$$

$$\therefore \quad P = I(IR) = I^2R$$

$$\text{Also,} \quad P = \left(\frac{V}{R}\right)V = \frac{V^2}{R}$$

Commonly used circuit symbols

Figure 2.2.2 shows a list of commonly used electrical circuit symbols that will be encountered in the chapters that follow.

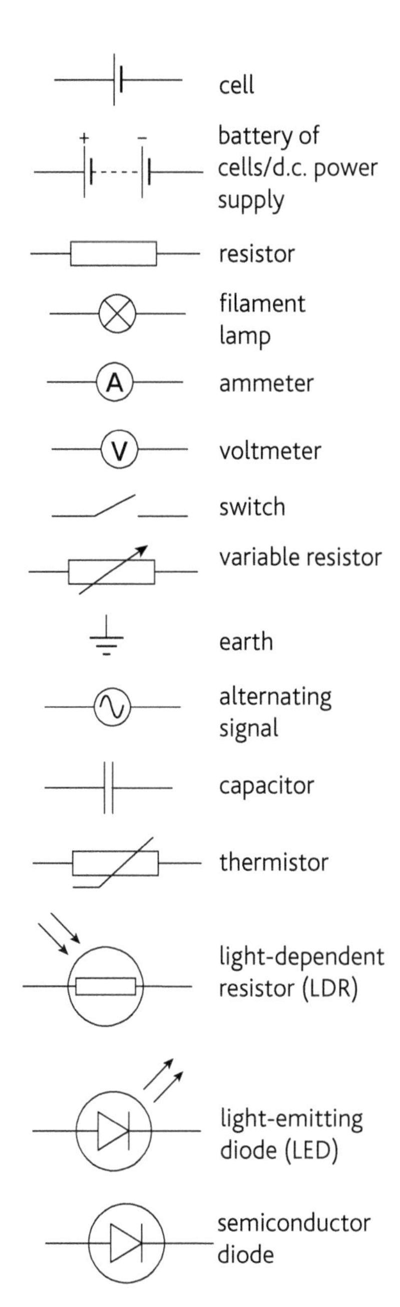

Figure 2.2.2 *Commonly used circuit symbols*

Key points

- The drift velocity is the mean value of the velocity of the electrons in a conductor when an electric field is applied.
- Power is the rate at which energy is converted.

Definition

1 watt is a rate of conversion of energy of 1 joule per second.

$1\,W = 1\,J\,s^{-1}$

2.3 Resistance

Learning outcomes

On completion of this section, you should be able to:

- define resistance and the ohm
- recall and use $V = IR$
- sketch I–V characteristics
- state Ohm's law
- define resistivity.

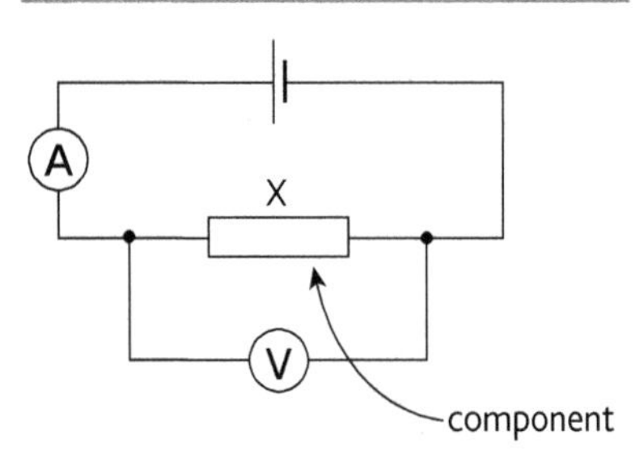

Figure 2.3.1 *Measuring electrical resistance*

Definition

1 ohm is the resistance of a conductor through which a current of 1 A flows when there is a potential difference of 1 V across it.

$1\,\Omega = 1\,\mathrm{V\,A^{-1}}$

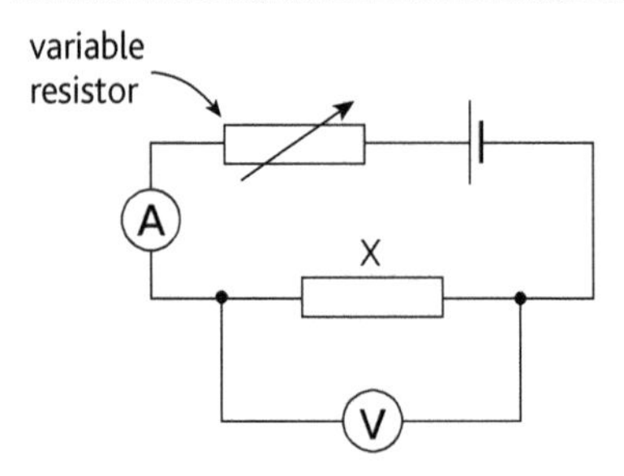

Figure 2.3.2 *Circuit used to obtain data to plot an I–V curve*

Definition

Ohm's law states that the current flowing through a conductor is directly proportional to the potential difference across it provided that there is no change in the physical conditions of the conductor.

Resistance

In a metal, there are free electrons throughout the structure. These electrons are free to move within the metal. When a cell is connected across the ends of a piece of metal, the electrons begin moving. The cell provides the necessary energy to allow the electrons to move. As the electrons move through the metal they collide with each other and the metal ions. These collisions restrict the flow of electrons. This property of the metal is known as **electrical resistance**.

The circuit in Figure 2.3.1 is used to determine the resistance of the component X. The component in this case is a resistor. The voltmeter reading gives the potential difference across the resistor. The ammeter reading gives the current flowing through the resistor. The ratio of the potential difference to the current flowing through component X is its resistance.

Definition

Resistance (R) is defined as the ratio of the potential difference (V) across the conductor to the current (I) flowing through it.

Equation

$$R = \frac{V}{I}$$

R – resistance/Ω
V – potential difference/V
I – current/A

The SI unit of resistance is the **ohm** (Ω).

Current–voltage graphs

In the circuit in Figure 2.3.2 a variable resistor is used to adjust the current flowing through the component X.

The variable resistor is adjusted and several values of V and the corresponding values of I are recorded. A graph of I against V is then plotted to obtain the I–V characteristic of the component.

Figure 2.3.3 shows the I–V characteristic for a metallic conductor at constant temperature. The graph is a straight line through the origin. The current I is directly proportional to the potential difference V. This relationship is known as **Ohm's law**.

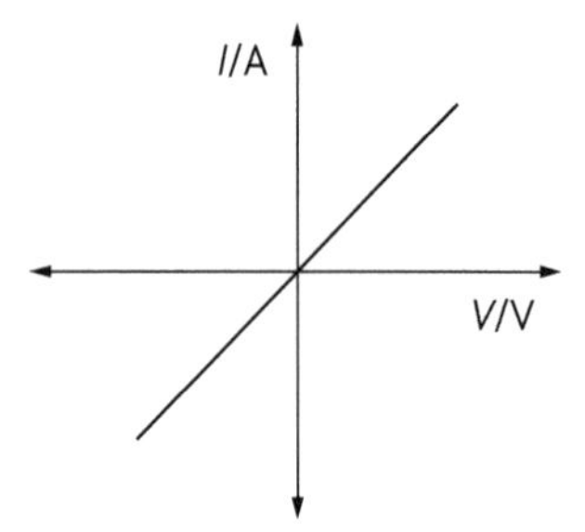

Figure 2.3.3 ***I–V*** *characteristic for an ohmic conductor*

A conductor that obeys Ohm's law is called 'ohmic'.

The physical conditions could be temperature or mechanical strain.

Figure 2.3.4 shows the I–V characteristic for a filament lamp.

The graph for the filament lamp does not obey Ohm's law. The resistance of the filament lamp is not constant. It varies with current.

From the I–V characteristics of the filament lamp it can be seen that the resistance of the lamp increases as the voltage increases. As more current flows through the lamp the temperature of the filament increases. The kinetic energy of the atoms inside it increases. The atoms vibrate rapidly about their mean positions. The moving electrons in the filament lamp collide with these vibrating atoms. As a result, the movement of electrons through the filament lamp becomes restricted. This explains why the resistance of the filament lamp increases.

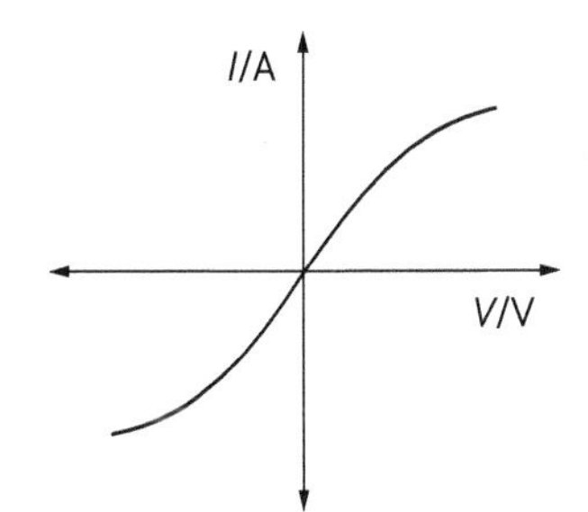

***Figure 2.3.4 I–V** characteristic for a filament lamp*

Figure 2.3.5 shows the I–V characteristic for a semiconductor diode.

The I–V graph for the semiconductor diode does not obey Ohm's law. When the semiconductor diode is connected so that it is reverse-biased (see 8.3) it does not allow any current to flow. The resistance is infinite when it is reverse-biased. This is the reason why the current is zero for negative voltages. If the semiconductor diode is now connected so that it is forward-biased (see 8.3), it begins to allow a current to flow when the voltage is approximately 0.6 V.

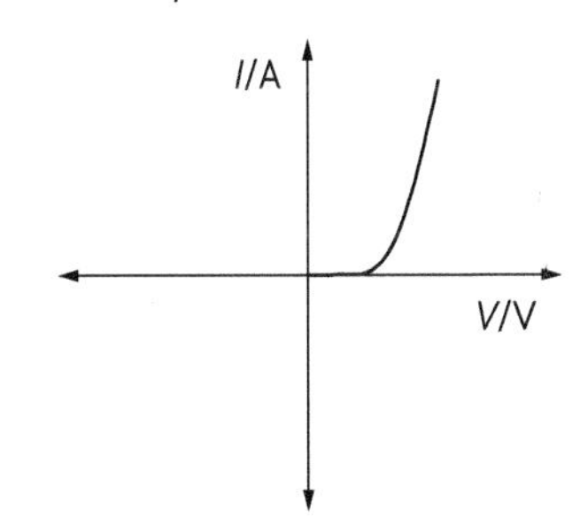

***Figure 2.3.5 I–V** characteristic for a semiconductor diode*

It may be desirable for the resistance of a device to vary with temperature. One such device is a thermistor. There are two kinds of thermistor. There is one type whose resistance decreases exponentially with increasing temperature. These thermistors are said to have a negative **temperature coefficient**. There is another kind whose resistance increases suddenly at a particular temperature. These thermistors are said to have a positive temperature coefficient.

Figure 2.3.6 shows how the resistance of a thermistor having a negative temperature coefficient varies with temperature.

Thermistors are used as temperature sensors in many electrical devices. They are used in aircraft wings to monitor external temperature.

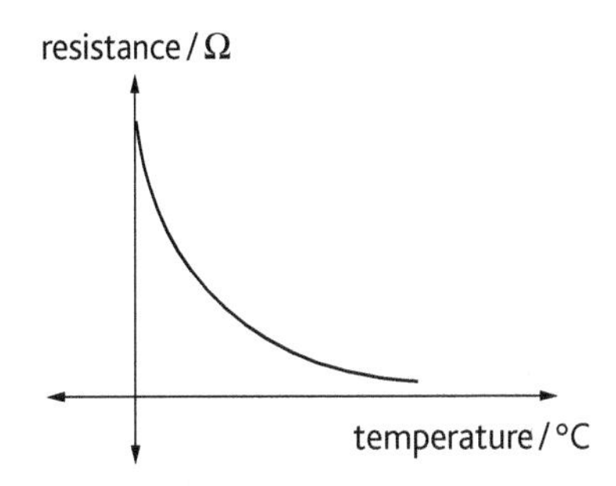

***Figure 2.3.6** Variation of resistance for a thermistor with negative temperature coefficient*

Resistivity

Resistance depends on several factors. It depends on the **resistivity** of the material, the length of the material, the cross-sectional area of the material and the temperature of the material.

The resistivities of silver and copper are $1.6 \times 10^{-8}\,\Omega\,\text{m}$ and $1.7 \times 10^{-8}\,\Omega\,\text{m}$ respectively. Copper and silver are good electrical conductors and therefore have low resistivities.

Definition

The following equation is used to define the resistivity of a material. The SI unit of resistivity is Ω m.

Equation

$$R = \frac{\rho L}{A}$$

R – resistance/Ω
ρ – resistivity/Ω m
L – length/m
A – cross-sectional area/m^2

Key points

- Resistance (R) is defined as the ratio of the potential difference (V) across the conductor to the current (I) flowing through it.
- 1 ohm is defined as the resistance of a conductor through which a current of 1 A flows when there is a potential difference of 1 V across it.
- Resistivity (ρ) is defined by $R = \frac{\rho L}{A}$.

2.4 Electromotive force and internal resistance

Learning outcomes

On completion of this section, you should be able to:

- understand the difference between potential difference and e.m.f.
- understand the concept of internal resistance.

Definition

The e.m.f. of a source is defined as the energy converted from chemical (or mechanical) energy into electrical energy per unit charge flowing through it.

Exam tip

1. e.m.f. is associated with active devices (e.g. batteries) that produce electrical power.
2. Potential difference is associated with passive devices (e.g. resistors).
3. Potential difference is also associated with electric fields (see 4.1).

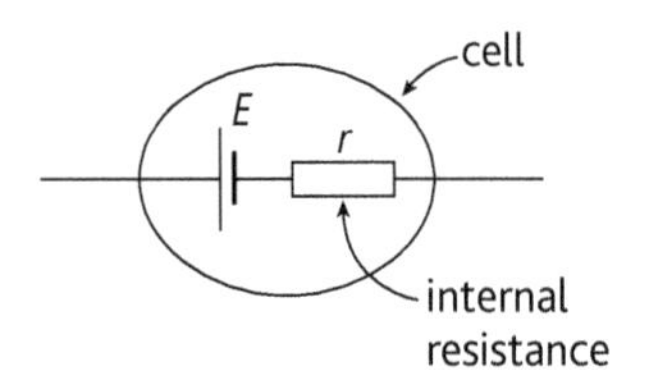

Figure 2.4.1 *e.m.f. and internal resistance*

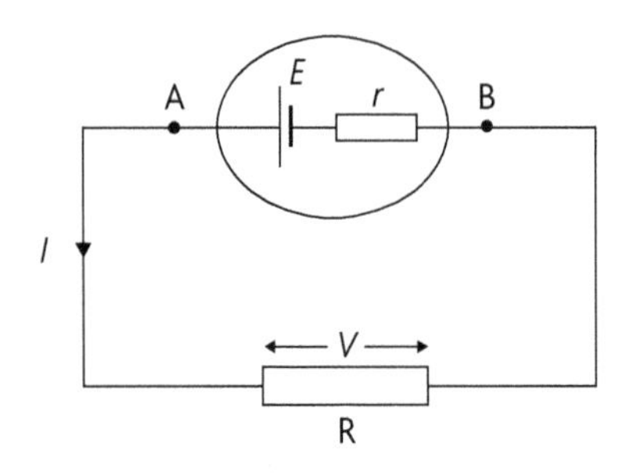

Figure 2.4.2

Electromotive force (e.m.f.) and internal resistance

In an electrical circuit, a source provides the energy required to drive an electric current in the circuit. The source has a positive and a negative terminal. Electrons are forced out of the negative terminal in a closed circuit. Examples of sources include cells, batteries, solar cells and dynamos.

In a cell, chemical energy is converted into electrical energy. The **electromotive force (e.m.f.)** of a cell is the energy converted from chemical energy to electrical energy per unit charge flowing through the cell. In 2.1, potential difference was defined as energy converted from electrical energy to other forms of energy per unit charge flowing between two points.

Equation

$$V = \frac{W}{Q}$$

V – electromotive force, e.m.f./V
W – work done/J
Q – charge/C

Definition

The potential difference between two points in a circuit is the energy converted from electrical energy to other forms of energy (e.g. heat) per unit charge flowing between the two points.

The chemicals inside a cell provide a resistance to the flow of current. The resistance that is internal to the cell is called the **internal resistance**. A cell having an e.m.f. of E and an internal resistance of r can be represented as shown in Figure 2.4.1.

Effect of increasing internal resistance

Suppose a cell is connected to an external load R and supplies a current I in the circuit as shown in Figure 2.4.2.

The e.m.f. can be written as follows:

$$E = I \times (R + r)$$

where $(R + r)$ is the total resistance in the circuit.

Therefore, $E = IR + Ir$, where IR is the potential difference across the external load and Ir is the potential difference across the internal resistance of the cell.

The potential difference across the external load is V.

$$\therefore \quad E = V + Ir$$

From the last equation, when I is equal to zero, $E = V$ (the potential difference between A and B). Therefore, when I is equal to zero, the **terminal potential difference** is equal to the e.m.f. of the cell.

The power dissipated in the external resistor, P_R, is given by:

$$P_R = I^2R$$

The total power generated by the cell P_T is given by:

$$P_T = I^2(R + r)$$

The fraction of the total power dissipated in the external resistor is given by:

$$\frac{P_R}{P_T} = \frac{I^2R}{I^2(R + r)}$$

$$\frac{P_R}{P_T} = \frac{R}{(R + r)}$$

After prolonged use, the internal resistance of a cell may increase. The increased internal resistance reduces the maximum current that the cell can supply. This reduces the total power P_T. This reduces the fraction of the total power supplied to the resistor R. As a result, the power supplied to R is reduced.

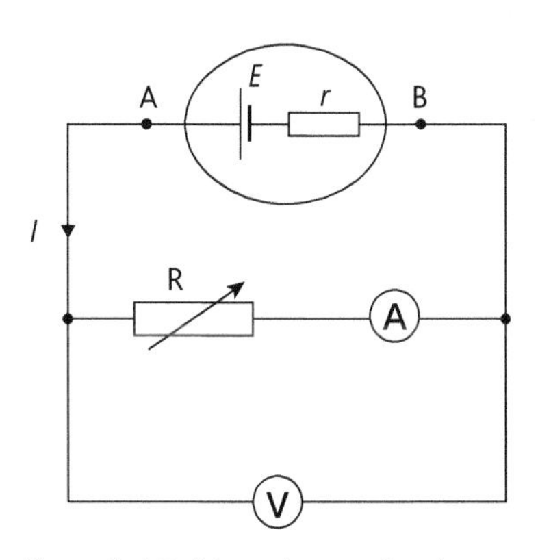

Figure 2.4.3 *Measuring e.m.f. and internal resistance*

Measuring e.m.f. and internal resistance

The circuit diagram in Figure 2.4.3 shows the circuit that can be used to measure the e.m.f. and internal resistance of a cell experimentally.

The resistance R is adjusted so that a series of readings (V) of the voltmeter V and the corresponding readings (I) of the ammeter A are recorded. A graph of V against I is then plotted.

The e.m.f. can be written as $E = I(R + r)$, where $(R + r)$ is the total resistance in the circuit.

The equation can be written as $E = IR + Ir$, where IR is the potential difference across the resistor R and Ir is the potential difference across the internal resistance of the cell.

The potential difference across the resistor R is V.

$$\therefore \quad E = V + Ir$$

$$\therefore \quad V = E - Ir$$

By comparing the equation above with $y = mx + c$ for a straight line, it can be seen that plotting a graph of V against I gives a straight line where

the y-intercept = e.m.f. of the cell

the gradient = negative the internal resistance of the cell.

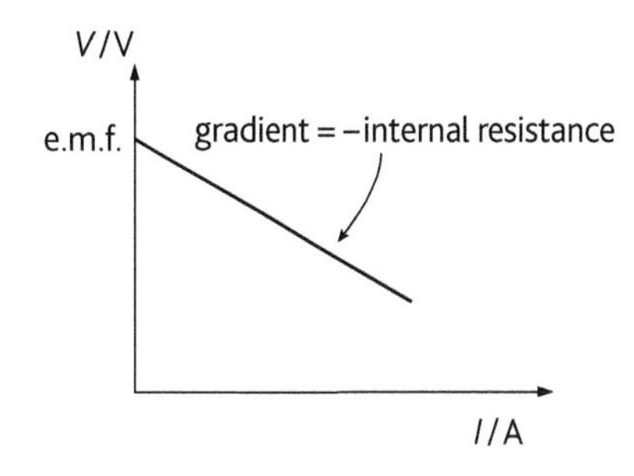

Figure 2.4.4 *Determining e.m.f. and internal resistance graphically*

Example

A cell has an e.m.f. of 1.32 V and internal resistance of r. The cell is connected across the terminals of a resistor of resistance 1.20 Ω. The cell provides a current of 0.65 A. (See Figure 2.4.5.)

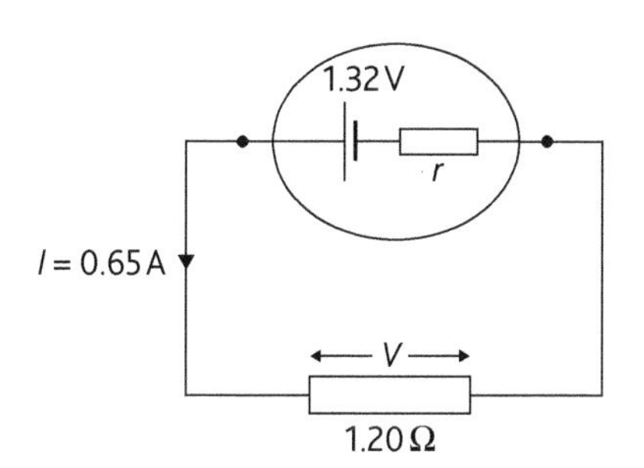

Figure 2.4.5

Calculate:

a the total resistance in the circuit

b the internal resistance r

c the potential difference across the terminals of the cell.

a The e.m.f. of the cell is given by $E = I(R + r)$

The total resistance in the circuit $= R + r = \frac{E}{I} = \frac{1.32}{0.65} = 2.03\,\Omega$

b The total resistance $= R + r = 2.03$

$$r = 2.03 - R$$
$$= 2.03 - 1.20$$
$$= 0.83\,\Omega$$

Therefore the internal resistance $r = 0.83\,\Omega$

c Potential difference across the terminals of the cell

$$V = IR = 0.65 \times 1.20 = 0.78\,\text{V}$$

Key points

- The e.m.f. of a source is defined as the energy converted from chemical (or mechanical) energy into electrical energy per unit charge flowing through it.
- The resistance that is internal to a cell is called the internal resistance.
- The e.m.f. and internal resistance of a cell can be determined experimentally.

3 d.c. circuits

3.1 Kirchhoff's laws

Learning outcomes

On completion of this section, you should be able to:

- state Kirchhoff's laws
- apply Kirchhoff's laws.

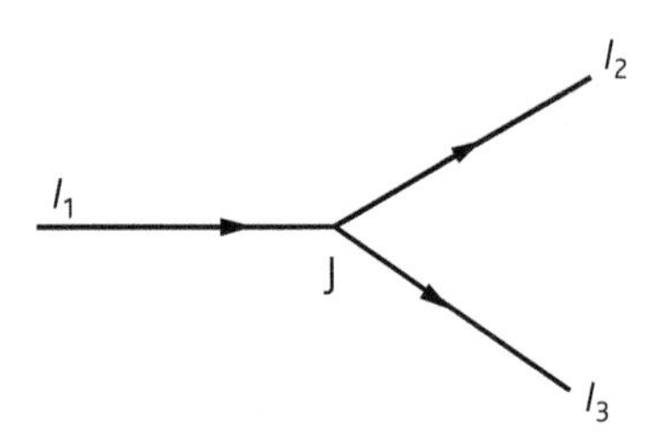

Figure 3.1.1 *Kirchhoff's first law*

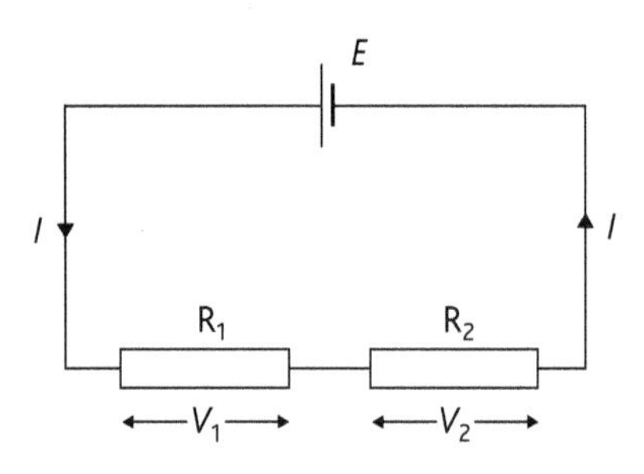

Figure 3.1.2 *Kirchhoff's second law*

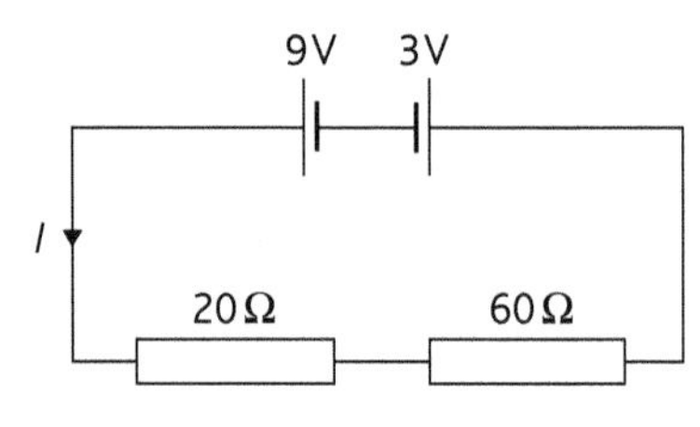

Figure 3.1.3

Kirchhoff's laws

Kirchhoff's first law

Kirchhoff's first law states that the sum of the currents flowing into any point in a circuit is equal to the sum of the currents flowing out of that point.

Consider a current I_1 flowing into a junction J. I_2 and I_3 are flowing out of the junction J.

According to **Kirchhoff's first law**:

$$I_1 = I_2 + I_3$$

The total charge flowing into the junction must equal the total charge leaving the junction in a given time. Therefore, Kirchhoff's first law is based on the conservation of charge.

Kirchhoff's second law

In the circuit in Figure 3.1.2 two resistors are connected in series with a cell of negligible internal resistance. The e.m.f. of the cell is E. The potential differences across resistors R_1 and R_2 are V_1 and V_2 respectively.

Kirchhoff's second law

Kirchhoff's second law states that the algebraic sum of the e.m.f.s around any loop in a circuit is equal to the algebraic sum of the p.d.s around the loop.

This can be stated as $E = V_1 + V_2$

If 1 coulomb of charge flows around a loop, it gains energy as it passes through the cell and loses energy as it passes through each resistor. If the charge moves around the loop and ends up at the same point at which it started, it will have the same energy at the end as at the beginning. Therefore, the energy gained by the charge passing through the cell is equal to the energy dissipated by the charge passing through resistors.

Kirchhoff's second law is based on the principle of conservation of energy.

Example

For the circuit in Figure 3.1.3, calculate the value of the current I.

Using Kirchhoff's second law

$$\text{sum of e.m.f.s in loop} = \text{sum of p.d.s in loop}$$

$$9 - 3 = 20I + 60I$$

Note the negative sign in front of the 3. This is because, the e.m.f. of the 3 V cell is opposite in direction to that of the 9 V cell.

$$80I = 6$$

$$I = \frac{6}{80} = 0.075\,\text{A}$$

Example

In the circuit in Figure 3.1.4, the batteries have negligible internal resistances and the resistance of the voltmeter is infinite. Determine the reading on the voltmeter.

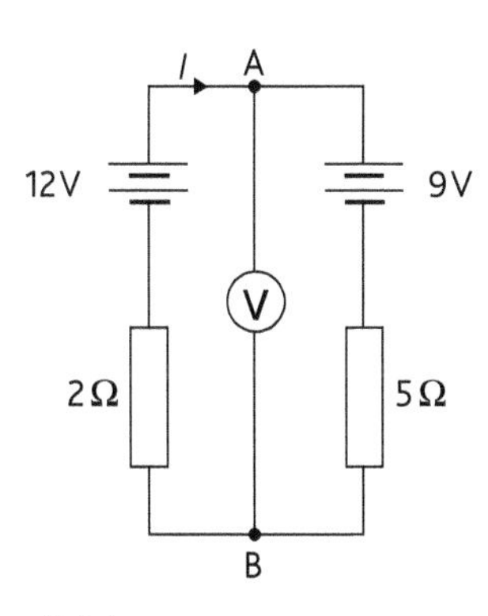

Figure 3.1.4

Using Kirchhoff's second law,

$$\text{sum of e.m.f.s in loop} = \text{sum of p.d.s in loop}$$
$$12 - 9 = 5I + 2I$$
$$7I = 3$$
$$I = \frac{3}{7} = 0.429\,\text{A}$$

Current always flows from a higher potential to a lower potential.

The potential difference across the 2 Ω resistor $= -(0.429 \times 2)$

$= -0.858\,\text{V}$

The voltmeter reading is therefore $= 12 - 0.858 = 11.1\,\text{V}$

Example

Calculate the currents in the circuit shown in Figure 3.1.5.

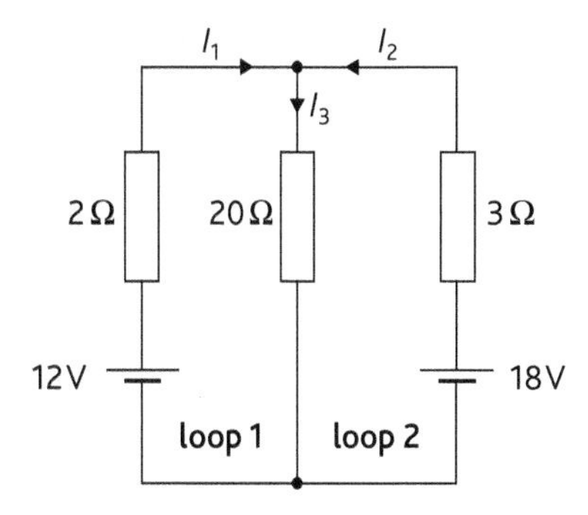

Figure 3.1.5

Using Kirchhoff's first law,

$$I_3 = I_1 + I_2 \qquad (1)$$

Using Kirchhoff's second law on loop 1,

$$12 = 2I_1 + 20I_3$$
$$12 = 2I_1 + 20(I_1 + I_2)$$
$$12 = 22I_1 + 20I_2 \qquad (2)$$

Using Kirchhoff's second law on loop 2,

$$18 = 3I_2 + 20I_3$$
$$18 = 3I_2 + 20(I_1 + I_2)$$
$$18 = 20I_1 + 23I_2 \qquad (3)$$

Equations (2) and (3) can be solved simultaneously as follows:

Multiplying equation (2) by 20,

$$240 = 440I_1 + 400I_2 \qquad (4)$$

Multiplying equation (3) by 22,

$$396 = 440I_1 + 506I_2 \qquad (5)$$

Equation (5) – Equation (4)

$$156 = 106I_2$$
$$I_2 = \frac{156}{106} = 1.47\,\text{A}$$

Substituting the value of I_2 into Equation (2)

$$12 = 22I_1 + 20 \times 1.47$$
$$22I_1 = 12 - 29.4$$
$$22I_1 = -17.4$$
$$I_1 = \frac{17.4}{22} = -0.791\,\text{A}$$

Substituting I_1 and I_2 into Equation (1)

$$I_3 = I_1 + I_2 = -0.791 + 1.47 = 0.679\,\text{A}$$

Exam tip

Practise many questions on Kirchhoff's laws to ensure that you grasp the concepts. Pay attention to the direction of the currents and the e.m.f.s.

Key points

- Kirchhoff's first law states that the sum of the currents flowing into any point in a circuit is equal to the sum of the currents flowing out of that point.
- Kirchhoff's first law is based on the conservation of charge.
- Kirchhoff's second law states that the algebraic sum of the e.m.f.s around any loop in a circuit is equal to the algebraic sum of the p.d.s around the loop.
- Kirchhoff's second law is based on the conservation of energy.

Learning outcomes

On completion of this section, you should be able to:

- derive an expression for resistors in series
- derive an expression for resistors in parallel.

Resistors in series

Consider two resistors R_1 and R_2 connected in series (Figure 3.2.1), with a current I flowing through them. The potential differences across R_1 and R_2 are V_1 and V_2 respectively. The potential difference across the two resistors is V.

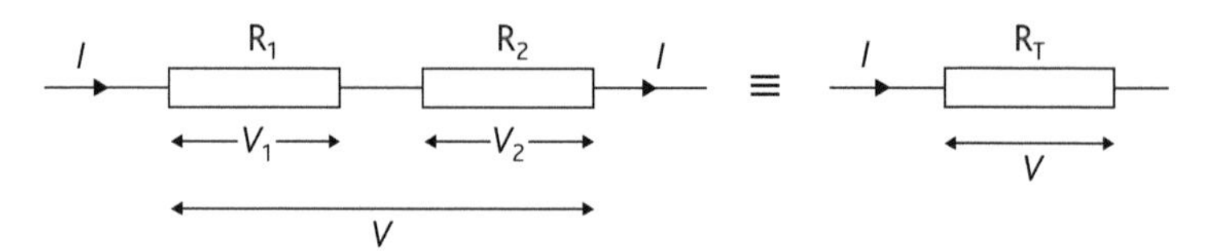

Figure 3.2.1 Resistors in series

From Kirchhoff's first law, the current I entering from the left is equal to the current I leaving on the right. This means that the same current flows through each of the resistors.

Also, the sum of the potential differences across each resistor is equal to V.

$$\therefore \qquad V = V_1 + V_2 \qquad (1)$$

From the definition of resistance,

$$V_1 = IR_1 \qquad (2)$$

$$V_2 = IR_2 \qquad (3)$$

$$V = IR_T \qquad (4)$$

where R_T is the combined resistance of the two resistors

Substituting Equations (2), (3) and (4) into Equation (1),

$$IR_T = IR_1 + IR_2$$

Dividing by I,

$$R_T = R_1 + R_2$$

Therefore, the combined resistance of resistors in series is found by summing the resistance of each resistor.

The expression can be extended to two or more resistors in series:

$$R_T = R_1 + R_2 + R_3 + \ldots$$

Resistors in parallel

Consider two resistors R_1 and R_2 connected in parallel (Figure 3.2.2). The potential difference across R_1 and R_2 is V.

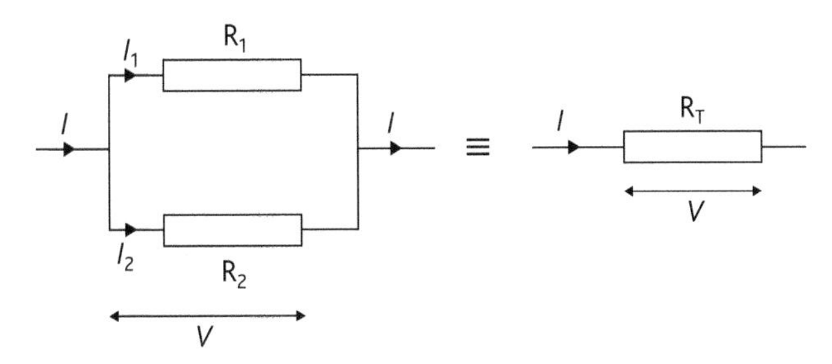

Figure 3.2.2 *Resistors in parallel*

From Kirchhoff's first law,

$$I = I_1 + I_2 \qquad (1)$$

Since the two resistors are in parallel, they each have a potential difference of V across them.

From the definition of resistance,

$$I_1 = \frac{V}{R_1} \qquad (2)$$

$$I_2 = \frac{V}{R_2} \qquad (3)$$

$$I = \frac{V}{R_T} \qquad (4)$$

where R_T is the combined resistance of the two resistors

Substituting Equations (2), (3) and (4) into Equation (1),

$$\frac{V}{R_T} = \frac{V}{R_1} + \frac{V}{R_2}$$

Dividing by V,

$$\frac{1}{R_T} = \frac{1}{R_1} + \frac{1}{R_2}$$

Therefore, the combined resistance of resistors in parallel is found by

$$\frac{1}{R_T} = \frac{1}{R_1} + \frac{1}{R_2}$$

The expression can be extended to two or more resistors in parallel:

$$\frac{1}{R_T} = \frac{1}{R_1} + \frac{1}{R_2} + \frac{1}{R_3} + \ldots$$

Exam tip

In the examination, if you are asked to derive an expression for the combined resistance for resistors in parallel:

1. Sketch a simple diagram to show the resistors, currents and potential differences.
2. Write out the steps in the derivation as clearly as possible.

Key points

- The combined resistance of two or more resistors in series is given by

  $R_T = R_1 + R_2 + R_3 + \ldots$

- The combined resistance of two or more resistors in parallel is given by

  $\frac{1}{R_T} = \frac{1}{R_1} + \frac{1}{R_2} + \frac{1}{R_3} + \ldots$

3.3 Worked examples on d.c. circuits

Learning outcomes

On completion of this section, you should be able to:

- analyse and solve problems involving d.c. circuits.

Exam tip

In an examination you may be given a diagram of a resistance combination that appears complicated. Take your time and break up the diagram into series and parallel resistors.

Example

Calculate the resistance of the network shown in Figure 3.3.1:

a between A and B

b between A and C.

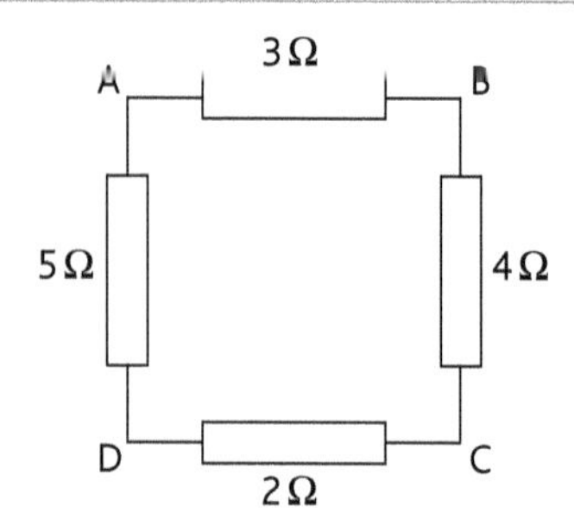

Figure 3.3.1

a Between A and B:

The first thing to do is to redraw the circuit as shown in Figure 3.3.2.

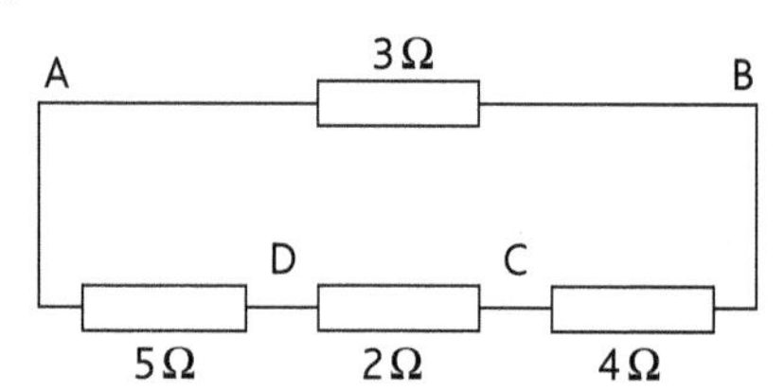

Figure 3.3.2

The combined resistance of the 5 Ω, 2 Ω and 4 Ω resistors is 11 Ω because they are in series.

This combined resistance is in parallel with the 3 Ω resistor.

$$\therefore \quad \frac{1}{R_{AB}} = \frac{1}{3} + \frac{1}{11}$$

$$\frac{1}{R_{AB}} = 0.424$$

$$R_{AB} = \frac{1}{0.424} = 2.36\,\Omega$$

b Between A and C:

The first thing to do is to redraw the circuit as shown in Figure 3.3.3.

The 3 Ω resistor and the 4 Ω resistor are in series, so their combined resistance is 7 Ω.

The 5 Ω resistor and the 2 Ω resistor are in series, so their combined resistance is 7 Ω.

This is equivalent to two resistors of value 7 Ω in parallel with each other.

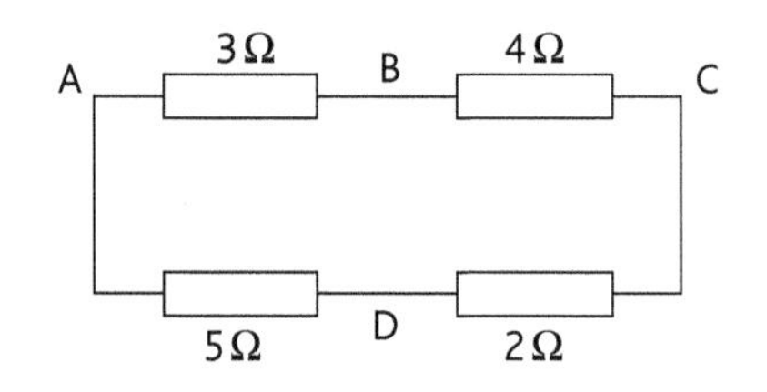

Figure 3.3.3

$$\therefore \quad \frac{1}{R_{AC}} = \frac{1}{7} + \frac{1}{7}$$

$$\frac{1}{R_{AC}} = 0.285$$

$$R_{AC} = \frac{1}{0.285} = 3.5\,\Omega$$

Example

A battery is connected to three resistors as shown in Figure 3.3.4.

The power dissipated in the 150 Ω resistor is 51.8 mW.

Calculate:

a the current flowing in the circuit
b the combined resistance in the circuit
c the e.m.f. of the battery
d the potential difference across the 25 Ω resistor
e the power dissipated in the 20 Ω resistor.

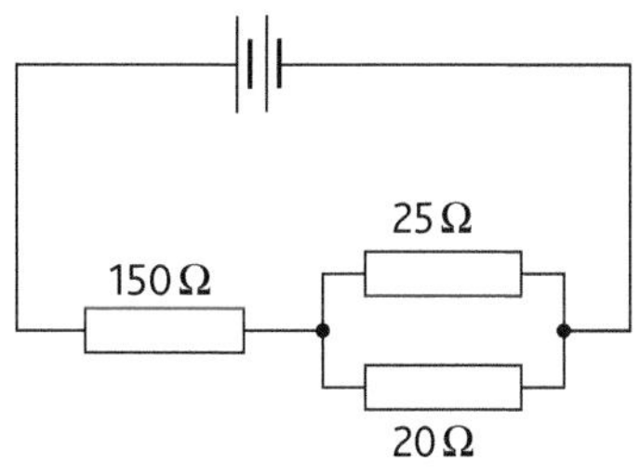

Figure 3.3.4

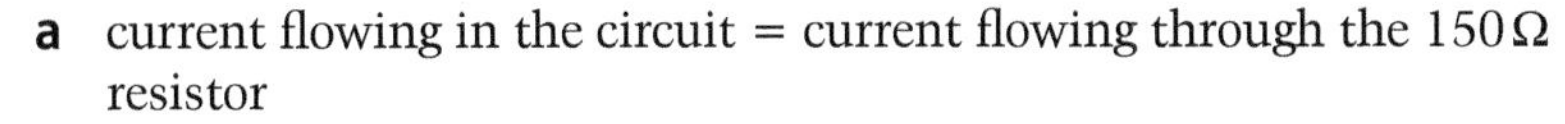

a current flowing in the circuit = current flowing through the 150 Ω resistor

$$P = I^2R$$
$$51.8 \times 10^{-3} = I^2(150)$$
$$I^2 = \frac{51.8 \times 10^{-3}}{150}$$
$$I^2 = 3.45 \times 10^{-4}$$
$$I = \sqrt{3.45 \times 10^{-4}} = 1.86 \times 10^{-2}\,\text{A}$$

b The 25 Ω resistor and the 20 Ω resistor are in parallel with each other.

$$\frac{1}{R} = \frac{1}{25} + \frac{1}{20}$$
$$\frac{1}{R} = 0.09$$
$$R = 11.1\,\Omega$$

This resistance is in series with the 150 Ω resistor.

Therefore, the combined resistance in the circuit is:

$$R_T = 150 + 11.1 = 161.1\,\Omega$$

c e.m.f. of battery = current flowing in circuit × combined resistance

$$= 1.86 \times 10^{-2} \times 161.1$$
$$= 3.00\,\text{V}$$

d Potential difference across the 25 Ω resistor

= e.m.f. of battery − potential difference across 150 Ω resistor

$$= 3.00 - (1.86 \times 10^{-2} \times 150)$$
$$= 0.21\,\text{V}$$

e Potential difference across the 20 Ω resistor

= potential difference across the 25 Ω resistor

$$= 0.21\,\text{V}$$

Power dissipated in 20 Ω resistor $= \frac{V^2}{R} = \frac{(0.21)^2}{20} = 2.21 \times 10^{-3}\,\text{W}$

3.4 Potential dividers

Learning outcomes

On completion of this section, you should be able to:

- describe how to use a potential divider as a source of fixed or variable potential difference.

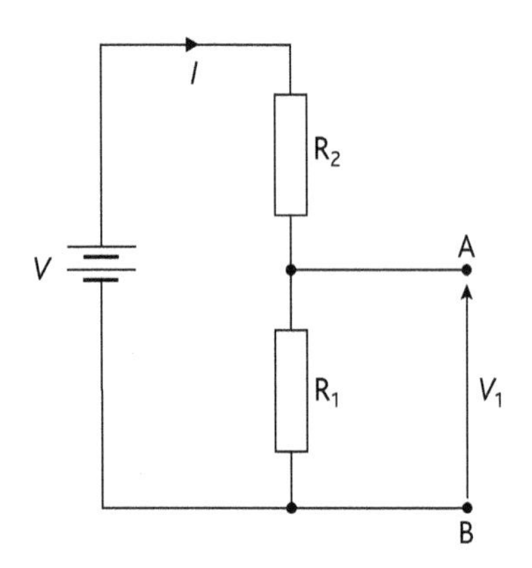

Figure 3.4.1 *A potential divider circuit*

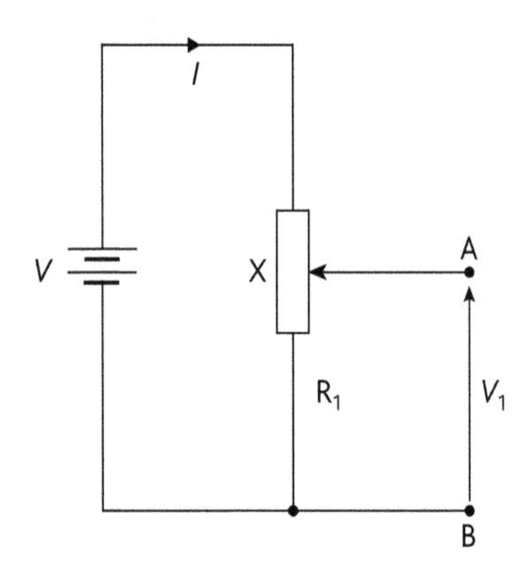

Figure 3.4.2 *A potential divider providing a variable p.d.*

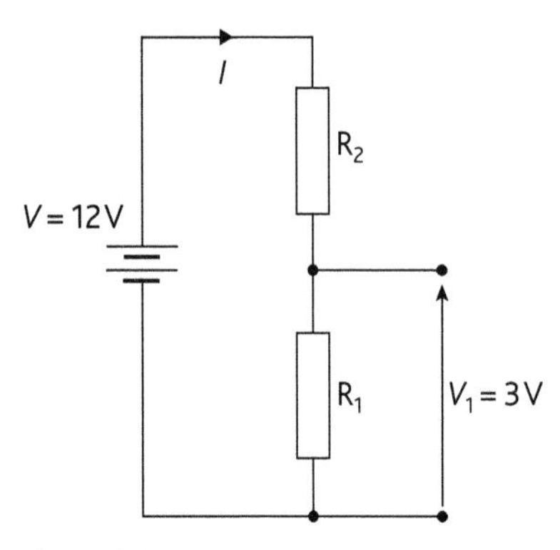

Figure 3.4.3

Potential dividers

A potential divider circuit is used to produce a small potential difference from a larger potential difference. The larger potential difference V is connected across two resistors in series, R_1 and R_2, as shown in Figure 3.4.1. This circuit is very useful when the only available power supply provides a greater p.d. than that required by some electrical circuit. The circuit can be supplied with the correct p.d. V_1 by connecting it across the terminals A and B.

The total resistance in the circuit is

$$R_T = R_1 + R_2 \quad (1)$$

From the definition of resistance,

$$I = \frac{V}{R_T} = \frac{V}{R_1 + R_2} \quad (2)$$

But,

$$I = \frac{V_1}{R_1} \quad (3)$$

Equation (2) is equal to (3)

$$\therefore \quad \frac{V}{R_1 + R_2} = \frac{V_1}{R_1}$$

$$V_1(R_1 + R_2) = VR_1$$

$$\therefore \quad V_1 = \left(\frac{R_1}{R_1 + R_2}\right)V$$

Apart from supplying a fixed voltage, the circuit can be made to supply a variable voltage by replacing the resistors with a single variable resistor X. This resistor is called a potentiometer. Therefore, the potential divider can supply any voltage between zero and V.

Example

Suppose a student has a 12 V d.c. supply. He wants to power a circuit that requires 3 V d.c. and he has no other power supply available. How can he accomplish this?

He can achieve this by using a potential divider circuit as shown in Figure 3.4.3.

$$V = 12\,\text{V and } V_1 = 3\,\text{V}$$

He can choose any combination of resistors, provided that the ratio:

$$\frac{V_1}{V} = \frac{R_1}{R_1 + R_2} = \frac{3}{12} = \frac{1}{4}$$

If he chooses $R_1 = 100\,\text{k}\Omega$, then:

$$\frac{100}{100 + R_2} = \frac{1}{4}$$

$$100 + R_2 = 400$$

$$R_2 = 400 - 100$$

$$R_2 = 300\,\text{k}\Omega$$

Example

A thermistor T has a resistance of 2500 Ω at 0 °C and a resistance of 950 Ω at 25 °C. The thermistor is used to monitor the temperature of a room as shown in the circuit in Figure 3.4.4.

a Assuming that the battery has a negligible internal resistance and the resistance of the voltmeter is infinite, calculate:

  i the value of R such that the reading on the voltmeter is 1.8 V when the temperature is 0 °C

  ii the reading on the voltmeter when the temperature in the room is 25 °C.

b The voltmeter is replaced with one having a resistance of 9 kΩ. Calculate the new voltmeter reading at 25 °C.

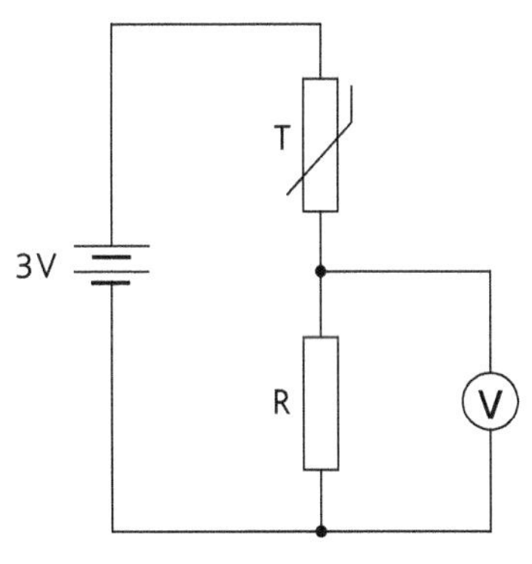

Figure 3.4.4

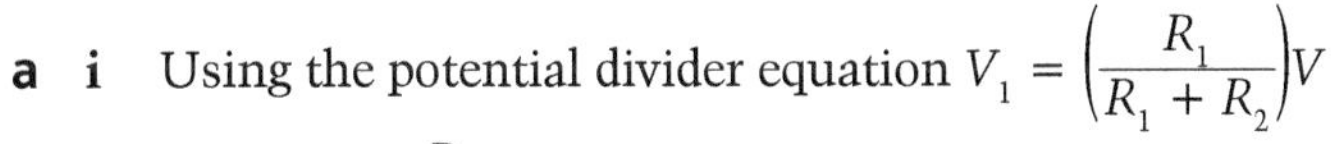

a **i** Using the potential divider equation $V_1 = \left(\dfrac{R_1}{R_1 + R_2}\right)V$

$$\frac{R}{R + 2500} \times 3 = 1.8$$

$$\frac{R}{R + 2500} = \frac{1.8}{3}$$

$$\frac{R}{R + 2500} = 0.6$$

$$0.6(R + 2500) = R$$

$$0.6R + 1500 = R$$

$$R - 0.6R = 1500$$

$$0.4R = 1500$$

$$R = \frac{1500}{0.4} = 3750\,\Omega$$

ii At 25 °C the resistance of the thermistor is 950Ω. Using the potential divider equation again

$$V_1 = \left(\frac{R_1}{R_1 + R_2}\right)V$$

$$V_1 = \frac{3750}{3750 + 950} \times 3$$

$$V_1 = 2.39\,\text{V}$$

Therefore, the reading on the voltmeter is $V_1 = 2.39$ V.

b Since the voltmeter has a resistance it needs to be taken into account. The combined resistance of the voltmeter and the resistor R is given by:

$$\frac{1}{R_T} = \frac{1}{3750} + \frac{1}{9000}$$

$$\frac{1}{R_T} = 3.78 \times 10^{-4}$$

$$R_T = \frac{1}{3.78 \times 10^{-4}} = 2647\,\Omega$$

Voltmeter reading $V_1 = \left(\dfrac{R_1}{R_1 + R_2}\right)V$

$$V_1 = \frac{2647}{2647 + 950} \times 3 = 2.21\,\text{V}$$

Key point

- A potential divider can be used as a source of fixed or variable potential difference.

3.5 The Wheatstone bridge

Learning outcomes

On completion of this section, you should be able to:

- describe how to use a Wheatstone bridge to compare two resistances
- understand that a Wheatstone bridge is a double potential divider.

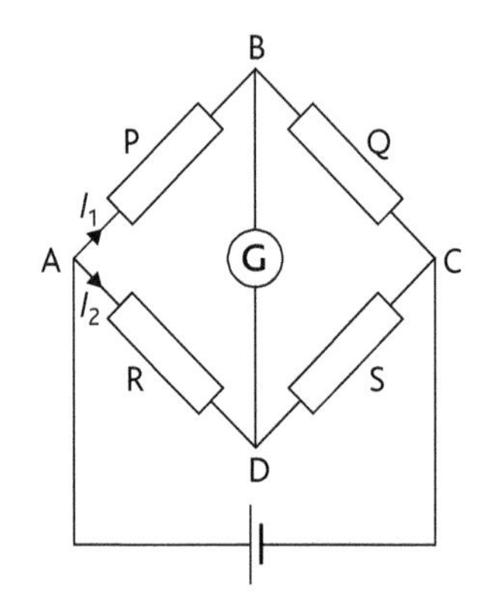

Figure 3.5.1 *The principle of the Wheatstone bridge*

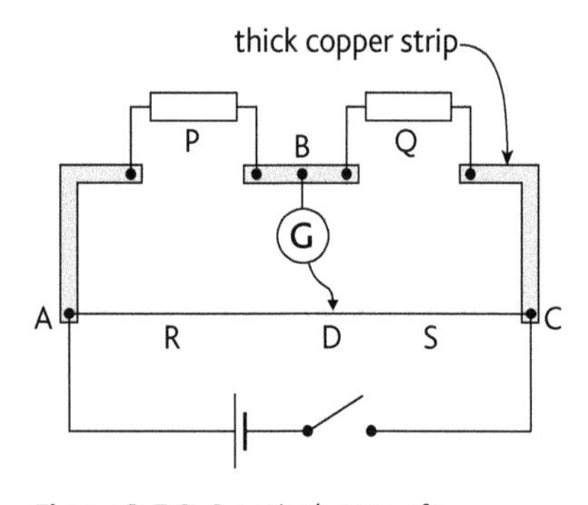

Figure 3.5.2 *Practical setup of a Wheatstone bridge*

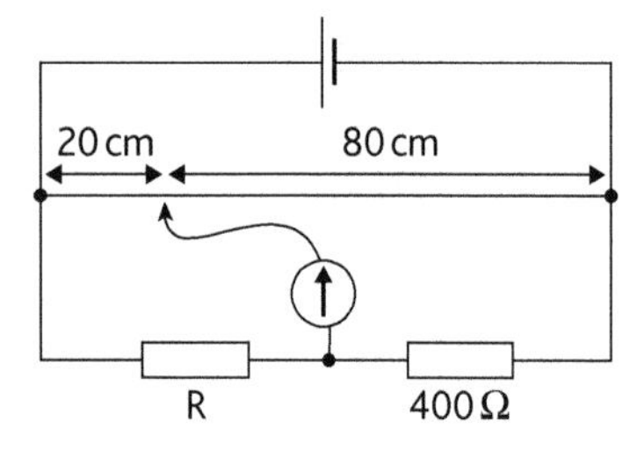

Figure 3.5.3

The Wheatstone bridge

A Wheatstone bridge is used to determine the resistance of an unknown resistive component accurately. It is essentially made up of two potential divider circuits. This technique uses a null method for determining the unknown resistance. The circuit diagram is shown in Figure 3.5.1. If R is an unknown resistance, it can be determined if the resistances P, Q and S are known. The relationship between P, Q, R and S is as follows:

$$\frac{P}{Q} = \frac{R}{S}$$

In practice, a fixed voltage is applied across a wire AC of uniform cross-sectional area, using a standard cell (e.g. a Leclanché cell) (Figure 3.5.2).

The slider D is adjusted until the galvanometer reading is zero. The distances AD and DC are measured using a metre rule. Since the wire is uniform, the resistance of the wire is proportional to the length.

$$\therefore \quad \frac{AD}{DC} = \frac{R}{S}$$

When the galvanometer reading is zero, the potential difference across the points B and D is zero.

∴ Potential difference across AB = potential difference across AD

The current flowing through P and Q is I_1.

The current flowing through R and S is I_2.

$$I_1P = I_2R \qquad (1)$$

$$I_1Q = I_2S \qquad (2)$$

Dividing the two equations we get:

$$\frac{I_1P}{I_1Q} = \frac{I_2R}{I_2S}$$

$$\frac{P}{Q} = \frac{R}{S}$$

Example

A slide-wire Wheatstone bridge is balanced when the uniform wire is divided as shown in Figure 3.5.3. Calculate the value of the unknown resistor R.

$$\frac{R}{400} = \frac{20}{80}$$

$$R = \frac{20 \times 400}{80} = 100\,\Omega$$

The potentiometer

A potentiometer is a device used to measure an unknown e.m.f. or a potential difference. It uses the principle of the potential divider to measure potential differences. A potential difference is applied across a wire AB of uniform resistivity and uniform cross-sectional area (Figure 3.5.4). The cell used is called the driver cell. The unknown potential difference is applied across the points X and Y. The point with the higher potential is connected at the point X. The jockey J (a metal contact) is moved along AB until the deflection on the sensitive galvanometer is zero. This technique is called a null deflection method. The point at which the deflection on the galvanometer is zero is called the balance point. At the balance point, no current flows through the galvanometer. The length of the wire AJ is measured using a metre rule.

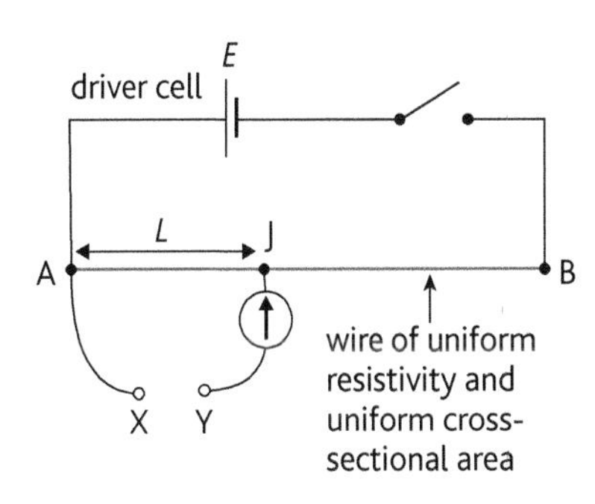

Figure 3.5.4 *A potentiometer*

A potentiometer is used to:

- compare e.m.f.s
- compare resistances
- measure currents.

Advantages of using a potentiometer to measure potential differences:

- No current is drawn from the p.d. being measured at the balance point. A moving coil voltmeter has a resistance and therefore draws some current from the p.d. being measured.
- Since a null deflection method is used to find the balance point, this can be found with a high degree of accuracy. The method does not depend on the calibration of the sensitive galvanometer.
- The results are dependent on lengths.

Disadvantages of using a potentiometer to measure potential differences:

- The resistivity and cross section of the wire must be uniform.
- It is difficult to measure changing potential differences, since it takes time to find the balance point and measure lengths.

Comparing e.m.f.s

A potentiometer can be used to compare two e.m.f.s. Figure 3.5.5 shows how this is done. A standard cell (a cell with an accurately known e.m.f.) E_s is initially connected as shown. The jockey is adjusted until a balance point l_1 is found. E_s is replaced by a second cell E_2 (of unknown e.m.f.). The jockey is adjusted until a balance point l_2 is found.

Therefore, $$\frac{E_s}{l_1} = \frac{E_2}{l_2}$$

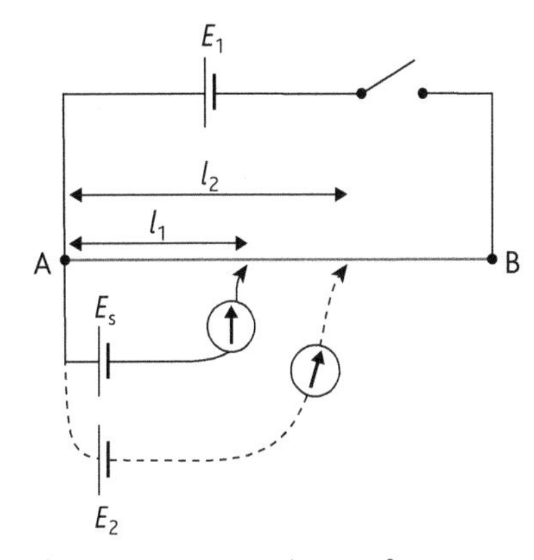

Figure 3.5.5 *Comparing e.m.f.s*

Comparing resistances

A potentiometer can be used to compare two resistances. Suppose a circuit consists of a cell E_2 and two resistors R_1 and R_2. The potentiometer is initially connected across R_1 and the balance point l_1 is found. The potentiometer is then connected across R_2 and the balance point l_2 is found.

Therefore, $$\frac{R_1}{l_1} = \frac{R_2}{l_2}$$

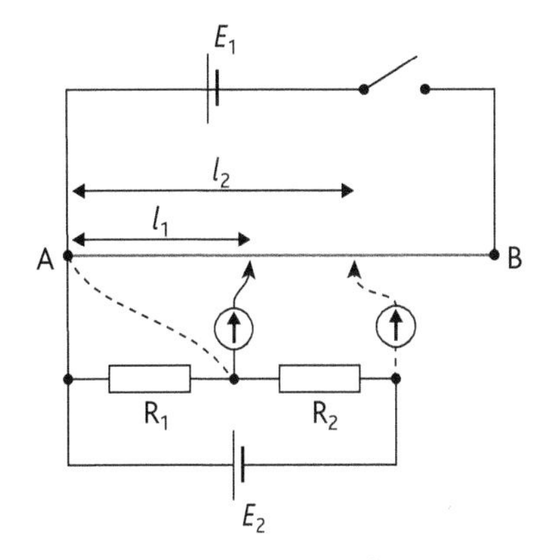

Figure 3.5.6 *Comparing resistances*

Key points

- A Wheatstone bridge is used to determine the value of an unknown resistance.
- A potentiometer is a device used to measure an unknown e.m.f. or a potential difference.
- A potentiometer can be used to compare two e.m.f.s and compare two resistances.

Revision questions 1

Answers to questions that require calculation can be found on the accompanying CD.

1 a Describe, in terms of an electron model, the difference between an electrical conductor and an electrical insulator. [2]

b Give an example of each of the following:

i An insulator [1]
ii A metal that is a conductor [1]
iii A non-metal that is a conductor [1]

2 a Discuss two hazards of electrostatics. [4]

b Discuss two applications of electrostatics. [6]

3 A 1.5 V cell delivers a constant charge of 420 C for a period of 2.8×10^4 s in a circuit.

Calculate:

a the current flowing through the circuit [3]
b the resistance in the circuit [2]
c the total number of electrons flowing during the period of 2.8×10^4 s. [2]

4 a Explain what is meant by the terms *electric current* and *potential difference*. [4]

b A student connects his cellular phone to charger for a period of 2 minutes for a quick charge. The charger delivers a constant current of 300 mA.

Calculate:

i the charge flowing during the period [2]
ii the number of electrons flowing during the period. [2]

5 a Define potential difference. [2]

b Define the unit of potential difference. [2]

c Using the definition of potential difference, show that the power dissipated in a resistor of resistance R is given by $P = \frac{V^2}{R}$, where V is the potential difference across the resistor. [3]

6 A potential difference of 12 V causes 2.4×10^{18} electrons to flow through a metallic conductor in 1.2 minutes. Calculate:

a the charge that flowed through the conductor [2]
b the electric current flowing through the conductor [2]
c the resistance of the conductor. [2]

7 a Define the terms:

i resistance [2]
ii resistivity. [2]

b A potential difference of 3 V is applied across a piece of copper wire of length 0.65 m and cross-sectional area 2.2×10^{-9} m². Calculate:

i the resistance of the copper wire [3]
ii the current flowing through the wire [2]
iii the power dissipated in the wire. [2]

(Resistivity of copper = 1.7×10^{-8} Ω m)

8 a Explain what is meant by the term *drift velocity*. [2]

b Derive the expression for the current in a conductor in terms of the drift velocity of the electrons flowing in the conductor. [4]

c Explain why in a semiconductor, the drift velocity of the charge carriers is much larger than in a metallic conductor of the same dimensions with the same current flowing in it. [3]

9 a Use energy considerations to distinguish between electromotive force (e.m.f.) and potential difference (p.d.). [3]

b Explain what is meant by the internal resistance of a cell. [1]

c Explain why the potential difference across the terminals of a battery is normally lower than the battery's e.m.f. [2]

d Under what condition is the potential difference across a battery's terminal equal to its e.m.f.? [1]

10 Two resistors having resistances of 1.5 kΩ and 4.5 kΩ are connected in series with a battery of e.m.f. 9.0 V and negligible internal resistance shown below.

a Calculate the potential difference across each of the resistors. [4]

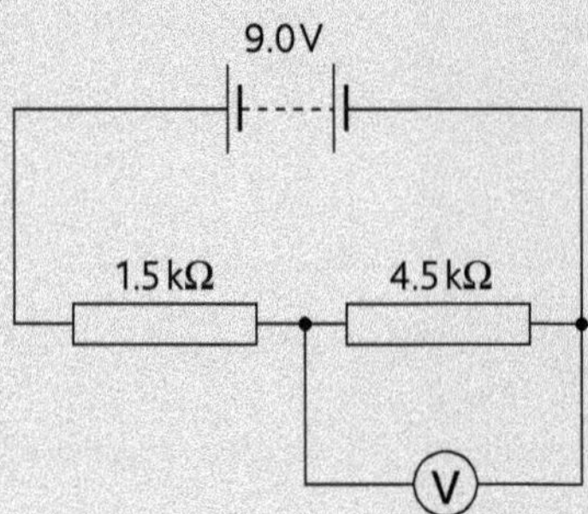

b A voltmeter of resistance R is placed across the 4.5 kΩ resistor and reads 5.92 V

i Calculate the resistance R of the voltmeter. [4]

ii What will the voltmeter read when placed across the 1.5 kΩ resistor? [4]

11 A household bulb is marked 125 V, 60 W. The bulb is switched on for 4 hours.

Calculate:

a the current flowing through the bulb [2]

b the charge that passes through the bulb for the 4 hours [2]

c the energy supplied to the bulb during the 4-hour period [2]

d the working resistance of the bulb. [2]

12 a Explain what is meant by the resistivity of a material. [2]

b Show that the unit in which resistivity is measured is Ω m. [2]

c The resistance of a piece of wire of length 12 cm and cross-sectional area of 2.0×10^{-8} m² is 3.66 Ω. Determine the resistivity of this wire. [3]

13 a Derive an expression for two resistors in series. [4]

b Derive an expression for two resistors in parallel. [4]

c Calculate the effective resistance between the points X and Y. [3]

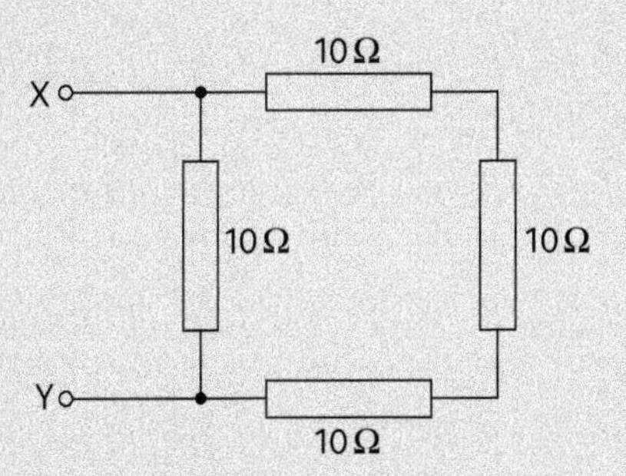

14 For the circuit below calculate:

a the value of I [4]

b the power dissipated in the 4 Ω resistor. [3]

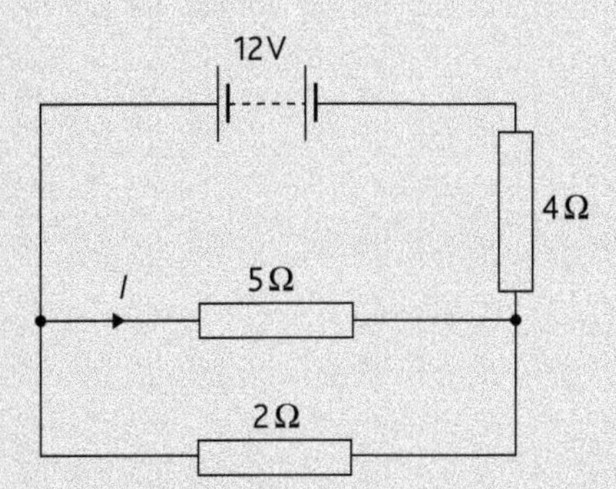

15 a Define electrical resistance and the ohm. [2]

b Sketch the I–V graph for a filament lamp. [2]

c State what happens to the resistance of the filament lamp as V increases.

Suggest an explanation for the shape of the I–V graph for a filament lamp. [3]

16 A car battery has an e.m.f. of 12 V and internal resistance of 0.10 Ω. When the ignition is turned, the battery delivers a current of 80 A to the starter motor.

Calculate:

a the resistance of the starter motor [3]

b the potential difference across the starter motor [2]

c the power dissipated in the starter motor [2]

d the power dissipated in the battery. [2]

17 A battery of e.m.f. 3.0 V and negligible internal resistance is connected in series with a resistor of 1.5 kΩ and a thermistor T as shown in the diagram below. At room temperature, the resistance of the thermistor is 2.0 kΩ.

a Calculate the potential difference across the thermistor at room temperature. [2]

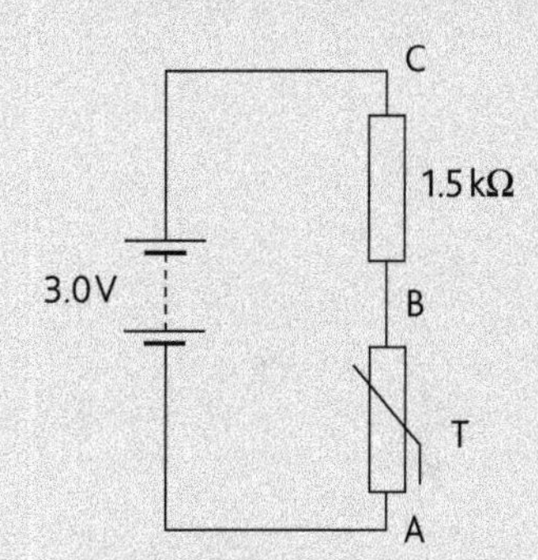

b A uniform resistance wire PQ of length 1.00 m is connected in parallel with the resistor and the thermistor as shown below. A sensitive voltmeter is connected between the point B and a movable contact M on the wire.

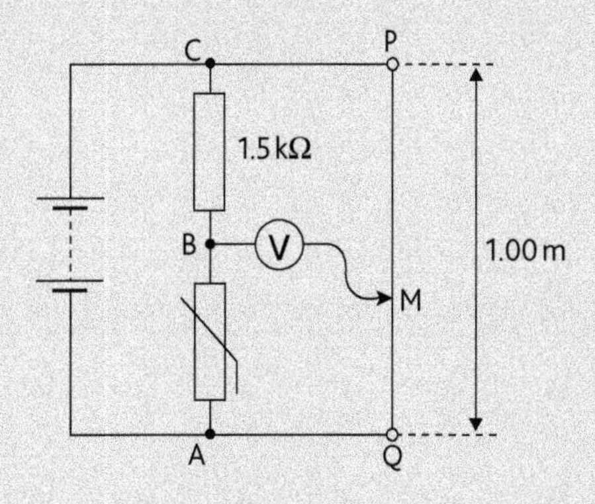

i Explain why, for a constant current in the wire, the potential difference between any two points on the wire is proportional to the distance between the two points. [2]

ii The contact M is moved along PQ until the voltmeter reading is zero.

1 State the potential difference between the contact at M and the point Q. [1]

2 Calculate the length of the wire MQ. [2]

iii The thermistor is cooled slightly. State and explain the effect on the length of the wire between M and Q for the voltmeter to remain at zero deflection. [2]

18 a Explain with the aid of a diagram how a slide-wire potentiometer could be used to measure the e.m.f. of a cell. [5]

b Explain why the slide-wire potentiometer gives a more accurate reading for the e.m.f. of the cell than a moving coil galvanometer. [3]

19 a State Kirchhoff's laws and state the physical principle upon which each is based. [6]

b Using Kirchhoff's laws, calculate the magnitude of the currents I_A, I_B and I_C. [8]

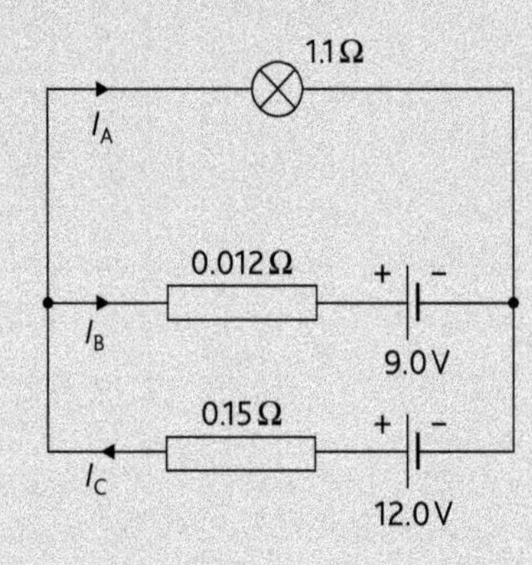

20 In the diagram below, a battery P has an e.m.f. of 9 V and negligible internal resistance. Battery Q has an e.m.f. of 6 V and its internal resistance is 1.2 Ω.

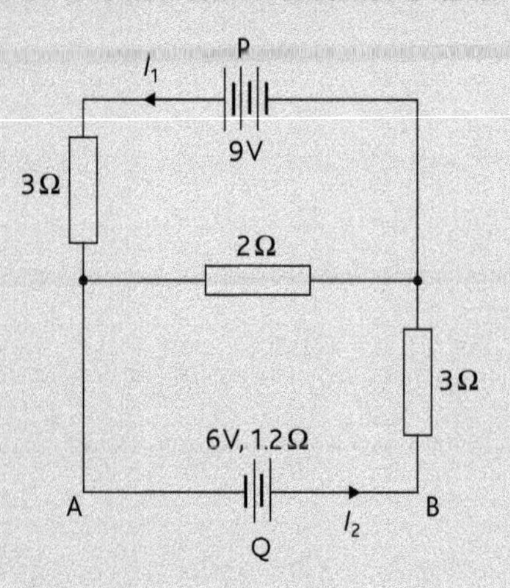

a Calculate the values of I_1 and I_2. [6]

b Determine the potential difference between the points A and B. [2]

21 An electric heater is made using nichrome wire. Nichrome has a resistivity of $1.0 \times 10^{-6}\,\Omega\,m$ at the operating temperature of the heater. The electric heater is designed to have a power dissipation of 75 W when a potential difference of 18 V is applied across the terminals of the heater.

When the heater is operating at 75 W, calculate:

a the current flowing through it [2]

b the resistance of the nichrome wire inside the heater [2]

c the length of nichrome wire of diameter of 0.75 mm required to make the heater. [3]

22 The diagram below shows a network of resistors and batteries. Each battery has an internal resistance of 0.1 Ω.

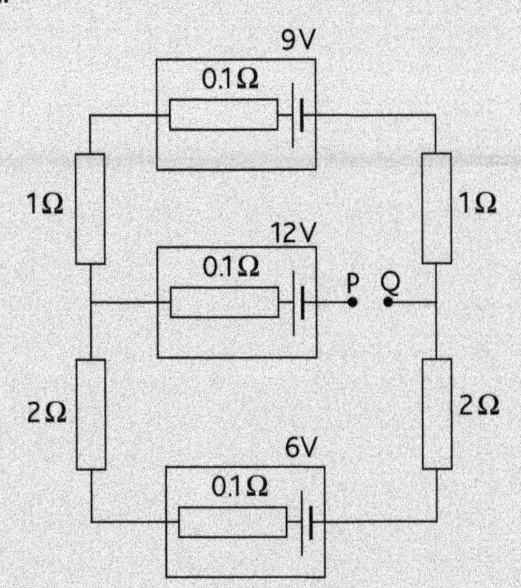

a Calculate the current flowing in the 9 V battery. [3]

b A thick wire of negligible resistance is used to connect the points P and Q. Calculate the new current flowing through the 9 V battery. [6]

23 For the circuit below calculate:

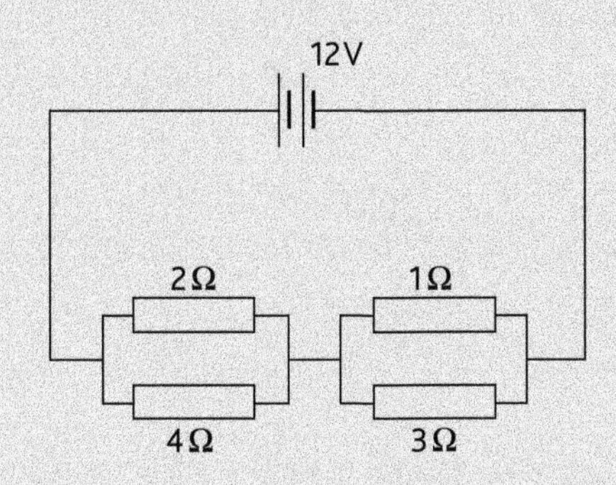

a the total resistance in the circuit [4]

b the current flowing through the 12 V d.c. supply [2]

c the potential difference across the 2 Ω resistor [2]

d the potential difference across the 1 Ω resistor [2]

e the current flowing through the 2 Ω resistor [2]

f the power dissipated in the 3 Ω resistor. [2]

24 Use Kirchhoff's laws to determine the currents I_1, I_2 and I_3 in the circuit below. [6]

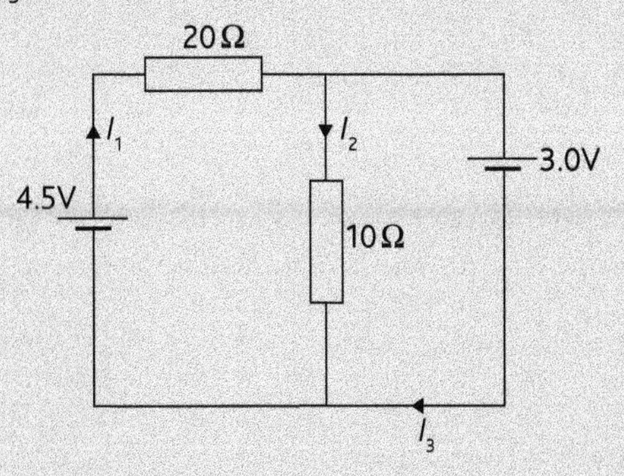

25 The following circuit is used to compare two e.m.f.s.

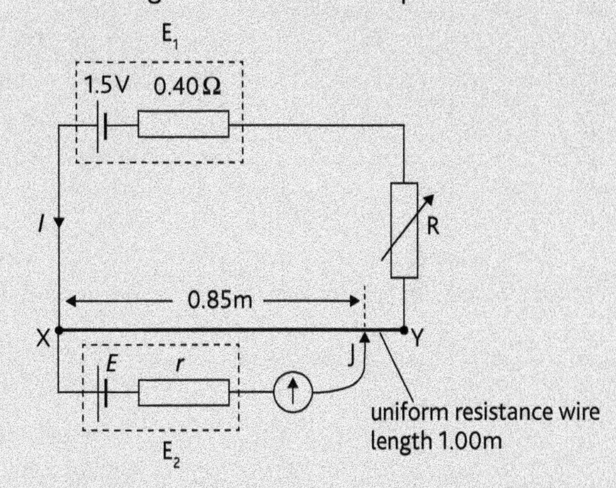

The uniform resistance wire XY has a length of 1.00 m and radius 0.55 mm. The resistivity of the material of the wire is 1.1×10^{-6} Ω m. E_1 has an e.m.f. of 1.5 V and internal resistance of 0.40 Ω. The current through E_1 is I.

E_2 has an e.m.f. of E and internal resistance of r.

The movable contact J is adjusted so that the current flowing through E_2 is zero. When this occurs, the length XJ is 0.85 m and the variable resistor has a resistance of 1.8 Ω.

a Calculate the resistance of wire XY. [3]

b Calculate the current I. [2]

c Calculate the potential difference across the length of wire XJ. [2]

d State the value of E. [1]

4 Electric fields

4.1 Electric fields

Learning outcomes

On completion of this section, you should be able to:

- explain what is meant by an electric field
- sketch electric field lines
- define electric field strength
- state and apply Coulomb's law.

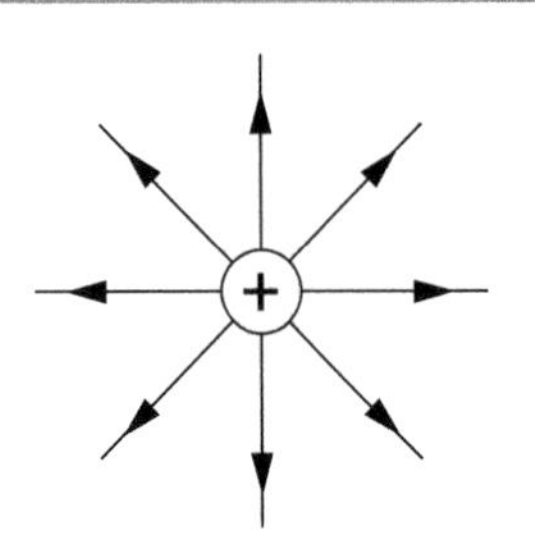

Figure 4.1.1 *Electric field due to an isolated positive charge*

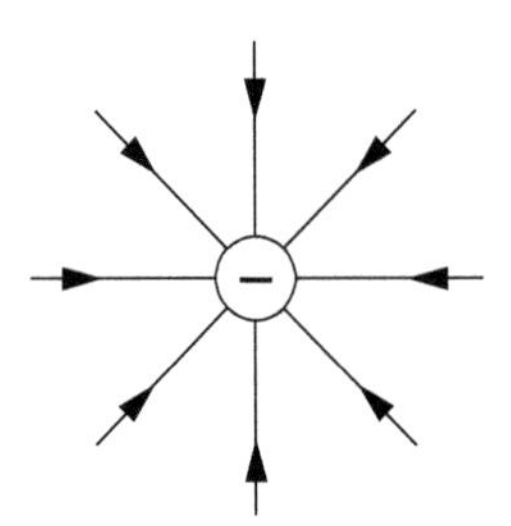

Figure 4.1.2 *Electric field due to an isolated negative charge*

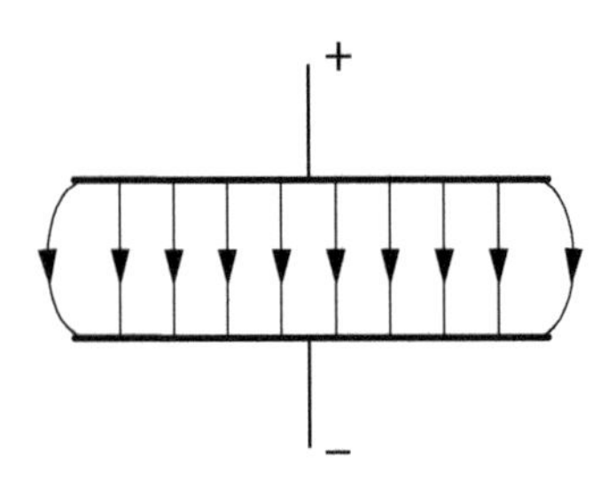

Figure 4.1.5 *Electric field due to a pair of parallel plates*

Electric fields

A charged body produces an **electric field** around itself. When another charged body is placed inside this electric field it experiences a force. The direction of the force is determined by the type of charge being placed in the field. For example, an isolated positive charge $+Q$ will have an electric field around it. When an object having a positive charge $+q$ is placed within that field it will experience a repulsive force. The direction of the field is therefore pointing away from the positive charge $+Q$.

Definition

An electric field is a region around a charged body where a force is experienced.

Definition

The direction of an electric field at any point is the direction of the force on a small positive charge placed at that point.

Figures 4.1.1–5 show various electric fields and their directions.

An isolated positive charge – the field is radial and pointing away from the isolated positive charge.

An isolated negative charge – the field is radial and pointing towards the isolated negative charge.

Two like charges

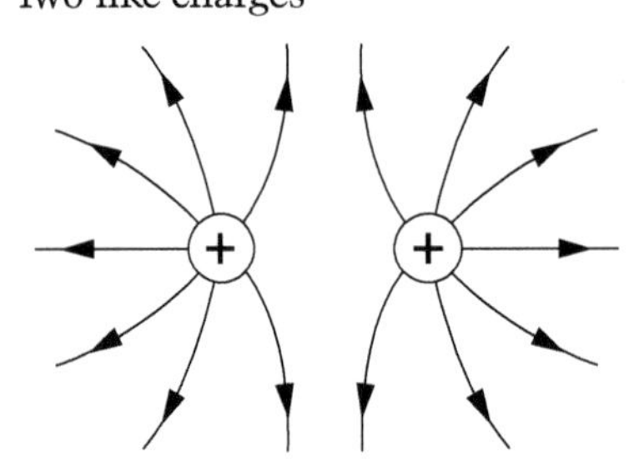

Figure 4.1.3 *Electric field due to two like charges*

Two unlike charges

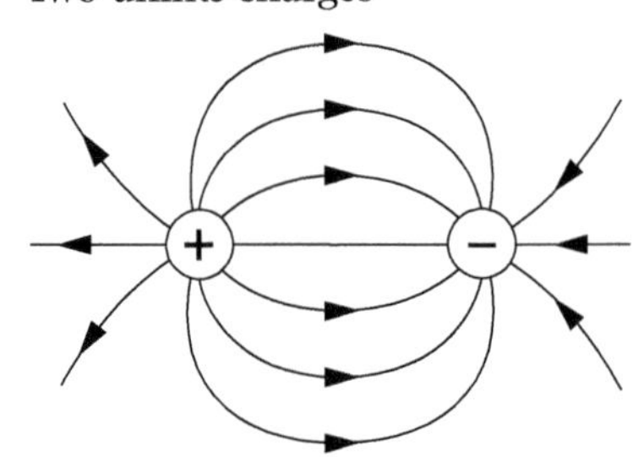

Figure 4.1.4 *Electric field due to two unlike charges*

A pair of parallel plates – the field between the plates is uniform. At the ends of the plates the field becomes curved.

Electric field strength *E*

Definition

The **electric field strength** *E* at a point in an electric field is the force acting per unit positive charge.

Equation

$$E = \frac{F}{Q}$$

E – electric field strength/$\mathrm{N\,C^{-1}}$
F – force/N
Q – charge/C

The SI unit of electric field strength is the newton per coulomb ($\mathrm{N\,C^{-1}}$). Electric field strength is a vector quantity.

Coulomb's law

The force experienced when two charged bodies are brought close to each other is dependent on the magnitude of their charges and the separation between them. The force is proportional to the product of their charges and is inversely proportional to square of the distance between them. This is **Coulomb's law**. Figure 4.1.6 illustrates two bodies having charges Q_1 and Q_2. They are separated by a distance d. The force between them is F.

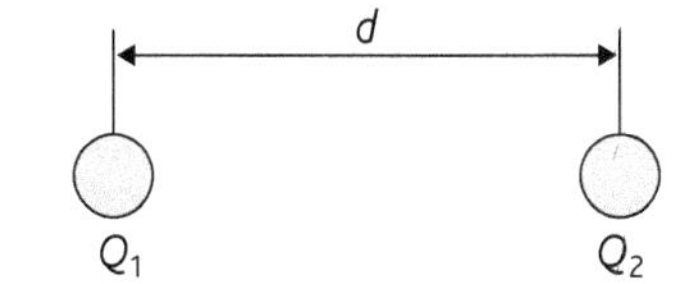

Figure 4.1.6

The equation below is used to calculate the force between the two charged bodies.

Equation

$$F = \frac{Q_1Q_2}{4\pi\varepsilon_0 d^2}$$

F – force/N
Q_1 – charge/C
Q_2 – charge/C
d – separation between charges/m
ε_0 – permittivity of free space, 8.85×10^{-12} F m^{-1}

ε_0 is a constant and is called the **permittivity of free space**. Its magnitude has been determined experimentally. It is a measure of how easy it is for an electric field to be transmitted through space. If the charges were positive and negative, the force calculated would be negative. A negative value for F indicates that the force is attractive. A positive value for F indicates that the force is repulsive. Remember, like charges repel and unlike charges attract each other.

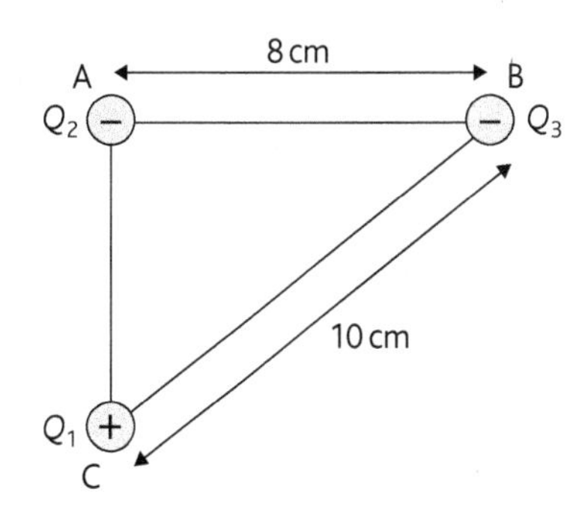

Figure 4.1.7

Example

Figure 4.1.7 shows three point charges located at the corners of a right-angled triangle BAC where:

$Q_1 = +4\,\mu C$, $Q_2 = -3\,\mu C$ and $Q_3 = -4\,\mu C$

Calculate:

a the magnitude of the force F_1 acting on Q_2, due to Q_1 alone

b the magnitude of the force F_2 acting on Q_2, due to Q_3 alone

c the magnitude of the resultant force acting on Q_2, due to Q_1 and Q_3.

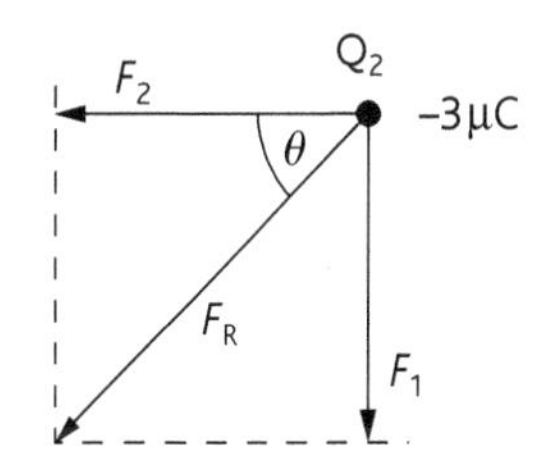

Figure 4.1.8

a $F_1 = \frac{Q_1Q_2}{4\pi\varepsilon_0 d^2} = \frac{(+4 \times 10^{-6})(-3 \times 10^{-6})}{4\pi(8.85 \times 10^{-12})(6 \times 10^{-2})^2} = -30.0\,N$

b $F_2 = \frac{Q_2Q_3}{4\pi\varepsilon_0 d^2} = \frac{(-3 \times 10^{-6})(-4 \times 10^{-6})}{4\pi(8.85 \times 10^{-12})(8 \times 10^{-2})^2} = 16.9\,N$

c Resultant force acting on $Q_2 = \sqrt{(-30.0)^2 + (16.9)^2} = 34.4\,N$

The angle the resultant force makes with the horizontal

$$= \tan^{-1}\frac{30.0}{16.9} = 60.6°$$

(See Figure 4.1.8.)

Key points

- An electric field is a region around a charged body where a force is experienced.
- The electric field strength E at a point in an electric field is the force acting per unit positive charge.
- Coulomb's law states the force between charges is proportional to the product of their charges and is inversely proportional to the square of the distance between them.

4.2 Electric field strength and electric potential

Learning outcomes

On completion of this section, you should be able to:

- recall and use $E = \dfrac{Q}{4\pi\varepsilon_0 r^2}$
- define electric potential
- recall and use $V = \dfrac{Q}{4\pi\varepsilon_0 r}$.

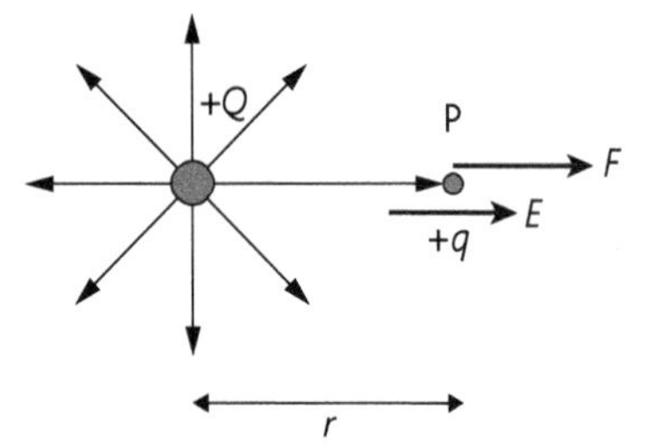

Figure 4.2.1 *Electric field strength due to a point charge*

Equation

Electric field strength due to a point charge is given by

$$E = \frac{Q}{4\pi\varepsilon_0 r^2}$$

E – electric field strength/$N\,C^{-1}$
Q – point charge/C
ε_0 – permittivity of free space, $8.85 \times 10^{-12}\,F\,m^{-1}$
r – distance from point charge/m

Definition

The electric potential at a point in an electric field is numerically equal to the work done in bringing unit positive charge from infinity to that point.

Equation

Electric potential at a point due to a point charge is given by

$$V = \frac{Q}{4\pi\varepsilon_0 r}$$

V – electric potential/V
Q – point charge/C
ε_0 – permittivity of free space, $8.85 \times 10^{-12}\,F\,m^{-1}$
r – distance from point charge/m

Electric field strength due to a point charge

Consider an isolated point charge $+Q$. The field lines around the point charge are shown in Figure 4.2.1. A small positive charge $+q$ is moved from an infinite distance to a point P, which is at a distance r from the point charge.

The electric field radiates outwards from the positive charge $+Q$. The direction of the field is outward, because the small positive charge $+q$ would be repelled.

The force exerted on the small positive charge at P can be found using Coulomb's law:

$$F = \frac{Qq}{4\pi\varepsilon_0 r^2}$$

Electric field strength E is defined as the force per unit positive charge:

$$E = \frac{F}{Q}$$

Therefore the electric field strength at the point P is given by

$$E = \frac{Qq/4\pi\varepsilon_0 r^2}{q} = \frac{Q}{4\pi\varepsilon_0 r^2}$$

Electric field strength is a vector quantity.

Electric potential due to a point charge

Consider an isolated point charge $+Q$. The field lines around the point charge are shown in Figure 4.2.2. A small positive charge $+q$ is moved from an infinite distance to a point P, which is at a distance r from the point charge.

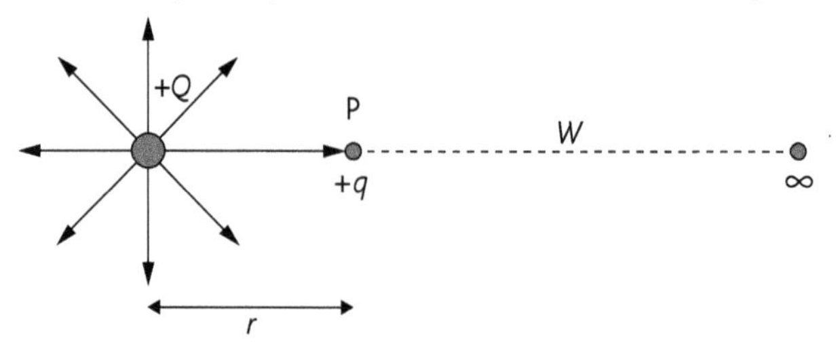

Figure 4.2.2 *Electric potential due to a point charge*

The **electric potential** V at the point P is defined as $V = \dfrac{W}{Q}$, where W is the work done in bringing unit positive charge from infinity to the point P. The SI unit of electric potential is the volt (V).

Electric potential is a scalar quantity. This means that the potential at a point due to several point charges is equal to the algebraic sum of potentials due to each charge.

Points that are equidistant from a point charge are at the same electric potential. A line drawn through these points is called an **equipotential line** (Figure 4.2.3).

Example

Figure 4.2.4 shows two point charges A and B.

Calculate:

a the force between the two point charges

b the electric potential at the point X due to the point charge A only

c the electric potential at the point X due to the point charge B only

d the electric potential at the point X due to the two point charges A and B

e the electric field strength at the point X due to the point charge A only

f the electric field strength at the point X due to the point charge B only

g the electric field strength at the point X due to the two point charges A and B.

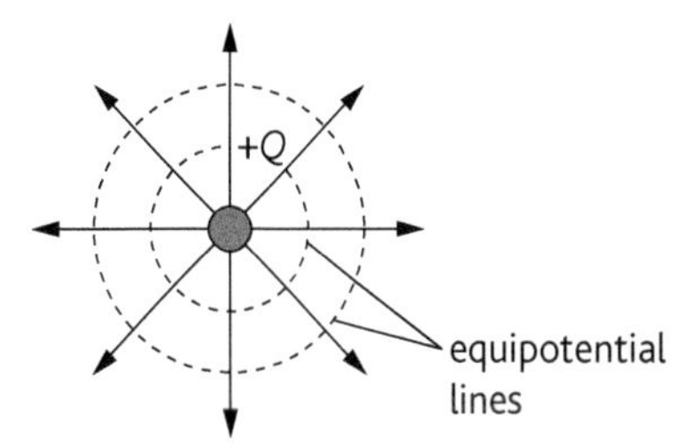

Figure 4.2.3 Equipotential lines

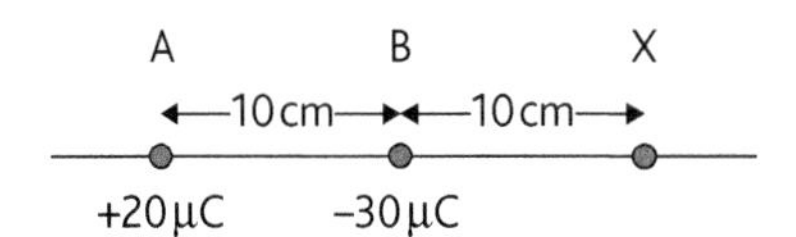

Figure 4.2.4

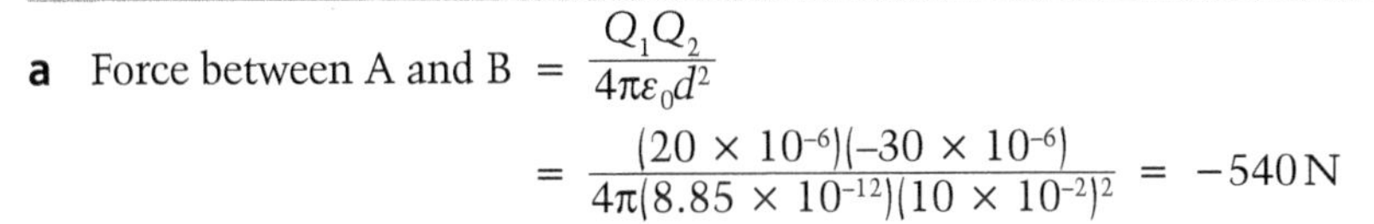

a Force between A and B $= \dfrac{Q_1Q_2}{4\pi\varepsilon_0 d^2}$

$$= \frac{(20 \times 10^{-6})(-30 \times 10^{-6})}{4\pi(8.85 \times 10^{-12})(10 \times 10^{-2})^2} = -540\,\text{N}$$

The negative sign indicates that the force is attractive.

b Electric potential at the point X due to the point charge A

$$= \frac{Q}{4\pi\varepsilon_0 r} = \frac{20 \times 10^{-6}}{4\pi(8.85 \times 10^{-12})(20 \times 10^{-2})} = 8.99 \times 10^5\,\text{V}$$

c Electric potential at the point X due to the point charge B

$$= \frac{Q}{4\pi\varepsilon_0 r} = \frac{-30 \times 10^{-6}}{4\pi(8.85 \times 10^{-12})(10 \times 10^{-2})} = -2.70 \times 10^6\,\text{V}$$

d Electric potential is a scalar quantity. Therefore, the electric potential at the point X due to the two point charges is equal to the algebraic sum of the electric potentials due to each charge.

$$\text{electric potential at the point X} = \text{electric potential at the point X due to A} + \text{Electric potential at the point X due to B}$$

$$= 8.99 \times 10^5 + (-2.70 \times 10^6) = -1.80 \times 10^6\,\text{V}$$

e Electric field strength at the point X due to charge A

$$= \frac{Q}{4\pi\varepsilon_0 r^2} = \frac{20 \times 10^{-6}}{4\pi(8.85 \times 10^{-12})(20 \times 10^{-2})^2} = 4.50 \times 10^6\,\text{NC}^{-1}$$

Electric field strength is a vector quantity. In order to determine its direction, a small positive charge is placed at the point X. It will be repelled by the point charge A. Therefore the direction of the electric field at the point X due to point charge A is to the right.

f Electric field strength at the point X due to charge B

$$= \frac{Q}{4\pi\varepsilon_0 r^2} = \frac{30 \times 10^{-6}}{4\pi(8.85 \times 10^{-12})(10 \times 10^{-2})^2} = -2.70 \times 10^7\,\text{NC}^{-1}$$

Electric field strength is a vector quantity. In order to determine its direction, a small positive charge is placed at the point X. It will be attracted to the point charge B. Therefore the direction of the electric field at the point X due to point charge B is to the left.

Figure 4.2.5

g Since, electric field strength is a vector quantity, the combined effect due to both point charges can be found by finding the vector sum of the individual field strengths.

The electric field strength at the point X due to both point charges is

$$= 2.70 \times 10^7 - 4.50 \times 10^6$$

$$= 2.25 \times 10^7\,\text{NC}^{-1}$$

The electric field strength at the point X due to B is larger than the electric field strength at the point X due to A. Therefore, the direction of the resultant electric field strength at the point X is acting to the left.

Key points

- Electric field strength is the force per unit charge.
- The electric potential at a point in an electric field is numerically equal to the work done in bringing unit positive charge from infinity to that point.
- Electric potential is a scalar quantity.
- Electric field strength is a vector quantity.
- An equipotential is a line drawn through points having equal potentials.

Learning outcomes

On completion of this section, you should be able to:

- recall that the field strength at a point in a field is numerically equal to the potential gradient at that point
- recall the formula for finding electric field strength between two charged parallel plates
- compare motion in an electric field with motion in a gravitational field.

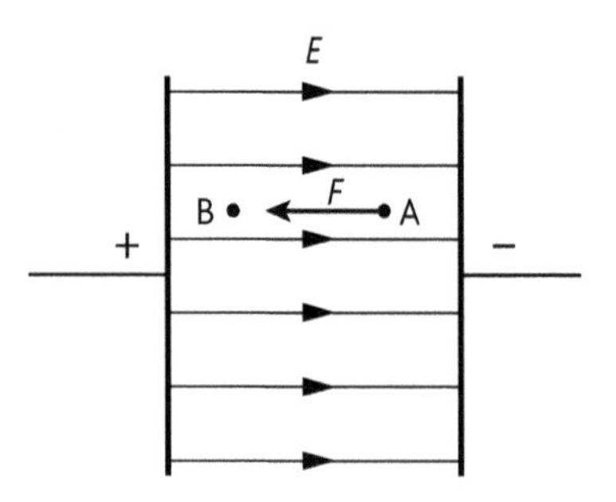

Figure 4.3.1

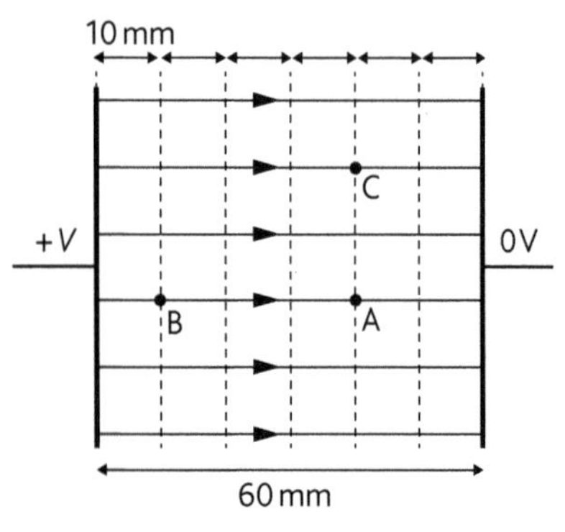

Figure 4.3.2

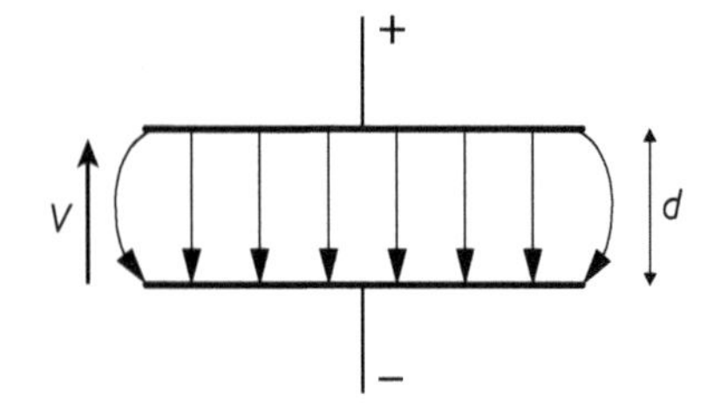

Figure 4.3.3

The relationship between *E* and *V*

Consider a uniform electric field between a pair of parallel metal plates. The electric field strength is E. Consider a charge $+Q$ being moved from point A to point B in the uniform electric field by a force F, as shown in Figure 4.3.1.

The distance between A and B (Δx) is very small so that the force F can be assumed to be constant. The work done by the force is

$$\Delta W = F\Delta x \quad (1)$$

But, by definition,

$$F = -EQ \quad (2)$$

The minus sign is required because work has to be done against the electric field in order to move the charge from A to B.

Substituting Equation (1) into Equation (2)

$$\therefore \quad \Delta W = -EQ\Delta x \quad (3)$$

The point B is at a higher electric potential than the point A.

Let the potential difference between A and B be ΔV.

$$\therefore \quad \Delta W = Q\Delta V \quad (4)$$

Equating Equation (3) and Equation (4)

$$-EQ\Delta x = Q\Delta V$$

$$\therefore \quad E = -\frac{\Delta V}{\Delta x} \quad (5)$$

$\frac{\Delta V}{\Delta x}$ is called the potential gradient. Since this expression can be used to find the electric field strength, another unit that can be used for electric field strength is the $\mathrm{V\,m^{-1}}$.

The electric field strength at a point in an electric field is therefore numerically equal to the potential gradient at that point.

Figure 4.3.2 shows two parallel plates separated by a distance of 60 mm. The potential difference between the plates is V. The dashed lines represent equipotentials. Points along any dashed line have the same potential. Therefore, the potential difference between A and C is zero.

The potential difference between each broken line is $\frac{V}{6}$ volts.

Therefore, the potential difference between A and B is $\frac{V}{6} \times 3 = \frac{V}{2}$ volts.

The potential difference between C and B is also $\frac{V}{2}$ volts.

The electric field between two parallel charged plates

Consider the electric field produced between two parallel plates as shown in Figure 4.3.3.

The separation between the plates is d and the potential difference across the plates is V. Along the middle of the plates the electric field is uniform. At the ends of the plate the electric field becomes non-uniform.

The electric field in the middle section of the plate is dependent on the voltage across the plates V and the separation of the plates, d.

The following equation is used to determine the electric field strength between two parallel plate conductors.

Equation

$$E = -\frac{V}{d}$$

E – electric field strength/V m^{-1}

V – potential difference across plates/V

d – separation between plates/m

The minus sign is necessary because the potential difference V and the electric force F are acting in opposite directions.

Exam tip

The minus sign is sometimes ignored in calculations, but keep in mind that the potential difference and the electric force act in opposite directions.

Suppose an electron is projected horizontally between a pair of parallel metal plates with a velocity v. As the electron enters the electric field it experiences a downward force. Just before entering the plates the vertical component of the electron's velocity is zero. (It was projected horizontally.) The downward force acting on the electron causes it to accelerate towards the positive plate. The horizontal component of the electron's velocity is unaffected because the force is not acting in a horizontal direction and v is therefore unchanged. The electron follows a curved path as it passes through the electric field. The curved path is parabolic. Once the electron is outside the electric field it continues moving in a straight line with the velocity it possessed coming out of the field.

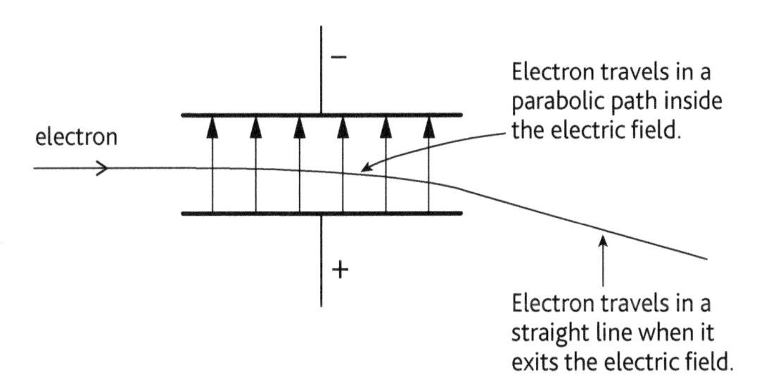

Figure 4.3.4 *A charged particle travelling in an electric field*

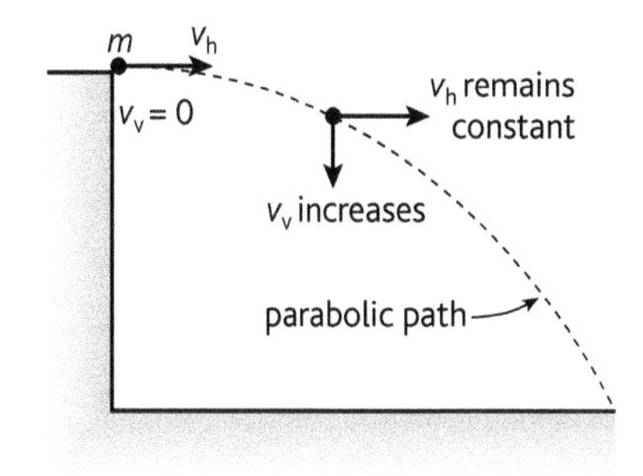

Figure 4.3.5 *A mass travelling in a gravitational field*

The path taken by a charged particle as it travels in a uniform electric field can be compared to the path taken by a mass travelling in a gravitational field. Suppose a mass m is projected horizontally with a velocity v_h. The initial vertical component of its velocity v_v is zero. The force of gravity acts vertically. It exerts a force of mg on the mass. This force causes the mass to accelerate vertically. The vertical component of the velocity v_v of the mass therefore increases. The horizontal component of its velocity v_h remains constant (assuming air resistance is negligible) since the force of gravity does not act horizontally. The path taken by the mass is parabolic.

Key points

- The electric field strength at a point in an electric field is numerically equal to the potential gradient at that point.
- The path taken by a charged particle travelling in a uniform electric field is parabolic.
- The path taken by a mass travelling in a uniform gravitational field is parabolic.

4.4 Worked examples on electric fields

Learning outcomes

On completion of this section, you should be able to:

- solve problems involving the motion of charge particles in uniform electric fields.

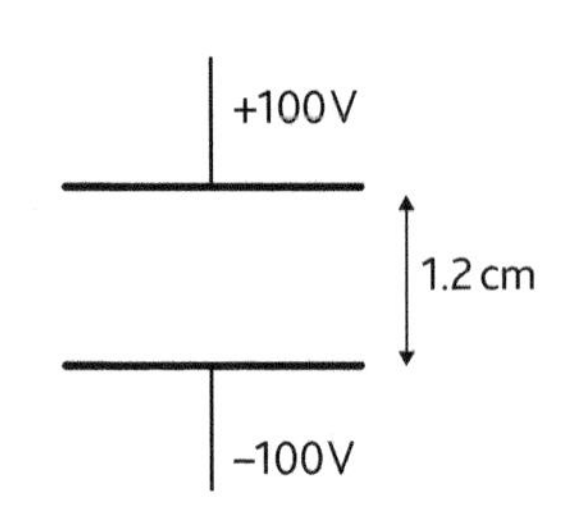

Figure 4.4.1

Example

Two flat metal plates are separated by a distance of 1.2 cm as shown in Figure 4.4.1.

Calculate the electric field strength between the plates.

Potential difference between the plates is 100 − (−100) = 200 V

(Note that the potential difference is not 0 V.)

$$\text{Electric field strength} \quad E = -\frac{V}{d} = -\frac{200}{1.2 \times 10^{-2}} = -1.67 \times 10^{4}\,\text{V m}^{-1}$$

Example

Two flat metal plates, each of length 4.0 cm, are separated by a distance of 1.5 cm as shown in Figure 4.4.2. A potential difference of 110 V between the plates provides a uniform electric field in the region between the plates. Electrons of speed $4.0 \times 10^{7}\,\text{m s}^{-1}$ enter this region at right angles to the field. Assume that the space between the plates is a vacuum and the field outside the plates is zero.

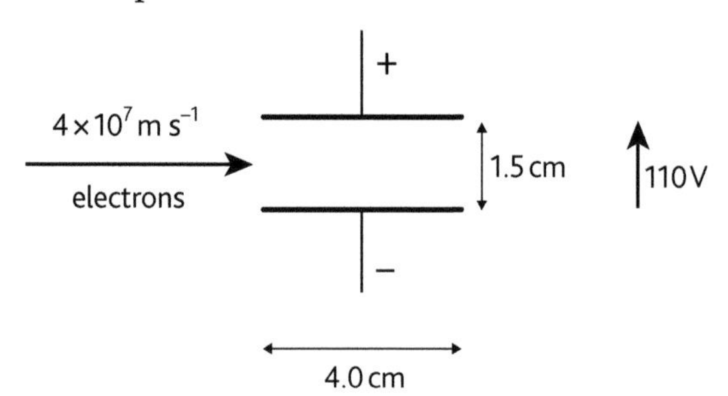

Figure 4.4.2

Calculate:

a the electric field strength between the plates

b the force on an electron due to the electric field

c the acceleration of the electron along the direction of the electric field

d the time taken for an electron to travel between the plates

e the speed of the electron at right angles to its original direction of motion as it leaves the region between the plates

f the velocity of an electron as it leaves the region between the plates.

(Charge on an electron = -1.6×10^{-19} C,
mass of an electron = 9.11×10^{-31} kg)

a Electric field strength $E = -\frac{V}{d}$

$$= -\frac{110}{1.5 \times 10^{-2}} = -7.33 \times 10^{3}\,\text{V m}^{-1}$$

b Force on an electron $F = EQ$

$= -7.33 \times 10^3 \times -1.6 \times 10^{-19}$

$= 1.17 \times 10^{-15}\,\text{N}$

c Using Newton's second law $F = ma$

Acceleration along the direction of the electric field $a = \dfrac{F}{m}$

$= \dfrac{1.17 \times 10^{-15}}{9.11 \times 10^{-31}}$

$= 1.28 \times 10^{15}\,\text{m}\,\text{s}^{-2}$

d The horizontal component of the electron's velocity is $4.0 \times 10^7\,\text{m}\,\text{s}^{-1}$. There is no force acting in this direction, hence the acceleration of the electrons in this direction is zero. The horizontal component of the velocity remains constant.

Using $s = ut + \frac{1}{2}at^2$

$s = ut$, since $a = 0$

$t = \dfrac{s}{u} = \dfrac{4 \times 10^{-2}}{4 \times 10^7} = 1 \times 10^{-9}\,\text{s}$

Time taken to travel between the plates is $1 \times 10^{-9}\,\text{s}$.

e The electron is accelerating vertically at a constant rate. The initial vertical component of the electron's velocity is zero.

Using $v = u + at$

$= 0 + (1.28 \times 10^{15} \times 1 \times 10^{-9})$

$= 1.28 \times 10^6\,\text{m}\,\text{s}^{-1}$

f The vertical component of the electron's velocity as it leaves the plates is $1.28 \times 10^6\,\text{m}\,\text{s}^{-1}$

Magnitude of electron's velocity as it leaves the plates

$= \sqrt{(1.28 \times 10^6)^2 + (4 \times 10^7)^2}$

$= 4.00 \times 10^7\,\text{m}\,\text{s}^{-1}$

Angle the velocity makes with the horizontal

$= \tan^{-1}\left(\dfrac{1.29 \times 10^6}{4 \times 10^7}\right)$

$= 1.85°$

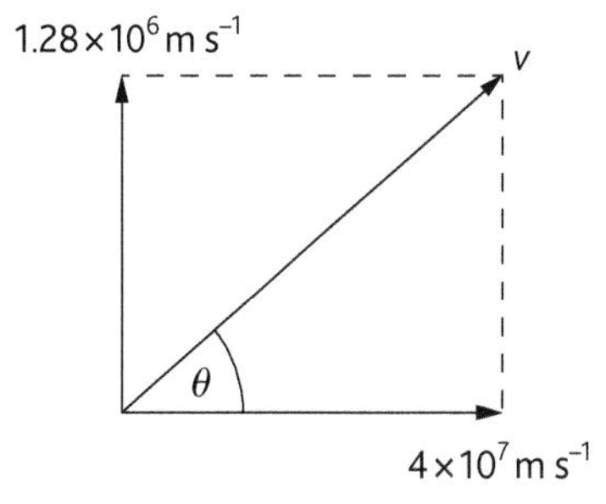

Figure 4.4.3

Example

An isolated conducting sphere of radius R has a positive charge $+Q$. Sketch a graph to show the variation of electric field strength E and electric potential V with distance r from the centre of the sphere.

Inside the sphere, there are no charges. The electric field strength E is zero everywhere inside the sphere. If the field lines are drawn from the surface of the sphere they would appear as if they were being produced from a point charge at the centre of the sphere.

For $r < R$, $E = 0$ and V is a constant.

For $r > R$, $E = \dfrac{Q}{4\pi\varepsilon_0 r^2}$ and $V = \dfrac{Q}{4\pi\varepsilon_0 r}$

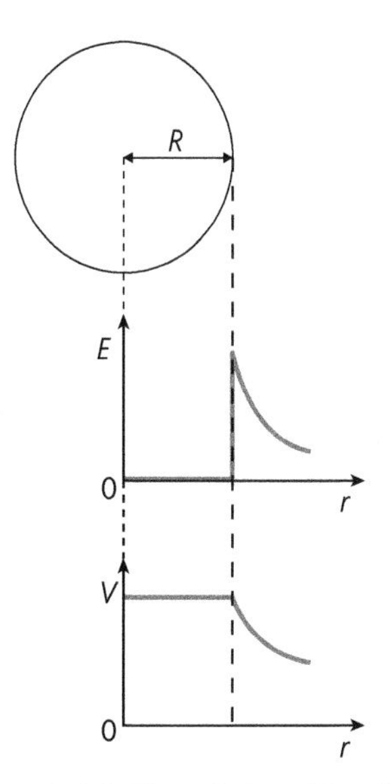

Figure 4.4.4 *The variation of E and V from the centre of an isolated charged hollow sphere*

5 Capacitance

5.1 Capacitance

Learning outcomes

On completion of this section, you should be able to:

- state the function of a capacitor
- define capacitance and the farad
- recall and solve problems using $C = \frac{Q}{V}$.

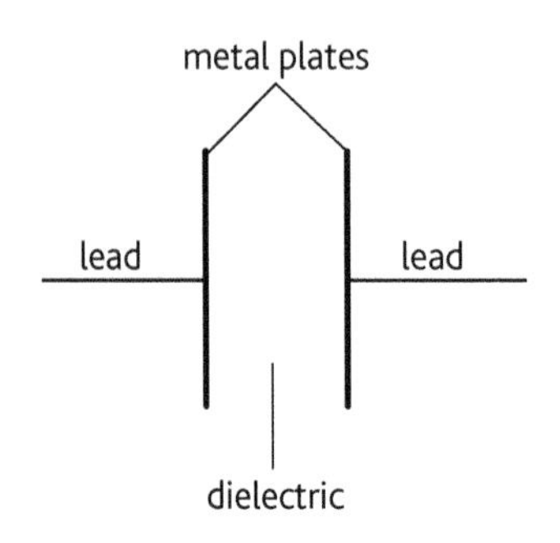

Figure 5.1.1 *The parts of a capacitor*

Figure 5.1.2 *The symbol for a capacitor*

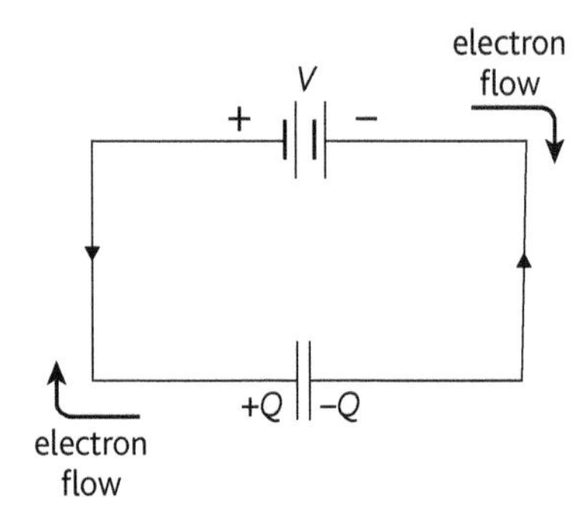

Figure 5.1.3

Capacitors

Capacitors are electrical components found in many electrical devices. They consist of two sheets of a conducting material separated by an insulator, called a **dielectric**. Commonly used dielectrics are air, oil and paper. The two conducting sheets are attached to leads (Figure 5.1.1). A capacitor is a device that stores electric charge. Capacitors are also used in electrical circuits to prevent the flow of direct currents (d.c.). Capacitors are widely used in digital cameras. A capacitor inside the camera is charged up using a battery. Once charged, it is made to discharge through a bulb and a bright flash is produced.

Electronic equipment such as cameras and television sets have the following warning message at the back of them:

'Danger: Even when switched off, it may be dangerous to remove the back of this equipment'.

Inside these types of electrical equipment, some of the capacitors store large amounts of electric charge. During normal operation of the equipment, these capacitors become charged. Even if the equipment is switched off, the capacitors may still have some electric charge stored in them. Opening the back of the equipment exposes the user to the capacitors. Accidentally touching the leads of the capacitor will cause the capacitor to discharge through the user's body, causing an electric shock.

The circuit symbol for a capacitor is shown in Figure 5.1.2.

In order for a capacitor to store charge, a potential difference must be applied across it. Consider the circuit shown in Figure 5.1.3. When a potential difference is applied across a capacitor an electric current flows momentarily in the circuit. Electrons flow towards one plate, while electrons flow away from the opposite plate. Charge is conserved in this process. If the amount of charge that flows to the plate on the right is $-Q$, then the charge that flows to the left plate is $+Q$. The total charge stored in the capacitor is Q.

In order to measure the ability of a capacitor to store charge, a quantity called **capacitance** is used. The greater the value of the capacitance, the more charge the capacitor is capable of storing. The amount of charge a capacitor can store depends on the voltage applied across it.

Definition	*Equation*
The capacitance of a capacitor is the charge stored per unit potential difference.	$C = \frac{Q}{V}$ C – capacitance/F Q – charge/C V – potential difference/V

The unit of capacitance is the **farad** (F). A capacitor has a capacitance of 1 farad if the charge stored is 1 coulomb when a potential difference of 1 volt is applied across it.

$$1\,F = 1\,C\,V^{-1}$$

The farad is a large unit and typical values of capacitors are of the order of microfarads (μF) or picofarads (pF).

Typically, there are two values printed on a capacitor. One value is the capacitance of the capacitor. The other value is a voltage. This voltage is the maximum allowable potential difference that can be applied across the capacitor. If this value is exceeded, the dielectric will break down and allow charge to flow between the plates.

Factors affecting capacitance

Experiments have shown that the capacitance of a capacitor depends on the cross-sectional area of the plates, the separation of the plates and the material used as the dielectric. The equation below shows how these quantities are related to capacitance.

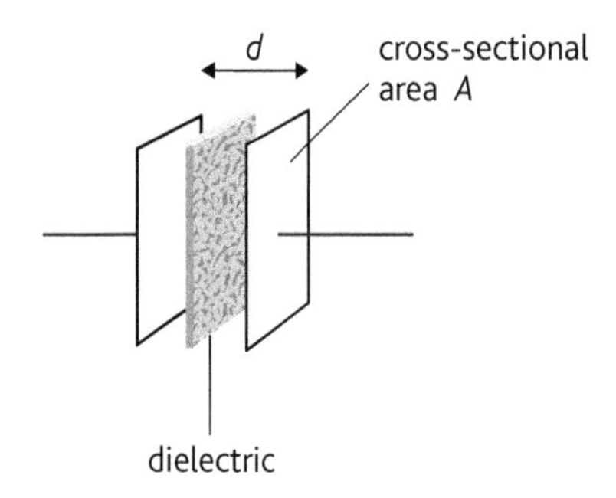

Figure 5.1.4 *Parameters affecting capacitance*

Equation

$$C = \frac{\varepsilon A}{d}$$

C – capacitance/F
ε – permittivity of the material/F m^{-1}
A – cross-sectional area/m^2
d – separation between plates/m

The permittivity of a material is a measure of its ability to transmit an electric field. The relative permittivity of a material ε_r (also known as its dielectric constant) is the ratio of the permittivity of the material to the permittivity of free space. It is a dimensionless quantity. The dielectric constant of vacuum is 1.0. Some common dielectrics are shown in Table 5.1.1.

Table 5.1.1 *Values of some dielectric constants*

Dielectric	Dielectric constant
Air	1.0006
Paper	2–5
Glass	4.5–8
Oil	2.5
Ceramic	45–6000

Equation

$$\varepsilon_r = \frac{\varepsilon}{\varepsilon_0}$$

ε – permittivity of the material/F m^{-1}
ε_0 – permittivity of free space/F m^{-1}
ε_r – relative permittivity of the material (no units)

When a dielectric is placed between the plates of an isolated capacitor, the molecules inside it become polarised. This causes the electric potential across the plates to decrease. As a result, the capacitance of the capacitor increases with the presence of a dielectric ($C \propto \frac{1}{V}$).

Key points

- A capacitor is a device that stores electric charge.
- The capacitance of a capacitor is the charge stored per unit voltage.
- A capacitor has a capacitance of 1 farad if the charge stored is 1 coulomb when a potential difference of 1 volt is applied across it.
- The capacitance of a capacitor depends on the cross-sectional area of the plates, the separation of the plates and the material used as the dielectric.

5.2 Charging and discharging a capacitor

Learning outcomes

On completion of this section, you should be able to:

- recall and use the equations for charging and discharging a capacitor
- understand the concept of time constant during the charging and discharging of a capacitor through a resistor.

Charging a capacitor through a resistor

A capacitor can be charged by connecting it to a power source in series with a resistor as shown in Figure 5.2.1. When the switch S is at position A, a current begins to flow and the capacitor C begins to charge.

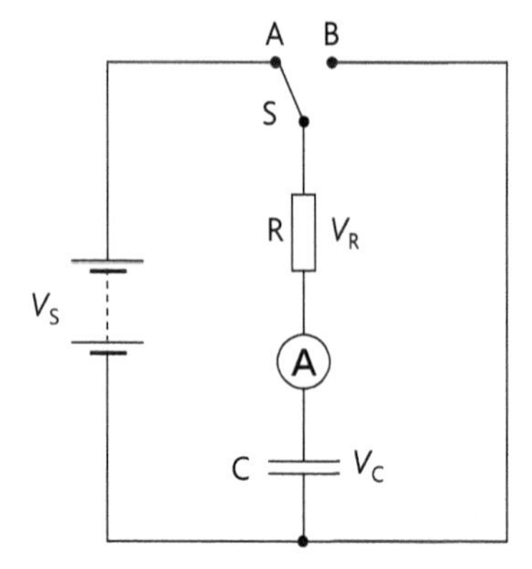

Figure 5.2.1 Charging and discharging a capacitor through a resistor

V_s is the e.m.f. of the d.c. power supply. V_C is the potential difference across the capacitor C and V_R is the potential difference across the resistor R. Figure 5.2.2 illustrates how the charge stored in the capacitor varies with time as the capacitor charges.

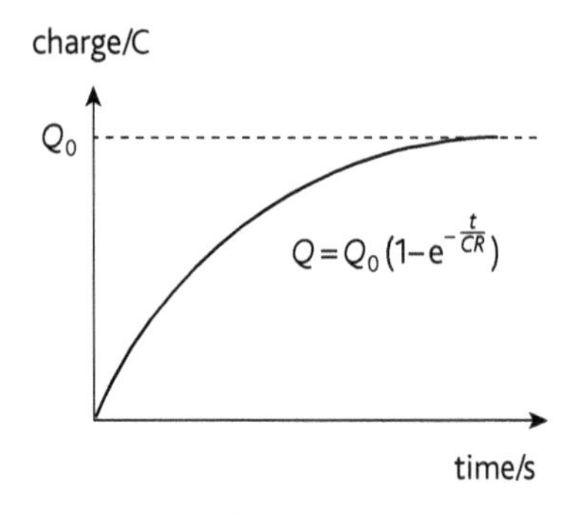

Figure 5.2.2 Charge Q against time

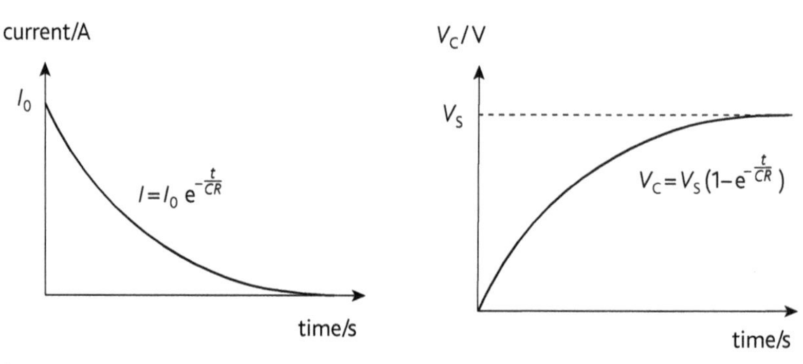

Figure 5.2.3 I against time

Figure 5.2.4 V_C against time

Figure 5.2.3 illustrates what happens to the current in the circuit over a period of time. The current approaches zero.

Figure 5.2.4 illustrates what happens to the potential difference across the capacitor as it charges.

Figure 5.2.5 illustrates what happens to the potential difference across the resistor as the capacitor is being charged.

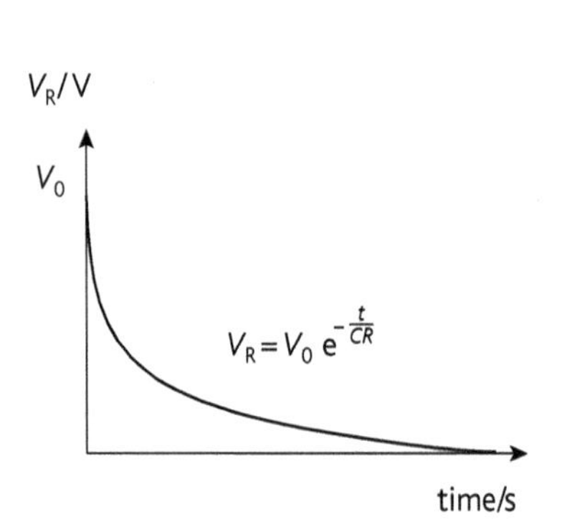

Figure 5.2.5 V_R against time

Discharging a capacitor through a resistor

When the capacitor C is fully charged, the switch S is moved to position B. The capacitor will discharge through the resistor R. Figure 5.2.6 shows how the charge in the capacitor varies with time as it discharges.

Figure 5.2.7 shows how the current flowing through the resistor varies with time.

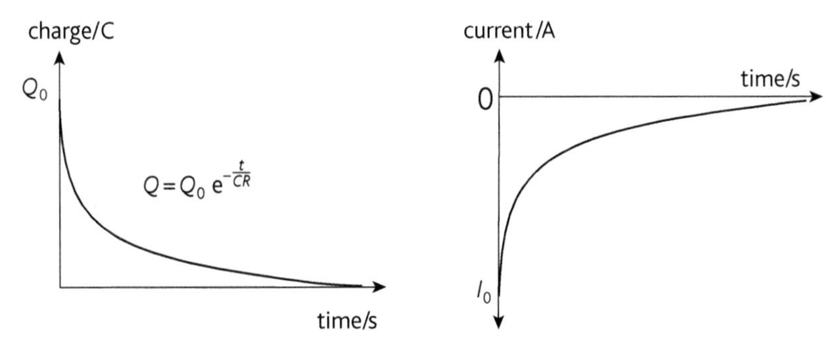

Figure 5.2.6 *Charge against time*

Figure 5.2.7 *I against time*

The graph shows the current as being negative during the discharge process. This is because the current now flows through the resistor in the opposite direction to that of the charging current.

Time constant

The **time constant** of a circuit is the time taken for the charge on a capacitor to fall to 1/e (0.368) of its initial value. The time constant is $\tau = CR$. The larger the value of R, the longer it takes for the capacitor to discharge. The time constant can be determined using a charging or discharging curve.

A charging curve is of the form $x = x_0(1 - e^{-t/CR})$

where x represents charge or potential difference.

A discharge curve is of the form $x = x_0 e^{-t/CR}$

where x represents charge, current or potential difference.

Figure 5.2.8 and Figure 5.2.9 show how the time constant is determined from a charging or discharging curve.

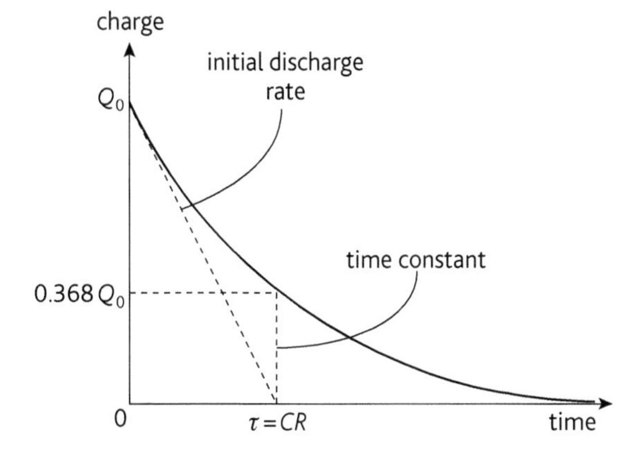

Figure 5.2.8 *Graphical determination of the time constant using a discharging curve*

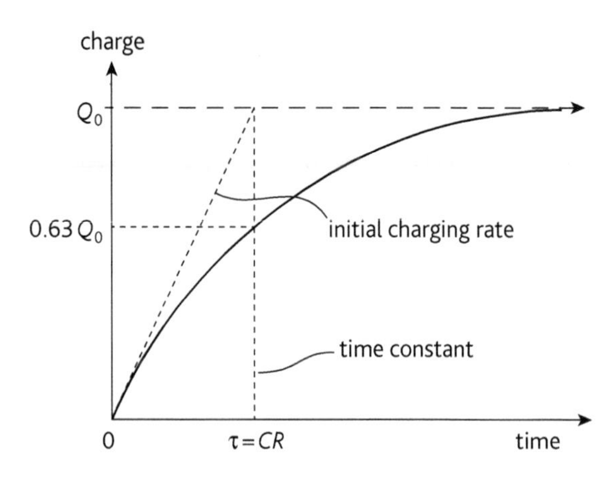

Figure 5.2.9 Graphical determination of the time constant using a charging curve

In laboratory experiments, currents and voltages are usually measured with respect to time during the charging or discharging of a capacitor. The time constant can be measured graphically using the same technique illustrated in Figures 5.2.8 and 5.2.9.

Example

A resistor of resistance 2.2 kΩ is connected in series with a capacitor of capacitance 12.0 μF and a battery of e.m.f. 20.0 V with negligible internal resistance (Figure 5.2.10).

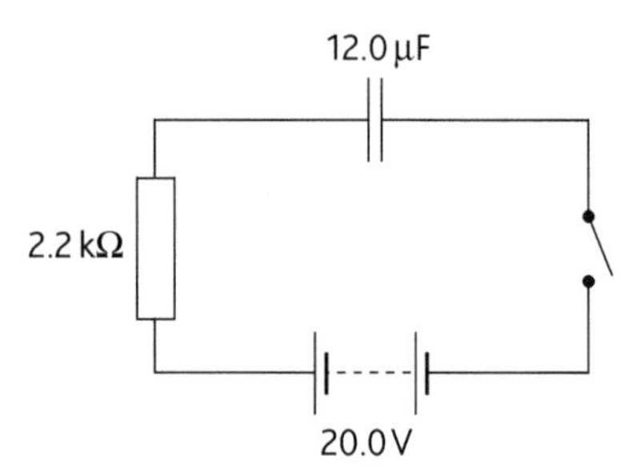

Figure 5.2.10

The capacitor is initially uncharged.

a Calculate:

i the current in the circuit immediately after the switch is closed

ii the final charge on the capacitor

iii the time constant of the circuit.

b The battery and switch are replaced by a voltage source that varies with time as shown in Figure 5.2.11.

[Assume that at time $t = 0$, the capacitor is uncharged.]

i Calculate the potential difference across the capacitor when $t = 0.04$ s.

ii Calculate the potential difference across the capacitor when $t = 0.08$ s.

iii Draw a sketch to show the variation of potential difference across the capacitor with time.

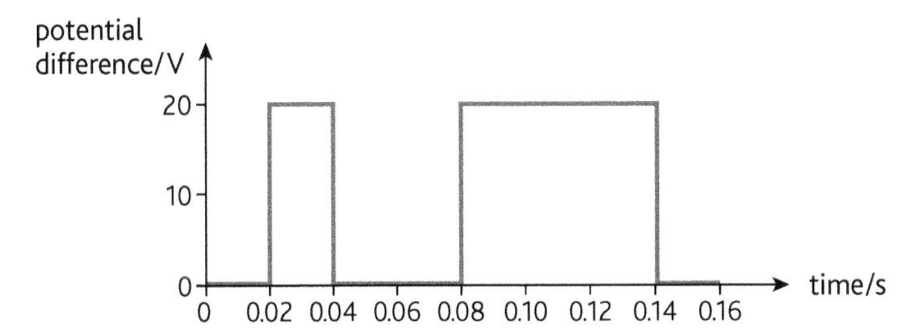

Figure 5.2.11

a **i** $I = \frac{V}{R} = \frac{20.0}{2.2 \times 10^3} = 9.1\,\text{mA}$

ii $Q = CV = 12.0 \times 10^{-6} \times 20 = 0.24\,\text{mC}$

iii $\tau = CR = 12 \times 10^{-6} \times 2.2 \times 10^3 = 0.026\,\text{s}$

b **i** $V = V_0(1 - e^{-t/CR}) = 20(1 - e^{-0.02/0.026}) = 10.7\,\text{V}$

ii $V = V_0 e^{-t/CR} = 10.7e^{-0.04/0.026} = 2.30\,\text{V}$

iii Between $t = 0$ s and $t = 0.02$ s, the capacitor is uncharged and the potential difference across it is zero.

Between $t = 0.02$ s and $t = 0.04$ s, the capacitor charges up and the potential difference across it reaches 10.7 V.

Between $t = 0.04$ s and $t = 0.08$ s, the capacitor discharges and the potential difference across it reaches 2.30 V.

Between $t = 0.08$ s and $t = 0.14$ s, the capacitor charges up.

Between $t = 0.14$ s and $t = 0.16$ s, the capacitor discharges.

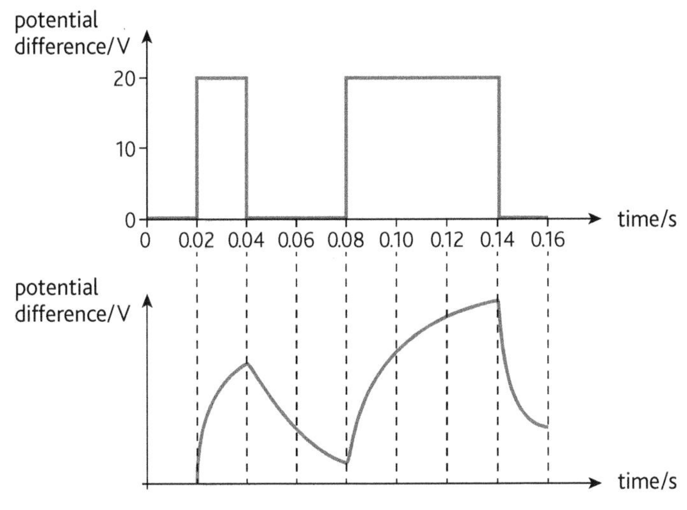

Figure 5.2.12

Key points

- A capacitor can be charged or discharged through a resistor.
- A charging curve is of the form $x = x_0(1 - e^{-t/CR})$ where x represents charge or potential difference.
- A discharging curve is of the form $x = x_0 e^{-t/CR}$ where x represents charge or potential difference.
- The time constant of a CR circuit is $\tau = CR$.
- The time constant τ of a circuit is the time taken for the charge on a capacitor to fall to 1/e (0.368) of its initial value.

5.3 Capacitors in series and in parallel

Learning outcomes

On completion of this section, you should be able to:

- derive and use an expression for capacitors in series
- derive and use an expression for capacitors in parallel
- derive and use an expression for the energy stored in a capacitor.

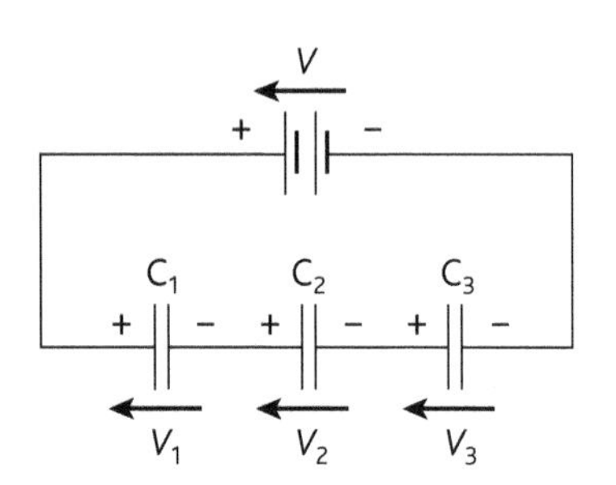

Figure 5.3.1 *Capacitors in series*

Equation

For capacitors in series:

$$\frac{1}{C_{total}} = \frac{1}{C_1} + \frac{1}{C_2} + \frac{1}{C_3}$$

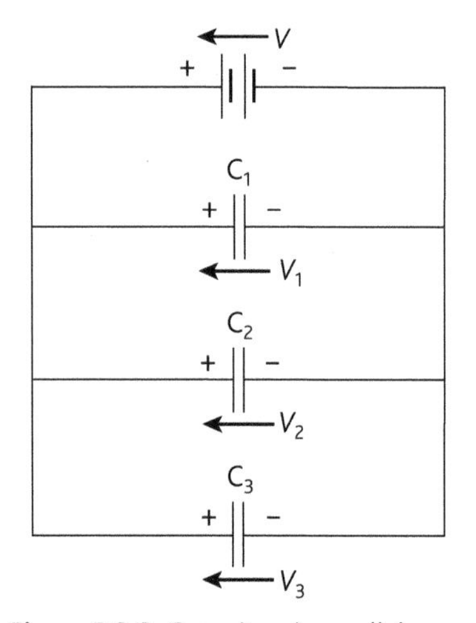

Figure 5.3.2 *Capacitors in parallel*

Equation

For capacitors in parallel:

$$C_{total} = C_1 + C_2 + C_3$$

Capacitors in series

Consider three capacitors, C_1, C_2 and C_3, in series with a d.c. supply having an e.m.f. of V (Figure 5.3.1). The potential differences across the capacitors are V_1, V_2 and V_3 respectively. Electrons from the d.c. supply flow towards the right-hand plate of capacitor C_3, while electrons flow away from the left-hand plate. At the same time, electrons flow towards the right-hand plate of capacitor C_2, while electrons flow away from the left-hand plate. The same process occurs for capacitor C_1. In this process, charge is conserved. Therefore, the charge stored in each capacitor is the same.

The charge stored in each capacitor is Q.

The three capacitors can be replaced by a single capacitor having a capacitance of C_{total} and a charge of Q_{total}.

Using Kirchhoff's second law we can write

$$V = V_1 + V_2 + V_3 \quad (1)$$

But $\quad V_1 = \frac{Q}{C_1},\ V_2 = \frac{Q}{C_2}$ and $V_3 = \frac{Q}{C_3}$

Therefore, $$\frac{Q_{total}}{C_{total}} = \frac{Q}{C_1} + \frac{Q}{C_2} + \frac{Q}{C_3} \quad (2)$$

where $\quad Q_{total} = Q$

Dividing Equation (2) by Q,

$$\frac{1}{C_{total}} = \frac{1}{C_1} + \frac{1}{C_2} + \frac{1}{C_3}$$

Capacitors in parallel

Consider three capacitors, C_1, C_2 and C_3, in parallel with a d.c. supply having an e.m.f. of V. The potential difference across each capacitor is therefore V. The charges stored in C_1, C_2 and C_3 are Q_1, Q_2 and Q_3 respectively.

The three capacitors can be replaced by a single capacitor having a capacitance of C_{total} and a charge of Q_{total}.

$$Q_1 = C_1V \qquad Q_2 = C_2V \qquad Q_3 = C_3V$$

The total charge $\quad Q_{total} = Q_1 + Q_2 + Q_3$

$$Q_{total} = C_1V + C_2V + C_3V = V(C_1 + C_2 + C_3)$$

Therefore, $$C_{total} = \frac{Q_{total}}{V} = C_1 + C_2 + C_3$$

Energy stored in a capacitor

When a capacitor is connected to a power supply, it begins to accumulate charge. Electrons flow towards one plate, while electrons flow away from the opposite plate. As more electrons move towards the negative plate, work has to be done against the repulsive forces of the electrons already on the plate. As more electrons move away from the positive plate, work has to be done against the attractive forces of the positive charges already on the plate. The work done is stored as electric potential energy. Figure 5.3.3 shows how the charge stored in a capacitor varies with potential difference across it. The energy, W, stored in the capacitor is numerically equal to the area under the graph.

Energy stored in a capacitor $W = \frac{1}{2}QV$

But, $Q = CV$, so

$$W = \frac{1}{2}(CV)V = \frac{1}{2}CV^2$$

Also, $V = \frac{Q}{C}$, so

$$W = \frac{1}{2}Q\left(\frac{Q}{C}\right) = \frac{1}{2}\frac{Q^2}{C}$$

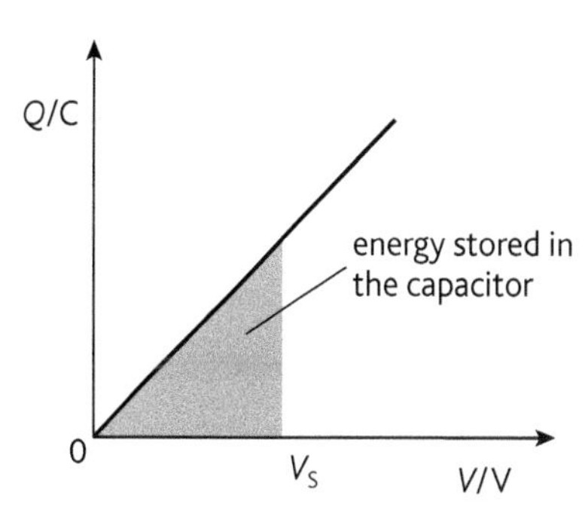

Figure 5.3.3

Equation

$$W = \frac{1}{2}QV$$

W – energy stored in capacitor/J
Q – charge stored in capacitor/C
V – potential difference across capacitor/V

Example

An isolated capacitor of capacitance 250 μF has a potential difference of 25 V across it.

a Calculate:

i the charge stored in the capacitor

ii the energy stored in the capacitor.

b An uncharged capacitor of capacitance 150 μF is then connected across the charged 250 μF capacitor. Calculate the total energy stored by the two capacitors after they have been connected and the potential difference across them.

---

a i Charge stored in capacitor

$Q = CV = 250 \times 10^{-6} \times 25 = 6.25 \times 10^{-3}\,\text{C}$

ii Energy stored in the capacitor $= \frac{1}{2}CV^2 = \frac{1}{2}(250 \times 10^{-6})(25)^2$

$= 7.81 \times 10^{-2}\,\text{J}$

b During the process of connecting the capacitors, charge is conserved. This means that the total charge before connection is equal to the total charge after the connection is made. Also, since the capacitors are now connected in parallel, the potential difference across each capacitor will be the same.

Effective capacitance after connection $C_{total} = C_1 + C_2$

$= 250 + 150 = 400\,\mu\text{F}$

Total charged stored after the connection $= 6.25 \times 10^{-3}\,\text{C}$

Total energy stored $= \frac{1}{2}\frac{Q^2}{C_{total}} = \frac{1}{2} \times \frac{(6.25 \times 10^{-3})^2}{400 \times 10^{-6}} = 4.88 \times 10^{-2}\,\text{J}$

Potential difference across the capacitors

$$V = \frac{Q}{C_{total}} = \frac{6.25 \times 10^{-3}}{400 \times 10^{-6}} = 15.6\,\text{V}$$

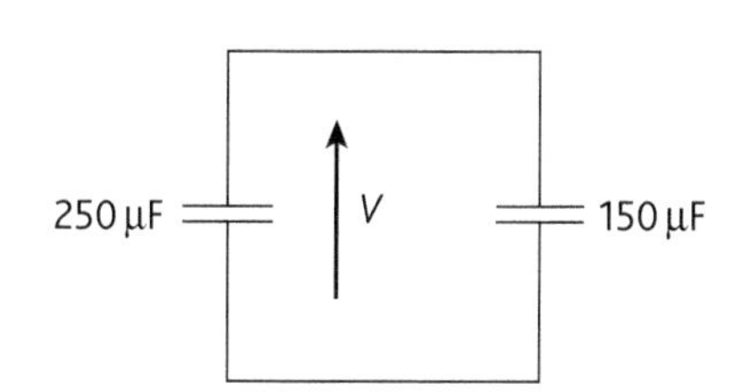

Figure 5.3.4

Key points

- The equivalent capacitance of several capacitors in series is given by:
  $\frac{1}{C_{total}} = \frac{1}{C_1} + \frac{1}{C_2} + \dots$
- The equivalent capacitance of several capacitors in parallel is given by:
  $C_{total} = C_1 + C_2 + \dots$
- The energy stored in a capacitor is given by: $W = \frac{1}{2}QV$.

Revision questions 2

Answers to questions that require calculation can be found on the accompanying CD.

1 **a** Explain what is meant by an electric field. [1]

b Explain what is meant by electric field strength E and electric potential V. [2]

c State the relationship between the two quantities. [2]

d Sketch the electric field:

i around an isolated positive point charge [2]

ii around an isolated negative point charge [2]

iii between a positive and negative point charge [2]

iv between two oppositely charged parallel metal plates. [2]

2 **a** State Coulomb's law. [3]

b The diagram below shows three point charges.

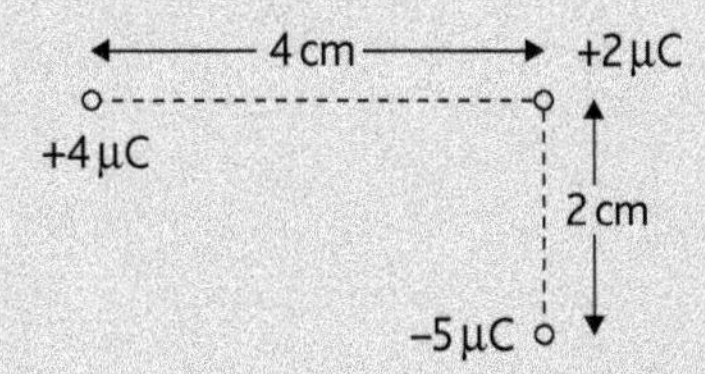

Calculate:

i the force acting on the +2 μC charge due to the +4 μC charge alone [2]

ii the force acting on the +2 μC charge due to the −5 μC charge alone [2]

iii the resultant force acting on the +2 μC charge due to the +4 μC charge and the −5 μC charge. [2]

3 **a** Explain what is meant by the terms *electric field* and *electric field strength*. [2]

b Sketch the field around two similar positive charges a distance *d* apart. [2]

c The diagram below shows two point charges A and B.

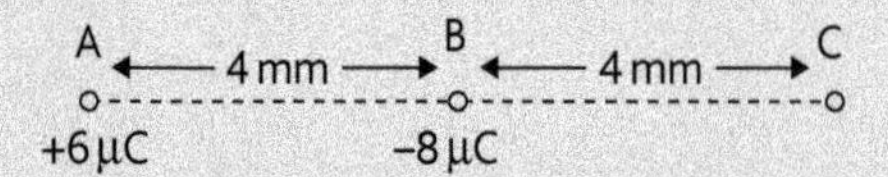

Calculate:

i the magnitude of the force between A and B and state whether the force is attractive or repulsive [3]

ii the electric field strength at the point C due to A alone [2]

iii the electric field strength at the point C due to B alone [2]

iv the electric field strength at the point C due to A and B [3]

v the electric potential at C due to A alone [2]

vi the electric potential at C due to B alone [2]

vii the electric potential at C due to A and B. [2]

4 **a** Two charges A and B are separated by a distance of 0.80 nm in a vacuum in the figure shown below. A has a charge of $+4.8 \times 10^{-19}$ C and B has a charge of -1.6×10^{-19} C. Calculate the force exerted by A on B. [3]

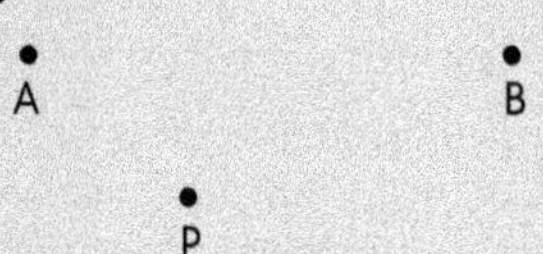

b Sketch the field lines between A and B, including the direction of the field. [3]

c Make a sketch of the points A, B and P. Use an arrow to show:

i the electric field strength at the point P due to the charge A only [1]

ii the electric field strength at the point P due to charge B only [1]

iii the resultant electric field strength at P due to both charges. [2]

5 **a** Define the term *capacitance*. [1]

b A capacitor with a capacitance of 2200 μF is charged until the potential difference between its plates is 5.0 V. Determine:

i the charge on one of the plates [2]

ii the energy stored in the capacitor. [2]

6 A capacitor is marked as having a capacitance of 220 μF. There is another marking indicating a voltage of 18 V.

a Explain what is meant by *a capacitance of 220 μF*. [1]

b Calculate the charged stored on the capacitor when a p.d. of 18 V is applied across it. [1]

c State the maximum charge that can be safely stored by this capacitor. [1]

d Calculate the energy stored in the capacitor. [2]

e State one reason why a maximum voltage is marked on a capacitor. [1]

7 a Define the terms *capacitance* and *farad*. [2]

b A capacitor of capacitance 12 μF is fully charged from a 20 V d.c. supply.

i Calculate the charge stored by the capacitor. [2]

ii Calculate the energy delivered by the d.c. supply. [1]

iii Calculate the energy stored in the capacitor. [2]

iv Account for the difference between your answers for **ii** and **iii**. [1]

c A 12 μF capacitor is charged from a 20 V d.c. supply through a resistor of 1.8 kΩ.

i Calculate the time constant for this circuit. [1]

ii Given that the capacitor is initially uncharged, calculate the time taken for the potential difference across the capacitor to reach 95% of its final value. [3]

8 a Derive an expression for the total capacitance for two similar capacitors connected in series. [4]

b A parallel-plate, air-filled capacitor consists of two plates, each having an area of $4.5 \times 10^{-3}\,\text{m}^2$. The separation of the plates is 1.2 mm. The p.d. across the plates of the capacitor is 450 V. Calculate:

i the capacitance of the capacitor [3]

ii the energy stored in the capacitor. [3]

9 a A parallel plate capacitor has a constant electric field strength of E between the plates. The area of each plate is A and the separation of the plates is d.

Write an expression for:

i the capacitance C of the capacitor in terms of A and d [1]

ii the energy per unit volume P of the capacitor in terms of E. [3]

b Each plate of a parallel-plate capacitor has an area of 18.0 cm². The separation between the plates is 4.0 mm. The electric field strength between the plates is $2.5 \times 10^6\,\text{V m}^{-1}$. Calculate:

i the capacitance of the capacitor [2]

ii the energy per unit volume P. [3]

10 A 6.0 μF capacitor is charged by a 12.0 V d.c. supply and is then discharged through a 1.8 MΩ resistor.

a Calculate the charge on the capacitor just before being discharged. [2]

b Calculate the time constant for the circuit. [1]

c After 6.0 s, what is:

i the charge on the capacitor [3]

ii the p.d. across the capacitor [3]

iii the current in the circuit? [3]

11 a State Coulomb's law for electrostatic charges. [3]

b Define the terms *electric field strength E* and *electric potential V*. [2]

c Write an equation to show the relationship between E and V. [2]

d Two point charges P and Q are situated in a vacuum. The charges are 18 cm apart. P has a charge of +45 μC and Q has a charge of −15 μC. The point X lies on the midpoint of a line joining P and Q.

Calculate:

i the electric field strength at the point X due to P only [2]

ii the electric field strength at the point X due to Q only [2]

iii the electric field strength at the point X due to P and Q [3]

iv the electric potential at the point X due to P only [2]

v the electric potential at the point X due to Q only [2]

vi the electric potential at the point X due to P and Q. [2]

12 Two horizontal metal plates are situated 1.5 cm apart. The electric field between the plates is $2.5 \times 10^4\,\text{N C}^{-1}$. Calculate:

a the potential difference between the plates [2]

b the acceleration of an electron between the plates, assuming there is a vacuum between them. [4]

13 An isolated metal sphere has a radius of 20 cm. All the charge on the surface may be considered as though it were concentrated at the centre of the metal sphere. The air around the metal sphere begins to conduct electricity when the electric field is greater than $2 \times 10^4\,\text{V cm}^{-1}$. Suppose that a spark is about to leave the sphere.

Calculate:

a the charge on the sphere [3]

b the potential of the sphere. [3]

14 a Define electric field strength. [1]

b Two flat parallel metal plates, each of length 10.0 cm, are separated by a distance of 1.2 cm. The potential difference between the plates is 200 V. An electron travels parallel to the plates along a line midway between the plates with a speed of $4.8 \times 10^7\,\text{m s}^{-1}$.

Calculate:

i the magnitude of the electric field strength between the plates [2]

ii the magnitude of the acceleration [4]

iii the time taken for the electron to travel the length of the plate (10.0 cm). [2]

15 Two flat metal plates, each of length 4.5 cm, are separated by a distance of 1.2 cm. A potential difference of 100 V between the plates provides a uniform electric field in the region between the plates. Electrons of speed $3.0 \times 10^7\,\text{m}\,\text{s}^{-1}$ enter this region at right angles to the field. Assume that the space between the plates is a vacuum and the field outside the plates is zero.

Calculate:

a the electric field strength between the plates [2]

b the force on an electron due to the electric field [2]

c the acceleration of the electron along the direction of the electric field [2]

d the time taken for an electron to travel between the plates [2]

e the speed of the electrons at right angles to its original direction of motion as it leaves the region between the plates [2]

f the velocity of an electron as it leaves the region between the plates. [2]

(Charge on an electron = -1.6×10^{-19} C, mass of an electron = 9.11×10^{-31} kg)

16 a Determine the equivalent capacitance of the network of capacitors in the circuit shown below. [4]

b What is the total charge stored by the network of capacitors? [2]

c What is the charge stored in the 6 μF capacitor? [2]

d What is the potential difference across the 6 μF capacitor? [2]

e What is the charge stored in the 14 μF capacitor? [2]

f What is the total energy stored by the network of capacitors? [2]

17 a State one function of a capacitor in simple circuits. [1]

b A capacitor is charged to a potential difference of 12 V and then connected in series with a switch, a resistor of resistance 10 kΩ and a sensitive ammeter as shown below.

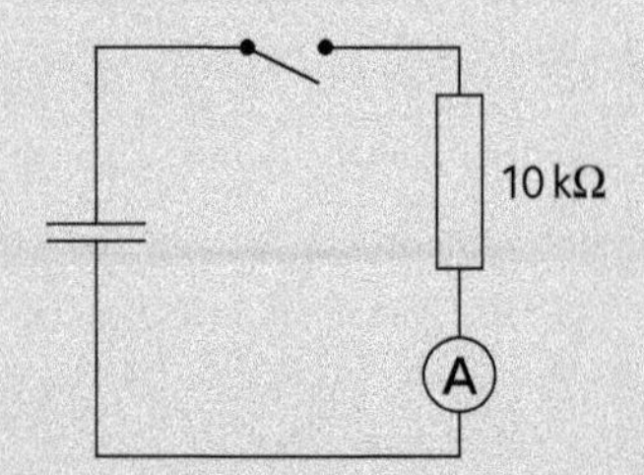

The switch is closed and the variation with time t of the current I in the circuit is shown below.

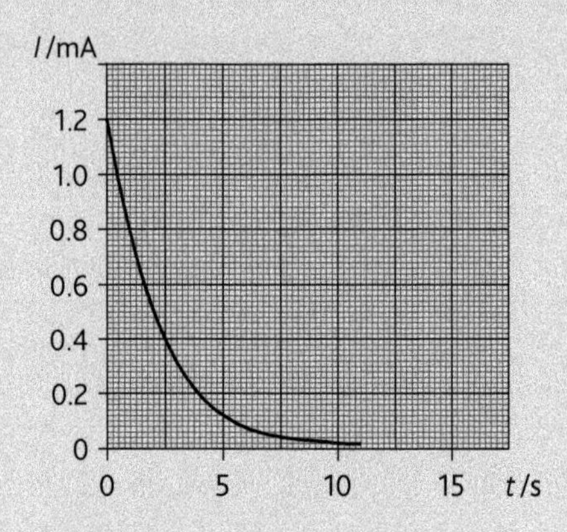

i State the relationship between the current in a circuit and the charge that passes a point in the circuit. [1]

ii The area under the graph above represents charge. Use the graph to estimate the initial charge stored in the capacitor. [4]

iii Calculate the capacitance of the capacitor. [2]

iv What is the time constant for the discharge circuit? [2]

v If the capacitor had discharged to one half of its initial energy, calculate the potential difference across the capacitor. [3]

18 An isolated conducting sphere of radius r is given a charge $+Q$. This charge may be assumed to act as a point charge situated at the centre of the sphere as shown below.

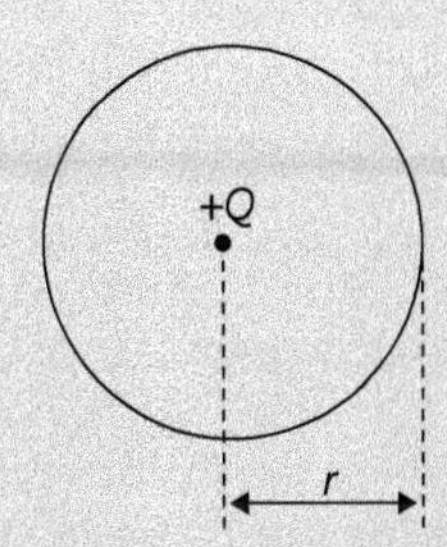

a Sketch a graph to show the variation with distance x from the centre of the sphere of the potential V due to the charge $+Q$. [2]

b State the relationship between electric field strength and potential. [1]

c Sketch a graph to show the variation with distance x of the electric field strength E due the charge $+Q$. [3]

19 An electron is accelerated from rest between a negative cathode and a positive anode in a vacuum. The potential difference between the anode and the cathode is 1.5 kV and their separation is 80 cm.

Calculate:

a the electric field strength E between the anode and cathode [2]

b the kinetic energy of the electron when it reaches the anode [2]

c the speed of the electron when it reaches the anode. [2]

20 Electrons in a cathode ray tube leave the cathode with negligible speed at a potential of −8 kV and are accelerated to an anode at a potential of −300 V. For an electron in this tube, calculate:

a the gain in electrical potential [1]

b the loss in potential energy [1]

c the gain in kinetic energy [2]

d the speed on reaching the anode. [2]

21 An oil droplet has a charge of 4.8×10^{-19} C. It is situated in a uniform electric field of strength 4.5×10^{5} V m^{-1}. Calculate the force experienced by the oil droplet. [2]

22 Two capacitors of capacitances 30 μF and 50 μF are connected in series with a 12 V supply. The capacitors are initially uncharged. Calculate:

a the total capacitance in the circuit [2]

b the charge delivered by the power supply [2]

c the charge stored in each capacitor [2]

d the potential difference across the 30 μF capacitor. [2]

6 Magnetic fields

6.1 Magnetic fields

Learning outcomes

On completion of this section, you should be able to:

- understand the concept of a magnetic field
- define the direction of a magnetic field
- sketch the flux patterns due to a long straight wire, a flat circular coil and a long solenoid.

Magnetic flux

Materials such as iron and steel are said to be **ferromagnetic**. These materials exhibit magnetic properties. A magnet is able to attract a paper clip at a distance. Experiments have shown that a bar magnet consists of two types of **magnetic poles**. There is a north pole and a south pole. Experiments have also shown that like poles repel each other and unlike poles attract each other.

The Earth behaves like a weak magnet. If a bar magnet is suspended from a string, it will align itself with the Earth's magnetic field.

The region around a magnet where a magnetic force is experienced is called a **magnetic field**. Lines are drawn around the magnet to represent the field. These lines are often referred to as lines of **magnetic flux**. The direction of the field line at a point in the field is the direction of the force on a north pole placed at that point. The closer the field lines are, the stronger the field.

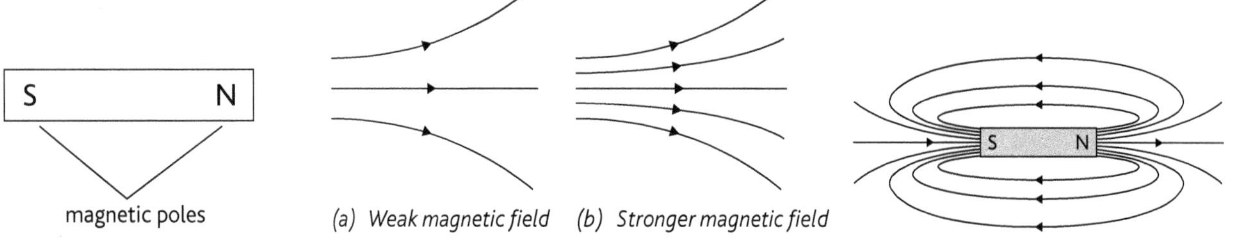

(a) Weak magnetic field *(b) Stronger magnetic field*

Figure 6.1.1 *Poles of a magnet* **Figure 6.1.2** *Magnetic flux* **Figure 6.1.3** *Field around a bar magnet*

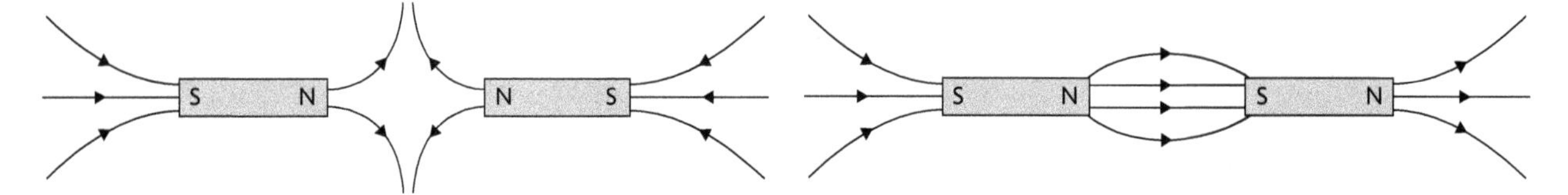

Figure 6.1.4 *Field between two north poles* **Figure 6.1.5** *Field between a north pole and a south pole*

Magnetic field produced by a current-carrying conductor

An electric current can also produce a magnetic field. The larger the current, the stronger is the field produced. The field around a long straight wire carrying a current is shown in Figure 6.1.6.

The direction of the field produced by the current flowing in the wire can be determined by using:

- Maxwell's corkscrew rule – imagine a corkscrew being driven into a cork on a wine bottle. The direction of the current is imagined as moving in the direction of the corkscrew. The direction of the field lines is the direction of the turning action of the corkscrew.
- Right-hand grip rule – imagine your right hand gripping a length of wire. Your thumb points in the direction of the current. The remaining fingers point in the direction of the field (Figure 6.1.7).

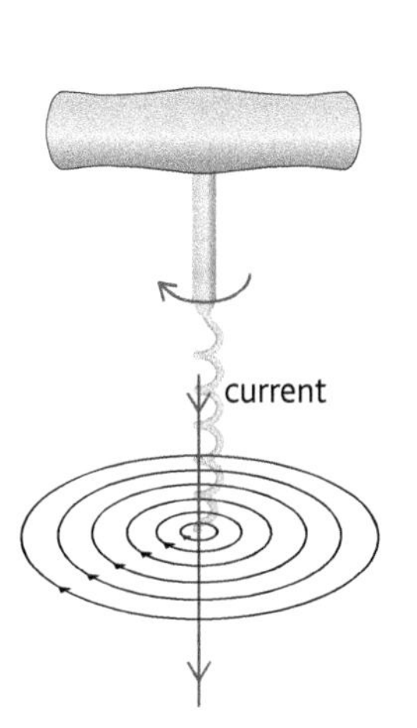

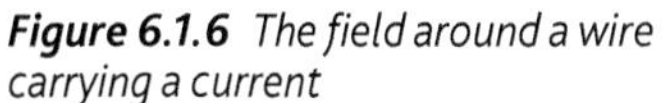

Figure 6.1.6 *The field around a wire carrying a current*

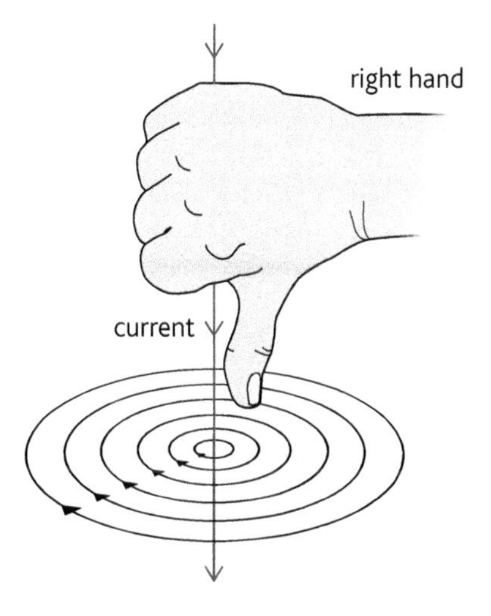

Figure 6.1.7 *Fleming's right-hand grip rule*

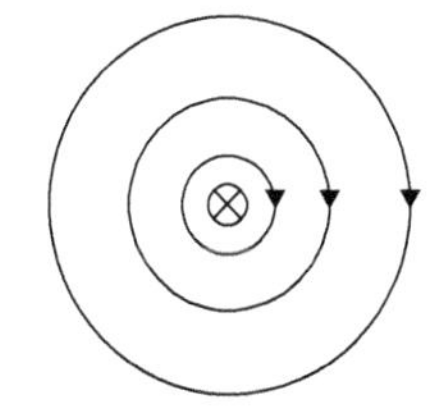

(a) Current flowing into the plane of the paper

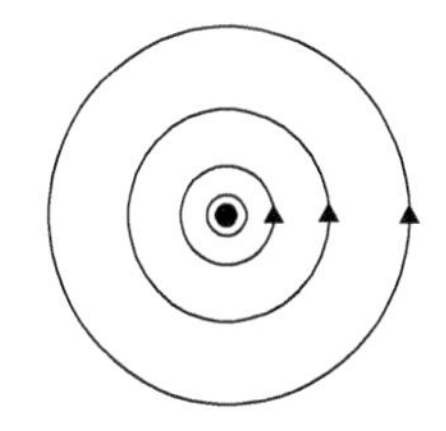

(b) Current flowing out of the plane of the paper

Figure 6.1.8 *The effect of changing the direction of current flow*

Figure 6.1.8 shows the effect of changing the direction of the current on the magnetic field. The symbols for representing current into and out of the plane are also shown.

Magnetic field produced by a flat circular coil

Figure 6.1.9 shows the magnetic field produced by a flat circular coil.

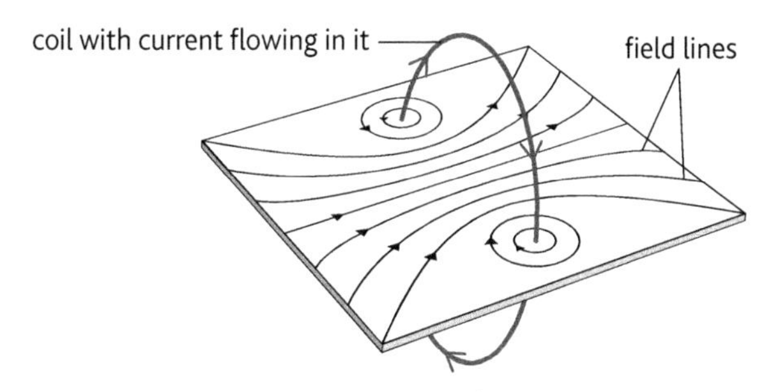

Figure 6.1.9 *Magnetic field produced by a flat circular coil*

Magnetic field produced by a long solenoid

Figure 6.1.10 shows the magnetic field produced by a long solenoid. The right-hand rule is used to determine the direction of the field. The right hand is used to grip the solenoid such that your four fingers point in the direction of the current flowing in the solenoid. The direction of the thumb points in the direction of the magnetic field produced by the solenoid.

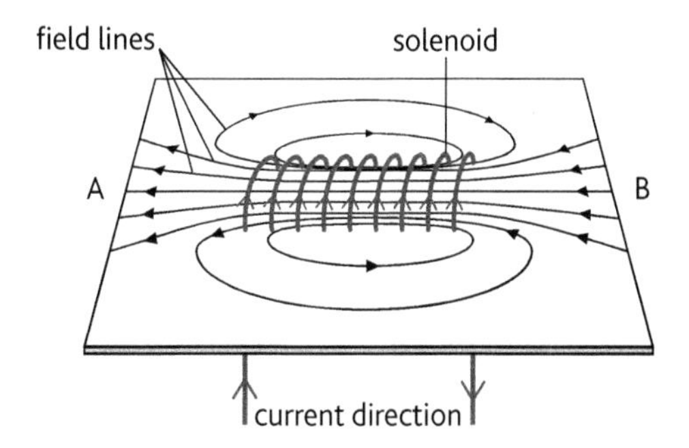

Figure 6.1.10 *Magnetic field produced by a long solenoid*

Key points

- A magnetic field can be produced by permanent magnets or current-carrying conductors.
- The region around a magnet where a magnetic force is experienced is called a magnetic field.
- Magnetic field lines are referred to as magnetic flux.
- The direction of the field line at a point in the field is the direction of the force on a north pole placed at that point.
- The field around a long straight wire is represented as concentric circles.
- The right-hand grip rule is used to determine the direction of a magnetic field.

6.2 Force on a current-carrying conductor

Learning outcomes

On completion of this section, you should be able to:

- appreciate that a force might act on a current-carrying conductor when placed in a magnetic field
- recall and use $F = BIl \sin\theta$
- understand how to use Fleming's left hand rule
- define magnetic flux density and the tesla.

Force acting on a current-carrying conductor placed in a magnetic field

When a current-carrying conductor is placed in a magnetic field it may experience a force. The magnetic field produced by the current flowing in the wire interacts with the external magnetic field to produce a force. Consider a current-carrying conductor placed at right angles to a magnetic field as shown in Figure 6.2.1. The permanent magnet produces a magnetic field. The current-carrying conductor also produces a magnetic field as shown in the diagram. These two fields interact to produce a downward force F acting on the wire. The direction of this force can be predicted using **Fleming's left hand rule**. Using the left hand:

- the index finger points in the direction of the magnetic field
- the second finger points in the direction of the current flowing in the conductor
- the thumb points in the direction of the force.

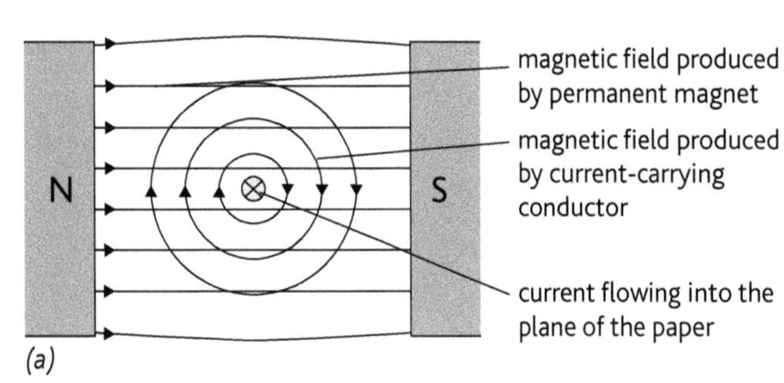

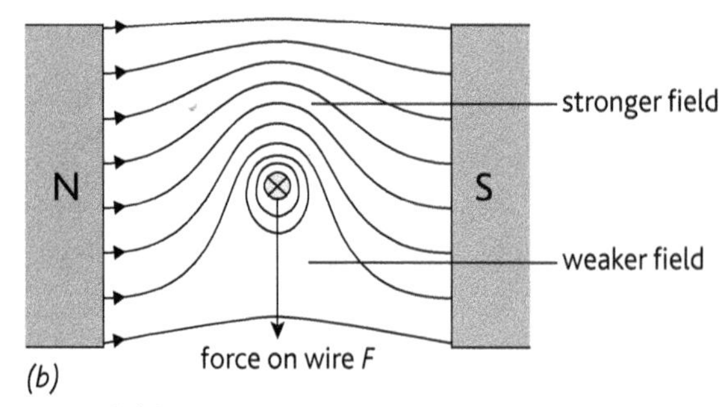

Figure 6.2.1 *The force acting on a current-carrying conductor placed in a magnetic field*

A magnetic field is represented by a vector quantity B. This is called the **magnetic flux density** and its SI unit is the **tesla** (T).

The magnitude of the force acting on a current-carrying conductor when placed in a magnetic field depends on the following:

- the magnitude of the current I flowing in the conductor
- the magnitude of the flux density B of the external magnetic field
- the length of the conductor l inside the external magnetic field
- the angle θ made between the current I and the direction of the magnetic field.

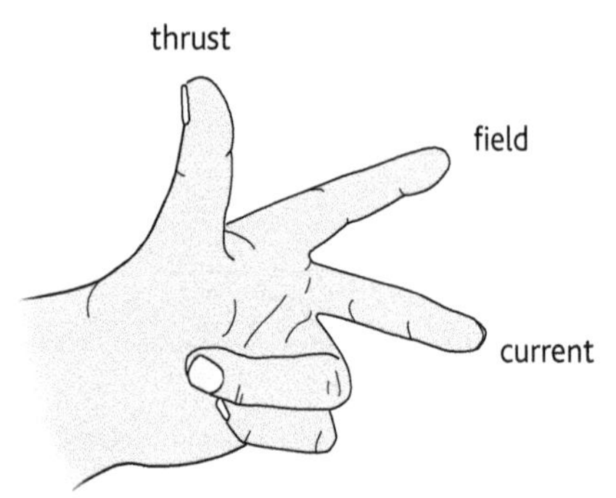

Figure 6.2.2 *Fleming's left hand rule*

Exam tip

The force F acts at right angles to the plane containing I and B.

Equation

$F = BIl \sin\theta$

F – force acting on conductor/N
B – magnetic flux density/T
I – current flowing in conductor/A
l – length of conductor in magnetic field/m
θ – angle made between the current and magnetic field

Figure 6.2.3 shows how the magnitude of the force acting on the conductor is affected by its orientation in the magnetic field.

Example

A conductor of length 1.5 m, carrying a current of 8.0 A, is placed in a magnetic field of flux density 0.12 T. Calculate the force acting on the wire when it is placed:

a at right angles to the magnetic field

b along the direction of the field

c at an angle of 30° to the field.

a $F = BIl\sin\theta = 0.12 \times 8.0 \times 1.5\sin 90° = 1.44\,\text{N}$

b $F = BIl\sin\theta = 0.12 \times 8.0 \times 1.5\sin 0° = 0\,\text{N}$

c $F = BIl\sin\theta = 0.12 \times 8.0 \times 1.5\sin 30° = 0.72\,\text{N}$

Magnetic flux density

Magnetic flux density is numerically equal to the force per unit length on a straight conductor carrying unit current normal to the field.

1 tesla is the magnetic flux density of a field producing a force of 1 N per metre on a wire carrying a current of 1 A normal to the field.

Equation

The magnetic flux density at a distance r from a straight conductor carrying a current I is given by:

$$B = \frac{\mu_0 I}{2\pi r}$$

μ_0 – permeability of free space/H m^{-1}
I – current/A
r – perpendicular distance from conductor/m

In the equations shown, the constant μ_0 is called the **permeability of free space**. It is a measure of the ability of a medium to transmit a magnetic field.

Its value is $4\pi \times 10^{-7}\,\text{H m}^{-1}$. The unit is the henry per metre.

Equation

The magnetic flux density at the centre of a flat circular coil of radius r, carrying a current I is given by:

$$B = \frac{\mu_0 N I}{2r}$$

μ_0 – permeability of free space/H m^{-1}
N – number of turns in the coil
I – current/A
r – radius of coil/m

Equation

The magnetic flux density at the centre of a long solenoid having n turns per unit length and carrying a current I is given by:

$$B = \mu_0 n I$$

μ_0 – permeability of free space/H m^{-1}
n – number of turns per unit length/m^{-1}
I – current/A

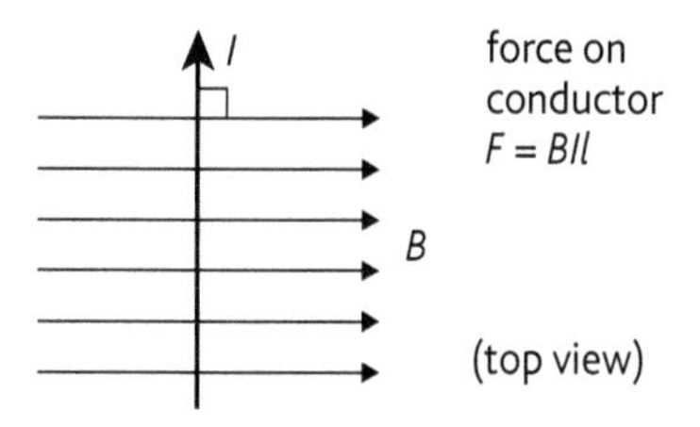

(a) Conductor is at right angles to the magnetic field

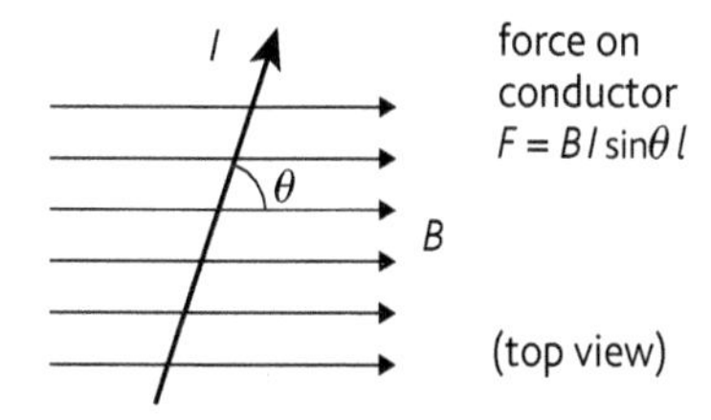

(b) Conductor makes an angle of θ with the magnetic field

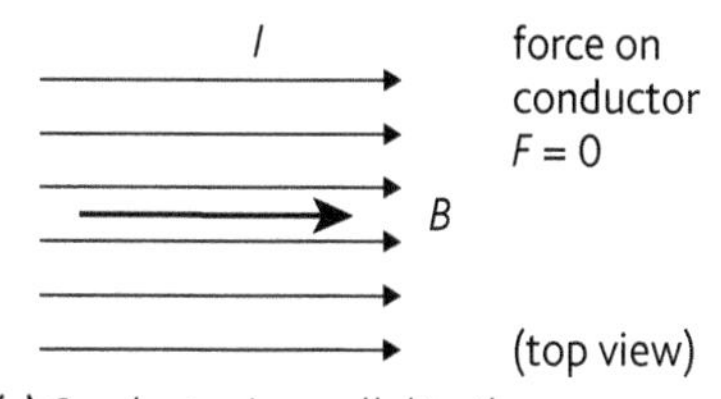

(c) Conductor is parallel to the direction of the magnetic field

Figure 6.2.3 *The effect of F when θ is changed*

Exam tip

If you cannot remember the definition for magnetic flux density, use the equation $F = BIl$ and state the meaning of the symbols.

Key points

- A current-carrying conductor may experience a force when placed in a magnetic field.
- Fleming's left hand rule is used to predict the direction of the force.
- Magnetic flux density is numerically equal to the force per unit length on a straight conductor carrying unit current normal to the field.
- 1 tesla is the magnetic flux density of a field producing a force of 1 N per metre on a wire carrying a current of 1 A normal to the field.

Force on a moving charge

Learning outcomes

On completion of this section, you should be able to:

- predict the direction of the force on a charge moving in a magnetic field
- recall and use $F = BQv\sin\theta$
- solve problems involving charged particles in mutually perpendicular electric and magnetic fields.

Equation

$F = BQv$

F – force on charge/N
B – magnetic flux density/T
Q – charge/C
v – speed/m s^{-1}

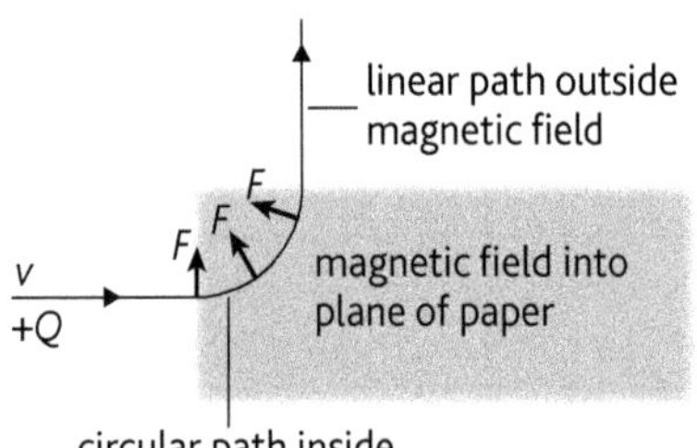

Figure 6.3.1 *Force acting on a positive charge travelling in a magnetic field*

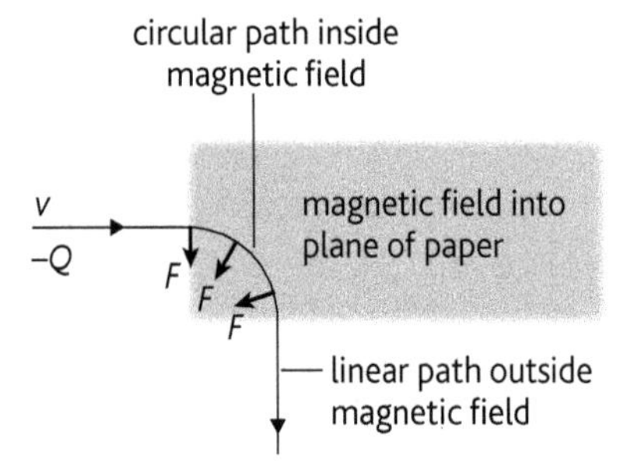

Figure 6.3.2 *Force acting on a negative charge travelling in a magnetic field*

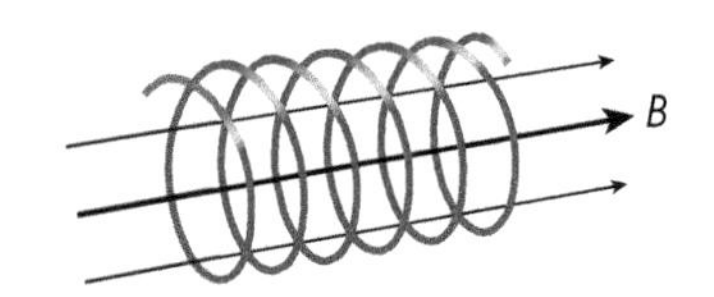

Figure 6.3.3 *Helical path*

Force acting on a charge moving in a magnetic field

An electric current was defined as the flow of charge. When a charged particle is in motion, it constitutes an electric current. If a positively charged particle is moving from left to right, the conventional current also flows from left to right. If, however, a negatively charged particle is moving from left to right, the conventional current flows from right to left.

An electric current produces a magnetic field. Therefore, the motion of a charged particle produces a magnetic field around itself. If a charged particle is moving in a magnetic field, it experiences a force because of the interaction of the two fields. A particle with charge Q, travelling with a speed v at right angles to a magnetic field experiences a force $F = BQv$.

Figures 6.3.1 and 6.3.2 show the effect of a magnetic field on the path of a positively charged and a negatively charged particle, respectively.

The force F acts at right angles to the plane containing v and B. The force changes the direction of the charged particle when it enters the magnetic field. The magnitude of v is unchanged because F does not act in the direction of motion of the charged particle. Since F is at right angles to v, no work is done by the force. This means that no energy is gained by the charge when it enters the magnetic field. The direction of F is predicted using Fleming's left hand rule.

In Figures 6.3.1 and 6.3.2 the charged particle with a speed v travels at right angles to the magnetic field B. As soon as the particle enters the magnetic field, a force F is experienced. The force acts at right angles to the path of the particle, causing the particle to change direction. As a force of magnitude F acts continuously on the particle, the particle follows a curved path. The motion of the particle is similar to that of circular motion. The force F provides the necessary centripetal force needed to maintain the circular motion of the charged particle.

Equation

A particle with charge Q, travelling with a speed v at an angle of θ to the magnetic field experiences a force F given by:

$F = BQv\sin\theta$

F – force on charge/N
B – magnetic flux density/T
Q – charge/C
v – speed/m s^{-1}
θ – angle between v and B

The path of the particle is helical (Figure 6.3.3).

Charged particles from the Sun, on approaching the Earth, may become trapped in the Earth's magnetic field near the poles. The trapped charged particles cause the sky to glow. The phenomenon is called the aurora borealis.

The Van Allen radiation belt is a torus of highly charged particles around the Earth. These charged particles are held in two distinct belts by the Earth's magnetic field. The inner belt is made up of protons and electrons, while the outer belt is made up of energetic electrons. These radiation belts pose a challenge for orbiting satellites.

The effect of speed on the curvature of a charged particle in a magnetic field

The force acting on a particle of mass m, having a charge Q, travelling with a speed v at right angles to a magnetic field B is given by the following equation:

$$F = BQv$$

This force provides the centripetal force required for the particle to follow a curved path. If the radius of curvature is r, the centripetal force acting on the particle is given by the following equation:

$$F = \frac{mv^2}{r}$$

Therefore, $$\frac{mv^2}{r} = BQv$$

$$r = \frac{mv}{BQ}$$

The radius of curvature is directly proportional to the speed with which the charged particle is travelling. Consider a stream of electrons of varying speeds travelling at right angles to a uniform magnetic field. The magnetic field acts into the plane of the paper as shown in Figure 6.3.4.

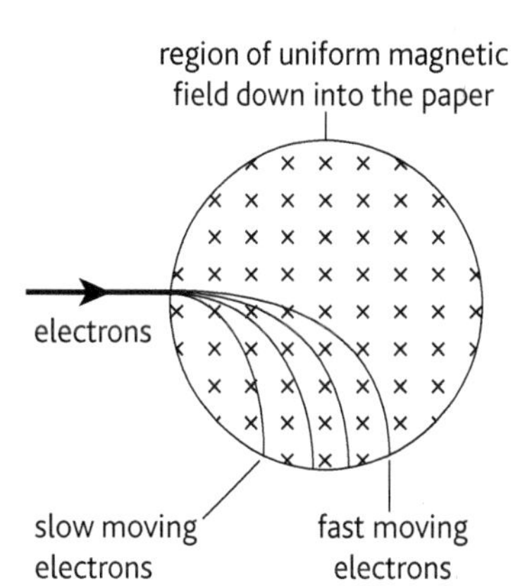

Figure 6.3.4 *Effect of speed on curvature*

Mutually perpendicular magnetic and electric fields

Consider a beam of particles having a charge $-Q$ travelling with a speed v through a region of space where the electric field strength is E. The force acting on the charged particles is given by $F = EQ$. The charged particles experience an upward force (Figure 6.3.5).

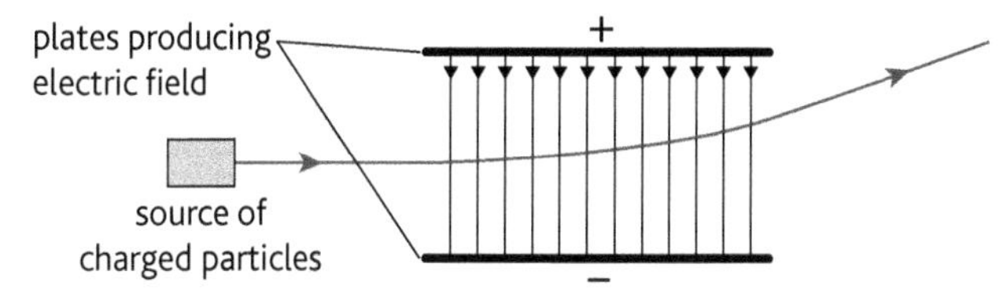

Figure 6.3.5 *Electric force*

Suppose a uniform magnetic field of magnetic flux density B is applied at right angles to the electric field. The magnetic field is in the plane of the page. The force acting on the charged particles is given by $F = BQv$ and acts downwards. In order for the charged particles to pass through both fields undeviated, the electric force and the magnetic force must be equal (Figures 6.3.6 and 6.3.7).

$$EQ = BQv$$

Therefore, $$\frac{E}{B} = v$$

Mutually perpendicular electric and magnetic fields can be used for velocity selection in a mass spectrometer.

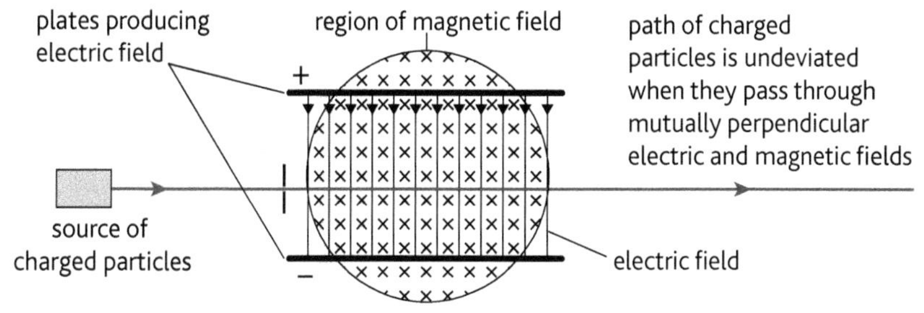

Figure 6.3.6 *Electric field and magnetic field at right angles to each other*

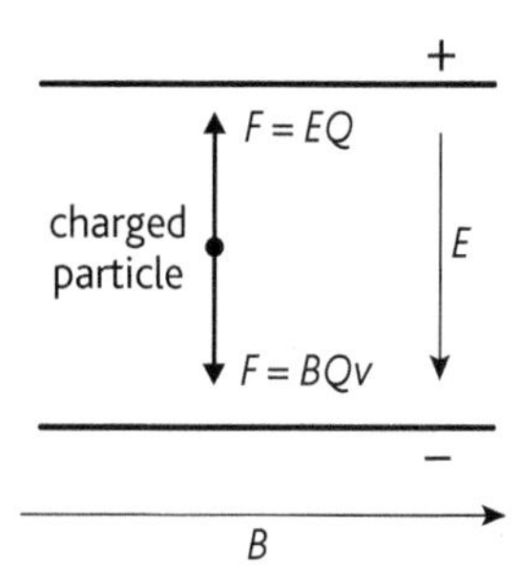

Figure 6.3.7 *Electric force and magnetic force are equal and opposite*

Key points

- A charged particle travelling at right angles to a magnetic field experiences a force at right angles to its motion.
- The path taken by the particle is circular.
- The radius of curvature is proportional to the speed of the particle.
- The speed of the particle does not change when in the magnetic field.
- If the particle is travelling such that it makes an angle θ with the magnetic field B, it travels in a helical path.
- Mutually perpendicular magnetic and electric fields can be used for velocity selection of charged particles.

6.4 Measuring magnetic flux density

Learning outcomes

On completion of this section, you should be able to:

- explain the Hall effect
- use a Hall probe to measure magnetic flux density
- describe the principle of operation of a current balance to measure magnetic flux density.

Definition

Hall effect

A potential difference is set up transversely across a current-carrying conductor when a perpendicular magnetic field is applied.

The Hall effect

Consider a metal conductor in which a small current I is flowing as shown in Figure 6.4.1. Inside the metal conductor there are free mobile electrons. The conventional current flows to the left. This means that the electrons are flowing from left to right. A magnetic field of flux density B acts at right angles to the face PQRS. The electrons have a mean drift velocity of v. Consider the force acting on a single electron. According to Fleming's left hand rule, the electron will experience a force F acting downwards. This force is at right angles to the plane containing v and B. The force F is given by:

$$F = Bev$$

where B is the magnetic flux density and e is the charge on an electron and v is the mean drift velocity on an electron.

Electrons begin collecting along the side RS, which therefore becomes negatively charged with respect to the side PQ. An electric field is therefore set up between sides RS and PQ. This creates a potential difference V_H between the sides RS and PQ. The potential difference is called the **Hall voltage**.

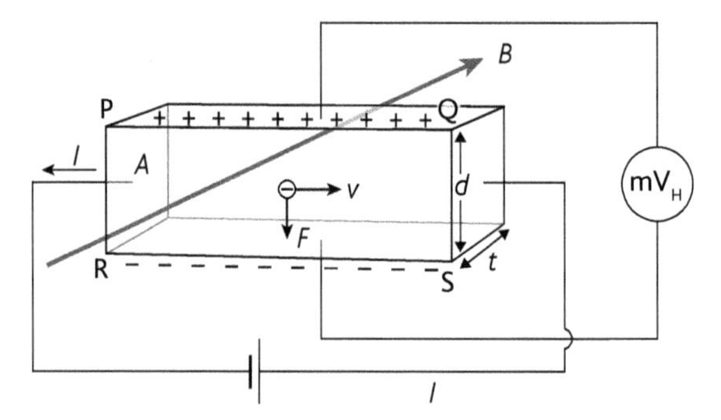

Figure 6.4.1 *Explaining the Hall effect*

The electric field strength is given by $E = \frac{V_H}{d}$. This causes an upward force Ee to act on the electron. As more and more electrons collect on the side RS, the electric field strength E increases. The voltage V_H becomes steady when the two forces acting on the electrons cancel each other.

Therefore, $\quad Bev = Ee$

But $\quad E = \frac{V_H}{d}, \quad$ so $\quad Bev = \frac{V_H}{d}e \quad$ and $\quad V_H = Bvd$

But $I = nevA$, where I is the current flowing in the conductor, n is number of electrons per unit volume, e is the charge on an electron, v is the mean drift velocity of electrons, and A is the cross-sectional area of the conductor.

So, $\quad v = \frac{I}{neA}$

Therefore, the Hall voltage is given by $V_H = \frac{BId}{neA}$

The cross-sectional area of the conductor is given by $A = dt$

$$\therefore \quad V_H = \frac{BI}{net}$$

The Hall probe

A device called a Hall probe is used to measure magnetic flux density. The Hall probe makes use of the Hall effect.

The Hall probe is made of a thin slice of a semiconductor. Semiconductors are used because the charge carriers inside them have larger drift velocities than those in pure metals. Since they have larger velocities, the measured Hall voltage is larger than that of a pure metal.

In order to measure magnetic flux density, the Hall probe is placed so that the semiconductor material is at right angles to the magnetic field. A small current I is passed through the probe and the Hall voltage is measured (Figure 6.4.2). The manufacturer of the probe normally supplies the value of net. The magnetic flux density B is calculated using the following equation:

$$B = \frac{V_H net}{I}$$

Note that the Hall probe is calibrated by measuring the Hall voltages in known magnetic fields.

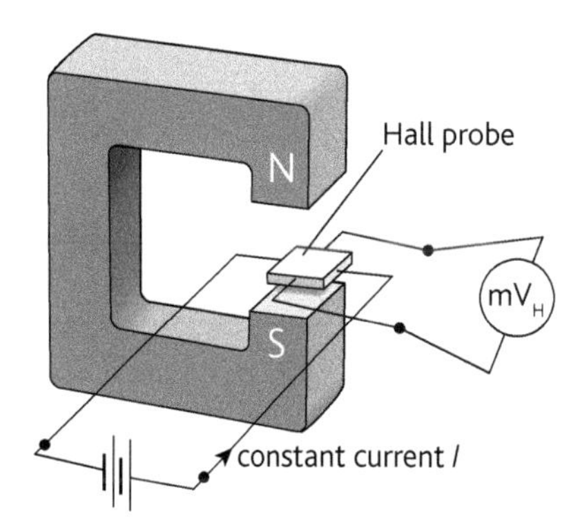

Figure 6.4.2 *Using a Hall probe to measure B*

The current balance

A current balance (Figure 6.4.3) is used to measure magnetic flux density. It consists of a wire frame ABCD pivoted about a horizontal axis (PQ). A current I enters one end of the frame through side AD and leaves through side BC.

Suppose the magnetic flux density at the centre of a solenoid needs to be measured. The side of the wire frame AB is placed such that it is at right angles to the magnetic field produced by the solenoid. The direction of the current I flowing through the frame ABCD is adjusted so that a downward force F acts on the side AB. A small mass (rider) of weight mg is adjusted so that the frame ABCD remains horizontal. This is verified by checking to see if the pointer lines up with the zero mark on the scale.

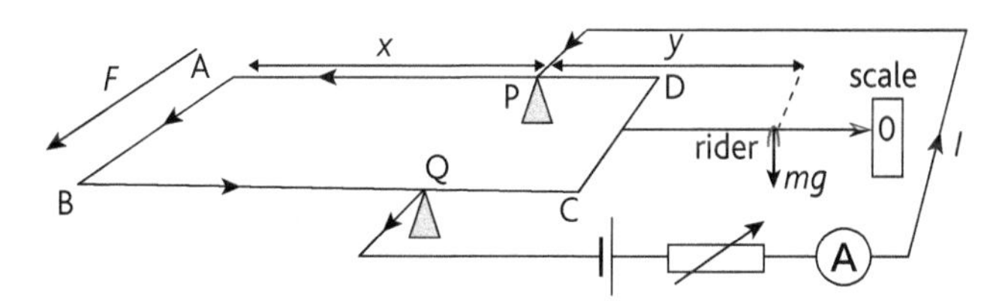

Figure 6.4.3 *A simple current balance*

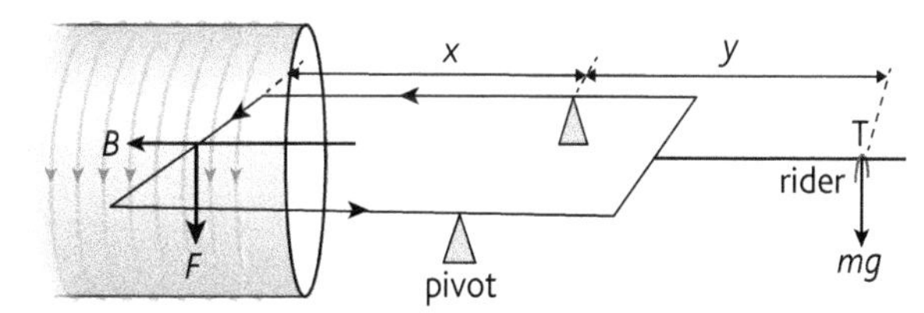

Figure 6.4.4 *Using a current balance to measure B*

The principle of moments is then applied to determine the magnitude of F.

$$F \times x = mg \times y$$

$$F = \frac{mgy}{x}$$

If the length of the side AB is l, then the magnetic flux density B is determined as follows:

Force acting on side AB is $F = BIl$

Therefore, magnetic flux density $B = \frac{F}{Il} = \frac{mgy}{xIl}$

Figure 6.4.4 shows how a current balance is used to measure the magnetic flux density inside a solenoid.

Key points

- In the Hall effect a potential difference is set up transversely across a current-carrying conductor when a perpendicular magnetic field is applied.
- A calibrated Hall probe and a current balance can be used to measure magnetic flux density.

6.5 Force between current-carrying conductors

Learning outcomes

On completion of this section, you should be able to:

- explain the forces between current-carrying conductors
- predict the direction of the forces between current-carrying conductors.

The force between current-carrying conductors

When two straight current-carrying conductors are placed parallel to each other a force is experienced between them. If the current flows in the same direction in the two conductors, the force is attractive. If the currents in the conductors flow in opposite directions, the force is repulsive. Figures 6.5.1 and 6.5.2 show the effect of passing a current through a thin sheet of aluminium foil.

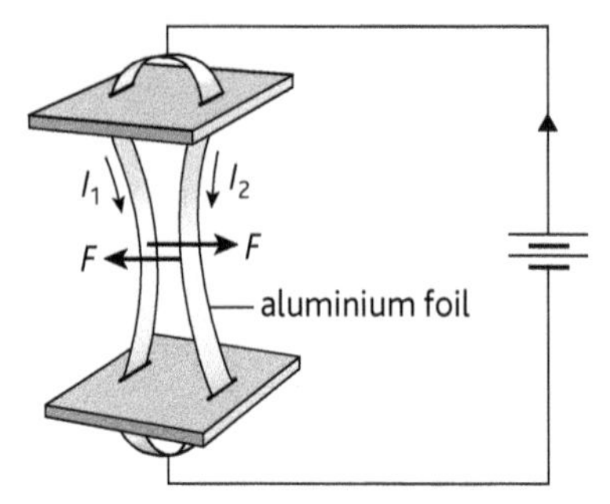

Figure 6.5.1 *Currents flowing in the same direction*

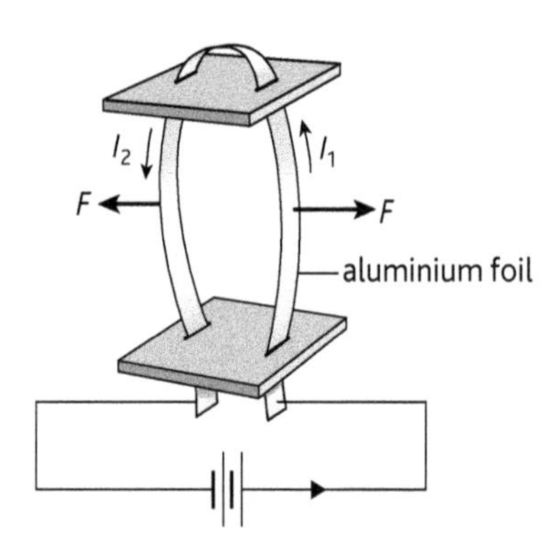

Figure 6.5.2 *Currents flowing in opposite directions*

Explaining the experiment

Currents flowing in the same direction

Figure 6.5.3 shows two wires X and Y that are at right angles to the plane of the paper. They carry currents I_1 and I_2 into the plane of the paper and they are separated by a distance r as shown.

Since wire X is carrying an electric current I_1, a magnetic field is created around it. This magnetic field is represented by concentric circles around the wire. The second wire Y is situated in this magnetic field. Since wire Y is carrying a current I_2, it experiences a force F as shown. The direction of F can be determined by Fleming's left hand rule. According to Newton's third law, an equal and opposite force will act on wire X.

Magnetic field produced by I_1 around wire X, $B_1 = \dfrac{\mu_0 I_1}{2\pi r}$

The force exerted on a section l of the wire Y, $F = BIl = \dfrac{\mu_0 I_1}{2\pi r} I_2 l$

The force per unit length of wire which I_1 causes on wire Y $= \dfrac{F}{l} = \dfrac{\mu_0 I_1 I_2}{2\pi r}$

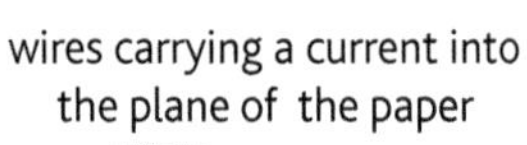

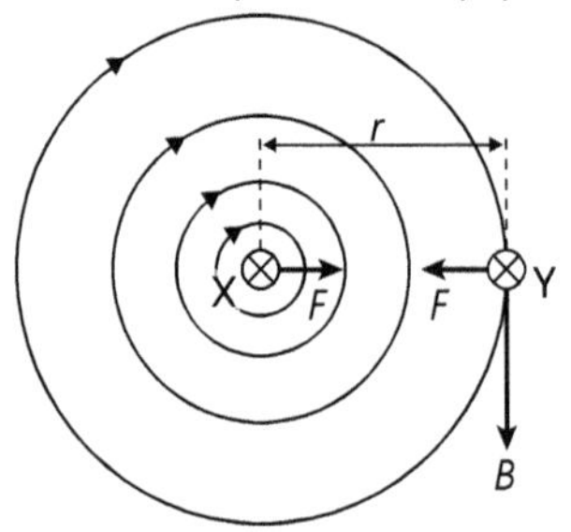

Figure 6.5.3 *Two currents acting in the same direction*

Currents flowing opposite directions

Figure 6.5.4 shows two wires X and Y which are at right angles to the plane of the paper, carrying currents I_1 and I_2. I_1 is flowing into the plane of the paper and I_2 is flowing out of the plane of the paper.

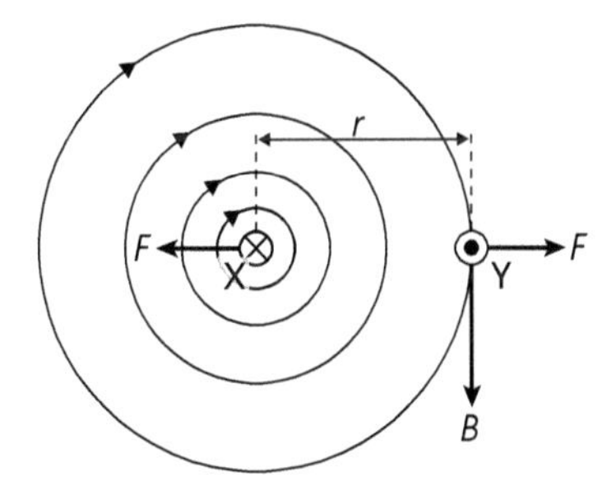

Figure 6.5.4 *Two currents acting in opposite directions*

Since wire X is carrying a current I_1, a magnetic field is created around it. This magnetic field is represented by concentric circles around the wire. The second wire Y is situated in this magnetic field. Since wire Y is carrying a current I_2, it experiences a force F whose direction, by Fleming's left hand rule, is as shown. According to Newton's third law, an equal and opposite force will act on wire X.

Magnetic field produced by I_1 around wire X, $B_1 = \frac{\mu_0 I_1}{2\pi r}$

The force exerted on a section l of the wire Y, $F = BIl = \frac{\mu_0 I_1}{2\pi r} I_2 l$

The force per unit length of wire which I_1 causes on wire Y $= \frac{F}{l} = \frac{\mu_0 I_1 I_2}{2\pi r}$

Example

Two long wires are mounted vertically and are 4 cm apart. A current of 2.8 A flows through each wire. Calculate the force acting on a 20 cm length of one of the wires.

Magnetic flux density produced by one of the wires:

$$B = \frac{\mu_0 I_1}{2\pi r} = \frac{4\pi \times 10^{-7} \times 2.8}{2\pi(4 \times 10^{-2})} = 1.4 \times 10^{-5}\,\text{T}$$

Force acting on a 20 cm length of wire:

$$F = BIl = 1.4 \times 10^{-5} \times 2.8 \times 20 \times 10^{-2} = 7.84 \times 10^{-6}\,\text{N}$$

Key points

- When two straight current-carrying conductors are placed parallel to each other, a force is experienced.
- If the currents are flowing in the same direction, the force is attractive.
- If the currents are flowing in opposite directions, the force is repulsive.

6.6 The electromagnet

Learning outcomes

On completion of this section, you should be able to:

- describe the principle of the electromagnet and appreciate its uses.

The electromagnet

An electromagnet makes use of the magnetic effect of a current. When a current flows through the solenoid shown in Figure 6.6.1, a magnetic field is produced. When a ferromagnetic material such as iron is introduced inside the solenoid, there is a concentration of magnetic field lines. The result is in an increase in the magnetic flux density. This is the principle by which an electromagnet works.

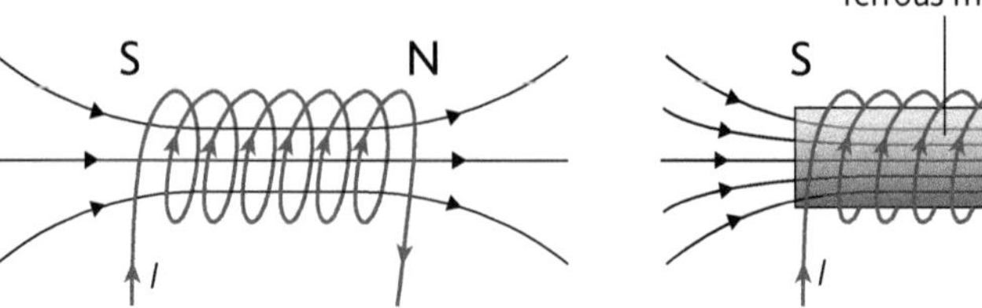

Figure 6.6.1 The magnetic field produced by a solenoid

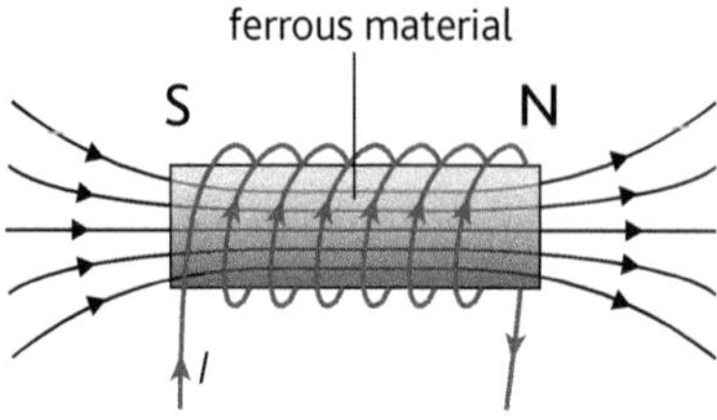

Figure 6.6.2 The effect of introducing a soft iron core

Soft iron is used to construct electromagnets because:

- it has a high permeability
- it is easily magnetised and demagnetised
- it is able to concentrate magnetic field lines.

Recall that the magnetic flux density at the centre of a solenoid is given by $B = \mu nI$. The relative permeability of iron μ_r is approximately 2000. Therefore, when a ferromagnetic material such as iron is introduced inside an air-filled solenoid the magnetic field density increases significantly. This can be compared with the effect of inserting a dielectric (e.g. ceramic) between the plates of an air-filled capacitor. The relative permittivity ε_r of ceramic can be as high as 3000. Recall that the capacitance of a capacitor is given by $C = \frac{\varepsilon A}{d}$.

The permeability of a material is a measure of its ability to transmit a magnetic field, whereas the permittivity of a material is a measure of its ability to transmit an electric field.

Electromagnets have many different practical applications:

- Magnetic resonance imaging (MRI)
- CAT scanners
- Used in the transportation industry to levitate trains
- Circuit breakers
- Scrapyards to separate and move metallic objects
- Magnetic door locks
- Relays
- Electric bells

Magnetic door locks

Electromagnets are used in modern locking systems. Powerful electromagnets are used as a locking mechanism for doors. A metal plate is attached to the door and the electromagnet is attached to the door frame. When the electromagnet is energised, the metal plate makes contact with one pole of the electromagnet. This type of operation is seen in many stores. A customer presses a button that alerts someone inside the store. The person inside the store presses a button to de-energise the electromagnet and release the door.

Relays

A magnetic relay is shown in Figure 6.6.3. A magnetic relay is a device that uses an electromagnet in one circuit to switch on a secondary circuit. The relay consists of two separate circuits. When the switch is closed in the first circuit a current flows through the solenoid and the soft iron core is magnetised. The soft iron armature is attracted to the electromagnet. The upper end of the armature moves upwards and closes the contacts in the second circuit. The main advantage of using a magnetic relay is that it allows a small current circuit to be used to switch on a large current circuit.

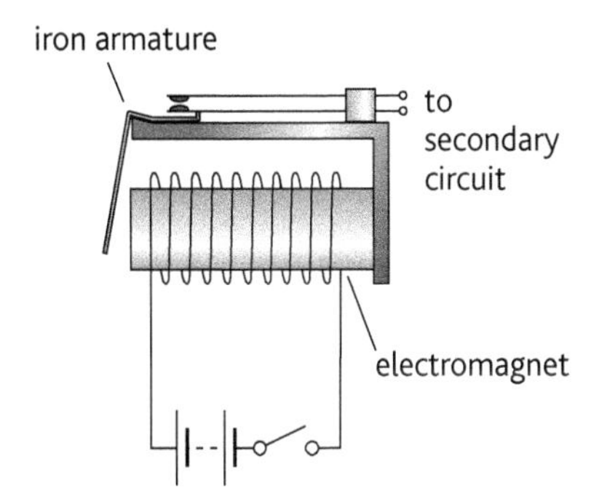

Figure 6.6.3 *A magnetic relay*

Electric bells

An electric bell is shown in Figure 6.6.3. When the switch is pressed, a current starts flowing and the electromagnet become magnetised. The electromagnet attracts the soft iron armature. The bell is struck by the hammer in this process and the electric circuit is broken. The electromagnet is demagnetised and the springy metal strip causes the armature to return to its original position. The circuit is closed again and the process repeats itself.

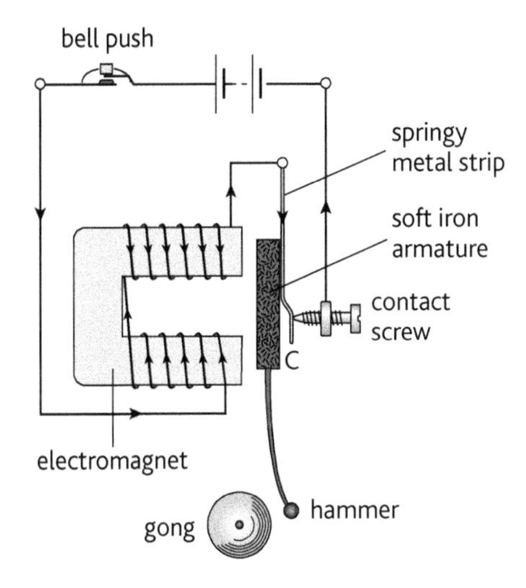

Figure 6.6.4 *An electric bell*

Key points

- An electromagnet makes use of the magnetic effect of a current.
- When a ferrous core is place inside a solenoid, it greatly increases the field.
- Electromagnets can be turned on and off, unlike permanent magnets.
- Electromagnets have many uses, including MRI machines, CAT scanners, electric locks, door bells and magnetic relays.

7 Electromagnetic induction

7.1 Faraday's and Lenz's laws

Learning outcomes

On completion of this section, you should be able to:

- state Faraday's law of electromagnetic induction
- use Faraday's law to determine the magnitude of an induced e.m.f.
- state Lenz's law and determine the direction of an induced e.m.f.
- discuss Lenz's law as an example of conservation of energy.

Electromagnetic induction

Experiment 1

Consider a solenoid attached to a sensitive galvanometer as shown in Figure 7.1.1.

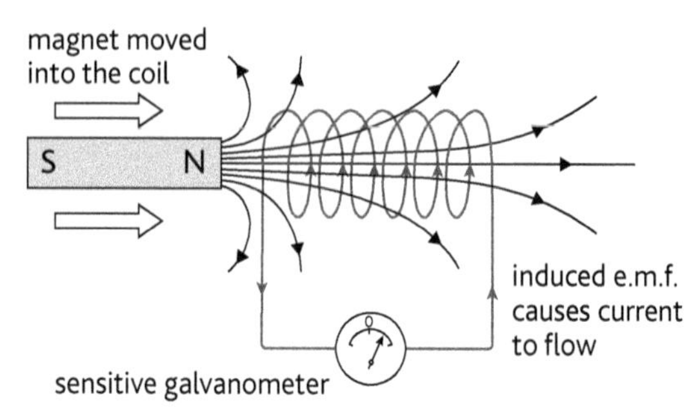

Figure 7.1.1

As the bar magnet is slowly moved towards the coil, a deflection is seen on the galvanometer. If the motion of the magnet stops, the deflection goes to zero. As the magnet passes through the coil and exits on the right, a deflection in the opposite direction is seen on the galvanometer.

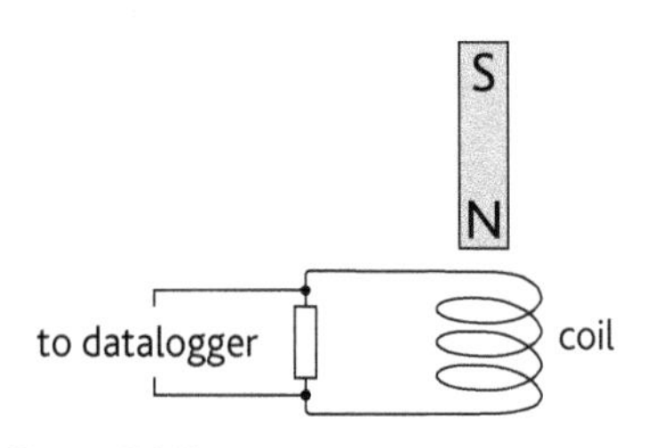

Figure 7.1.2

Experiment 2

Consider another experiment in which a solenoid is connected to a resistor of known resistance (Figure 7.1.2).

A datalogger is connected across the resistor. The datalogger takes samples of the voltage across the resistor over a period of time. Since the resistance is known, the current in the circuit can be determined. The bar magnet is dropped from a height above the solenoid. The magnet passes through the solenoid and exits through the bottom. Figure 7.1.3 shows the variation of the current in the resistor with time as the magnet passes through the solenoid.

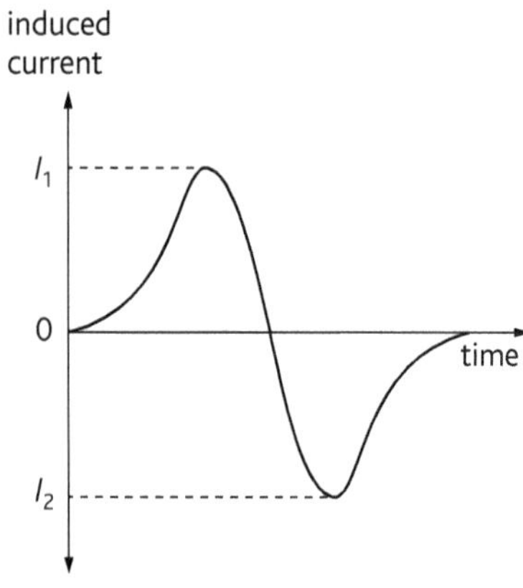

Figure 7.1.3

Experiment 3

In this experiment a galvanometer is connected to the ends of a stiff piece of copper wire AB. The wire is moved up and down between the poles of two magnets as shown in Figure 7.1.4. As the wire moves downwards the galvanometer deflects in one direction. When the wire is moved upwards the galvanometer deflects in the opposite direction. The faster the movement of the wire, the greater is the magnitude of the deflection on the galvanometer. If the wire is moved horizontally between the poles of the magnets, no deflection on the galvanometer is observed.

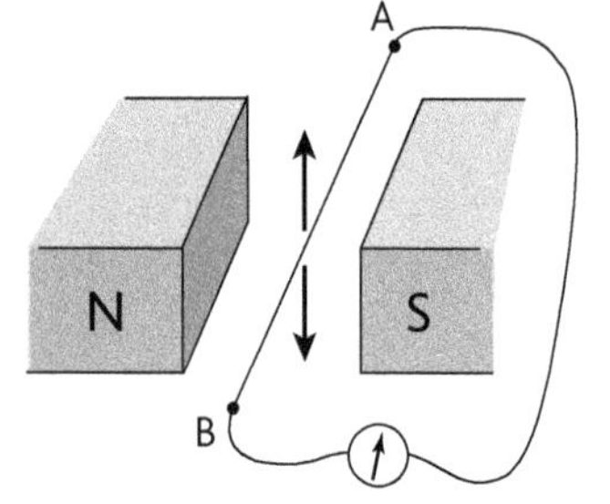

Figure 7.1.4

In order to explain the experiments above you need to see the link between them. In each there is relative motion between a conductor and a magnetic field. A magnetic field exists around a bar magnet. In 6.1 it was seen that magnetic fields are represented by lines called magnetic flux.

Magnetic flux

Definition

The total amount of magnetic flux ϕ through an area A at right angles to a magnetic field is given by

$$\phi = BA$$

where B is the magnetic flux density of the magnetic field.

The unit of magnetic flux is the **weber** (Wb). 1 weber is defined as the magnetic flux when a flux density of 1 T passes perpendicularly through an area of 1 m^2.

If the magnetic field acts at an angle of θ to the area A, then the total magnetic flux is given by $\phi = BA\cos\theta$.

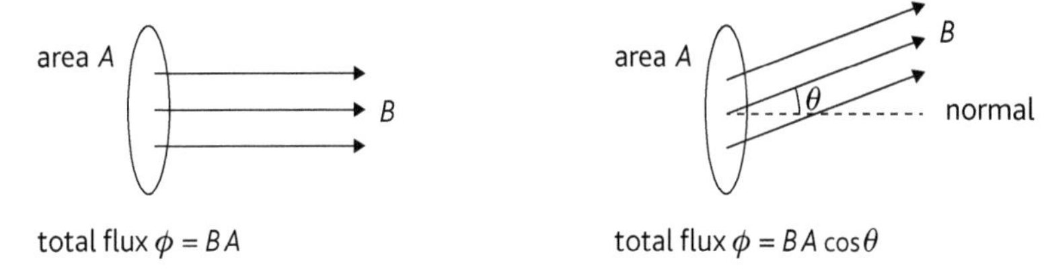

Figure 7.1.5 *Defining magnetic flux*

In Experiment 3, as the wire AB cuts magnetic flux, electrons in the wire experience a force that causes them to move to one end of the wire.

Recall, that when a charged body moves in a magnetic field, it experiences a force $F = BQv$, where B is the magnetic flux density of the magnetic field, Q is the charge on the body and v is the speed with which the charged body is moving in the magnetic field.

Each electron in the wire AB carries a tiny negative charge. When the wire moves downwards, the electrons inside it move downwards. This means that the conventional current moves upwards. Using Fleming's left hand rule, we see that the electrons experience a force that pushes them towards end A of the wire. This end of the wire becomes negatively charged and the other end becomes positively charged. The result is that an e.m.f. is induced between the ends of the wire. If the wire were to form part on an electric circuit, as when the galvanometer is connected, an electric current flows. This current is called an **induced current**.

In order to predict the direction of the induced current, **Fleming's right hand rule** is used. The thumb points in the direction of the force acting on the conductor. The index finger points in the direction of the field. The second finger points in the direction of the induced current.

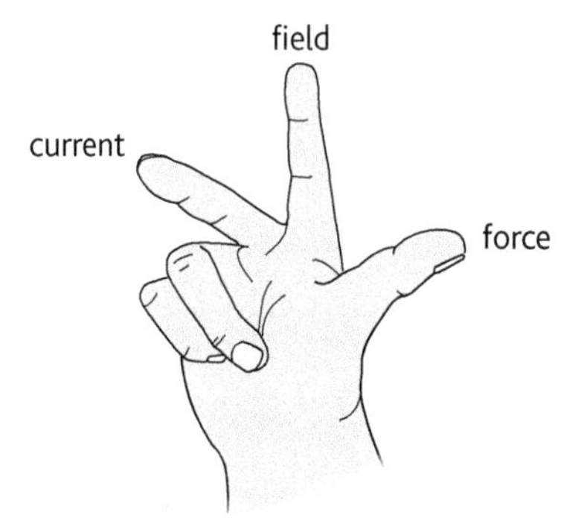

Figure 7.1.6 *Fleming's right hand rule*

When wire AB moves upwards, the electrons inside it move upwards. This means that the conventional current moves downwards. Using Fleming's left hand rule, the electrons experience a force that pushes them towards end B of the wire. This end of the wire becomes negatively charged and the other end becomes positively charged.

When wire AB is moved horizontally between poles of the magnet, the force experienced by the electrons in the wire is zero. Therefore no e.m.f. is induced in the wire.

The effect of producing an electric current using magnetism is called **electromagnetic induction**.

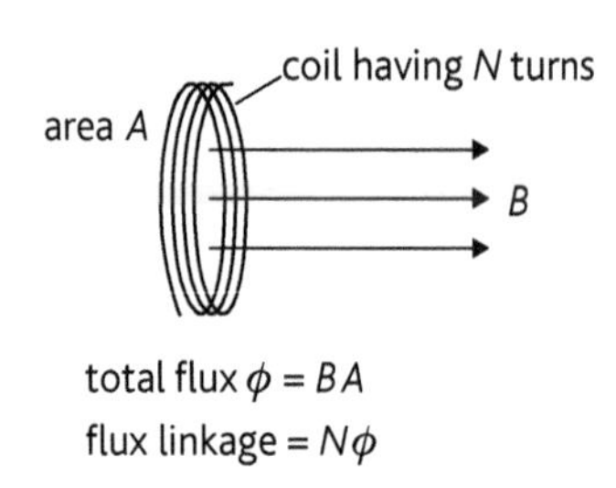

Figure 7.1.7 Defining magnetic flux linkage

Magnetic flux linkage

Definition

Magnetic flux linkage refers to the flux linking or passing through a coil and is numerically equal to $N\phi$.

Flux linkage $= N\phi$

N – number of turns of the coil
ϕ – flux passing through the coil

Definition

Faraday's law of electromagnetic induction states that the magnitude of the induced e.m.f. is proportional to the rate of change of magnetic flux linkage.

Definition

Lenz's law states that the induced e.m.f. (or current) acts in such a direction to produce effects to oppose the change causing it.

Faraday's and Lenz's laws

Faraday's law is expressed as $E = -N\frac{\Delta\phi}{\Delta t}$ where E is the induced e.m.f., N is thc number of turns of the coil, $\Delta\phi$ is the change in magnetic flux in a time Δt.

Faraday's law is also expressed as $E = -NA\frac{\Delta B}{\Delta t}$, where N is the number of turns of the coil, A is the cross-sectional area of the coil, ΔB is the change in magnetic flux density in a time Δt.

The magnitude of the induced e.m.f. in a conductor moving through a magnetic field can be increased by:

- increasing the length of conductor situated in the magnetic field
- increasing the speed at which the conductor is moving through the magnetic field
- increasing the strength of the magnetic field.

The magnitude of the induced e.m.f. in a coil of wire situated in a magnetic field can be increased by:

- increasing the cross-sectional area of the coil
- increasing the number of turns of wire in the coil
- increasing the rate at which the magnetic field changes
- ensuring that the magnetic field is perpendicular to the plane of the coil.

The negative sign (–) in the equation for Faraday's law implies that the induced e.m.f. opposes the rate of change of magnetic flux linkage. Lenz's law is a direct consequence of the principle of conservation of energy.

Lenz's law is used to predict the direction of the induced current.

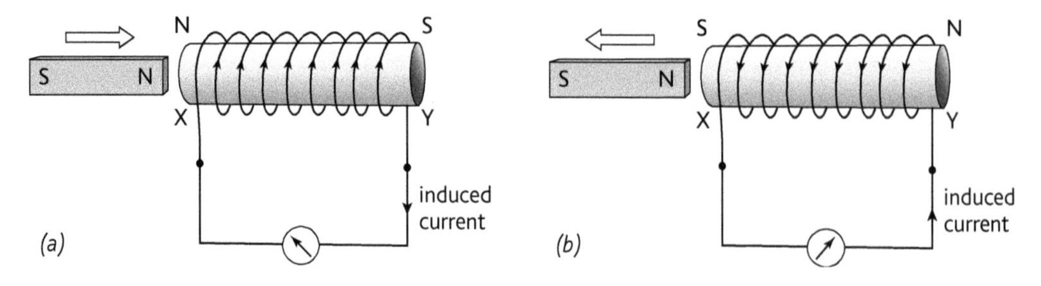

Figure 7.1.8 Applying Lenz's law

When the north pole of a magnet approaches a solenoid, as in Figure 7.1.8(a), lines of magnetic flux are cut. An e.m.f. is induced in the solenoid. The induced current in the solenoid flows in a direction to produce a north pole to oppose the motion of the magnet coming towards the solenoid.

When a north pole moves away from the solenoid, as in Figure 7.1.8(b), lines of magnetic flux are again cut, and an e.m.f. is induced in the solenoid. This time the induced current in the solenoid flows in a direction to produce a south pole to attract the magnet moving away from the solenoid.

Figure 7.1.9 shows a metal sheet oscillating freely in a plane at right angles to a magnetic field. The metal sheet comes to rest after a few oscillations. According to Faraday's law, an induced e.m.f. is proportional to the rate of change of flux linkage. As the metal sheet moves between the poles of the magnet, magnetic flux is cut and an e.m.f. is induced in it. The magnetic field between the north pole and south pole is not uniform at the edges. As a result, different e.m.f.s are induced in different parts of the metal sheet. Since the metal sheet is a conductor, eddy currents begin flowing in it. The direction of flow of the induced currents produces magnetic fields that oppose the motion of the metal sheet. Hence the amplitude of the oscillations decreases.

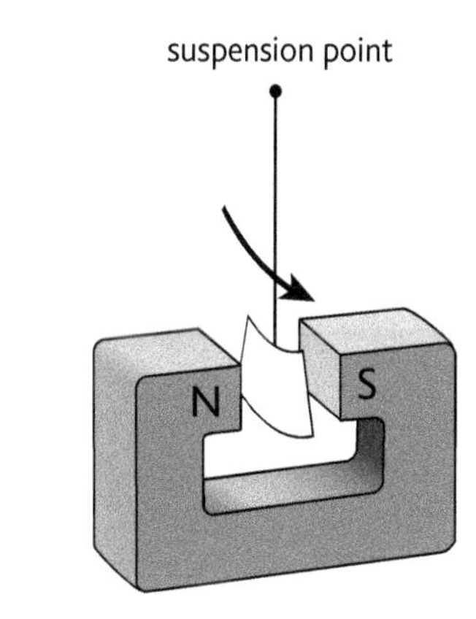

Figure 7.1.9

The reduction in amplitude of oscillations can also be explained in terms of energy considerations. The eddy currents dissipate thermal energy in the metal sheet. This energy is derived from the oscillation of the metal sheet. Since energy is being dissipated, the amplitude of the oscillations decreases.

Induced e.m.f. in a straight conductor

Consider a straight conductor of length l moving with a speed v at right angles to a magnetic field of field strength B (Figure 7.1.10).

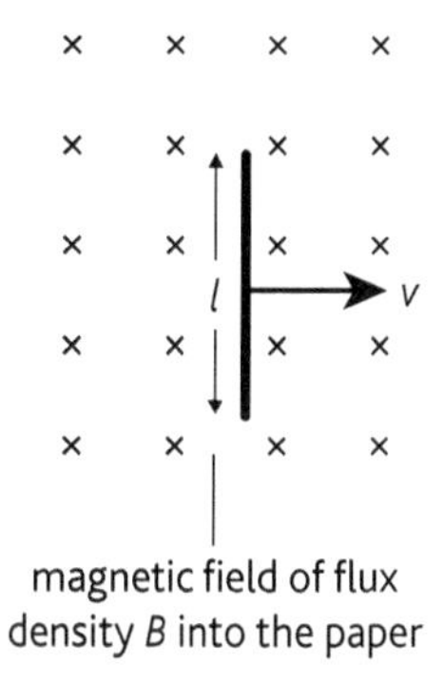

Figure 7.1.10

In a time t, the distance travelled by the conductor $= vt$

Change in magnetic flux linked with the conductor $= \phi = BA$

$$= B(vt \times l) = Bvtl$$

Induced e.m.f. E = rate of change of magnetic flux $= \dfrac{Bvtl}{t} = Blv$

Key points

- The total amount of magnetic flux ϕ through an area A at right angles to the magnetic field is given by $\phi = BA$.
- Magnetic flux linkage refers to the flux linking or passing through a coil and is numerically equal to $N\phi$.
- Faraday's law of electromagnetic induction states that the magnitude of the induced e.m.f. is proportional to the rate of change of magnetic flux linkage.
- Lenz's law states that the induced e.m.f. (or current) acts in such a direction to produce effects to oppose the change causing it.

Learning outcomes

On completion of this section, you should be able to:

- describe and explain simple applications of electromagnetic induction:
  - a simple d.c. motor
  - a simple a.c. motor
  - a d.c. generator
  - an a.c. generator.

A simple d.c. motor

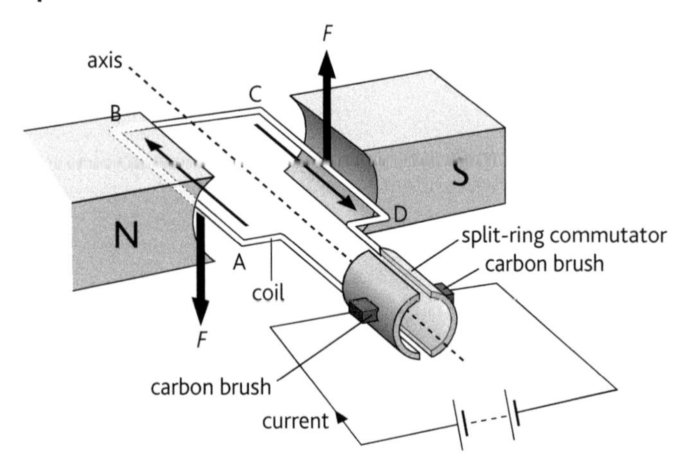

Figure 7.2.1 *A simple d.c. motor*

An electric motor is an application of electromagnetic induction. Figure 7.2.1 shows a diagram of a simple d.c. motor. It consists of a rectangular coil of wire situated in a magnetic field. The coil is mounted on pivot. Each end of the coil of wire is connected to half of a split ring of copper. This is called the **split-ring commutator**. Two rectangular carbon blocks, called brushes are pressed tightly against the commutator by springs. The brushes are connected to a d.c. power supply. Fleming's left hand rule can be used to determine the direction of the forces acting on the coil. When the current flows through the section AB, it experiences a downward force. The section CD experiences an upward force. These two forces produce a couple and cause the coil to rotate in an anticlockwise direction until the coil is vertical. When the coil is vertical, the brushes line up with the gap in the commutator and the current stops flowing through the coil. As a result of its inertia, the coil overshoots the vertical position, and then the commutator halves change from one carbon brush to the other. The current in the coil reverses direction, so that the section of the coil AB is now on the right and experiences an upward force. The section of the coil CD is now on the left and experiences a downward force. As a result, the coil continues to rotate in an anticlockwise direction.

A simple a.c. motor

Figure 7.2.2 shows a diagram of simple a.c. motor. It consists of a rectangular coil of wire situated in a magnetic field. The coil is mounted on a pivot. Each end of the coil of wire is connected to **slip rings** of copper. Two brushes are pressed tightly against the slip rings. The brushes are connected to an a.c. power supply. An a.c. power supply changes direction regularly with time. Suppose that initially a current is flowing in the coil from A to B. According to Fleming's left hand rule a downward force is experienced. The section CD will experience an upward force. The two forces produce a couple and the coil rotates in an anticlockwise direction. The combination of the slip rings and the a.c. power supply function similarly to the split-ring commutator in the d.c motor. As a result, the coil continues to rotate in an anticlockwise direction.

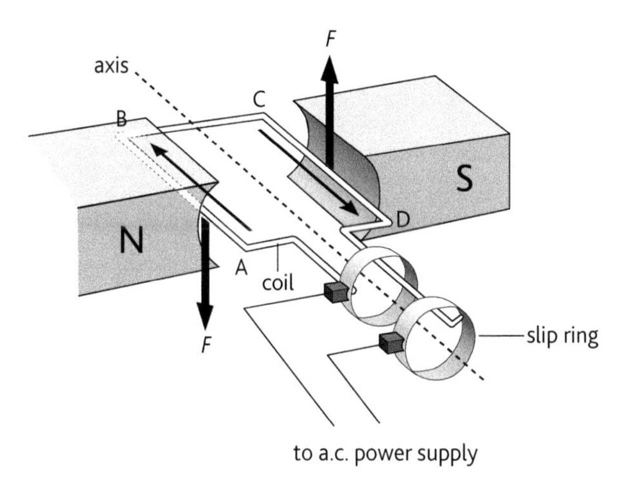

Figure 7.2.2 *A simple a.c. motor*

A simple d.c. generator

A generator is another application of electromagnetic induction. According to Faraday's law of electromagnetic induction, when a coil is rotated in a magnetic field, an e.m.f. is induced in it. The magnitude of the induced e.m.f. depends on the speed of rotation and the number of turns in the coil.

The construction of a simple d.c. generator (Figure 7.2.3) or dynamo is similar to that of a d.c. motor. The main difference is that there is no d.c. power supply in the circuit. The coil is made to rotate by some mechanical means. As the coil rotates in the magnetic field, an e.m.f. is generated. If the output of the generator is connected to a resistor, a current is produced.

The direction of the current can be determined by using Fleming's right hand rule. Suppose the coil is rotated in an anticlockwise direction. Then the side AB moves downwards and the side CD moves upwards. By Fleming's right hand rule, an induced current will flow from D to C to B to A. When the coil is vertical, the current flowing in the coil is zero. The split-ring commutator causes the current to flow in the same direction all the time. Figure 7.2.4 shows how the output voltage of a d.c. generator varies with time.

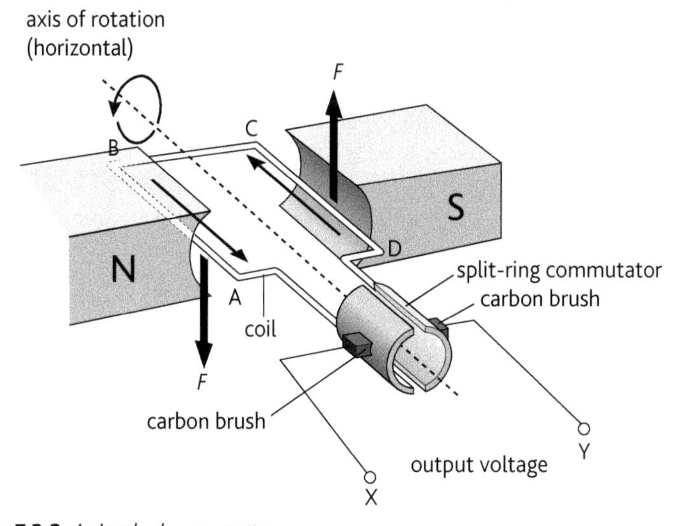

Figure 7.2.3 *A simple d.c. generator*

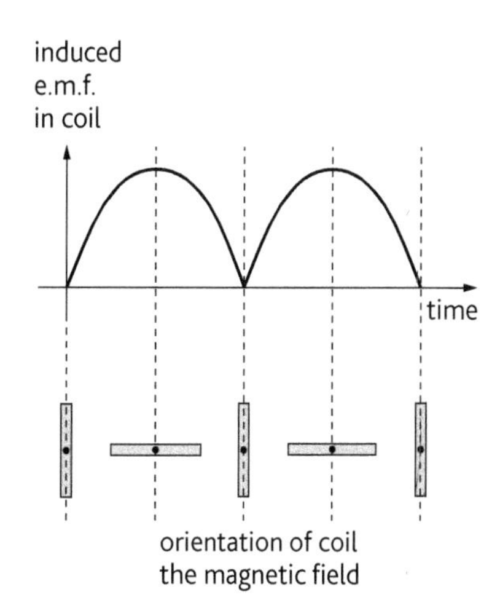

Figure 7.2.4 *The output voltage from a d.c. generator*

A simple a.c. generator

The construction of a simple a.c. generator or dynamo is similar to that of an a.c. motor. Figure 7.2.5 illustrates a simple a.c. generator. The main difference is that there is no a.c. power supply in the circuit. The coil is made to rotate by some mechanical means. As the coil rotates in the magnetic field, an e.m.f. is generated. If the output of the generator is connected to a resistor, a current is produced. The direction of the current can be determined by using Fleming's right hand rule. Suppose the coil is rotated in an anticlockwise direction. Then the side AB moves downwards and the side CD moves upwards. By Fleming's right hand rule, an induced current will flow from D to C to B to A. When the sides AB and CD change positions, the direction of flow of current in the coil changes. Figure 7.2.6 shows how the output voltage of the a.c. generator varies with time.

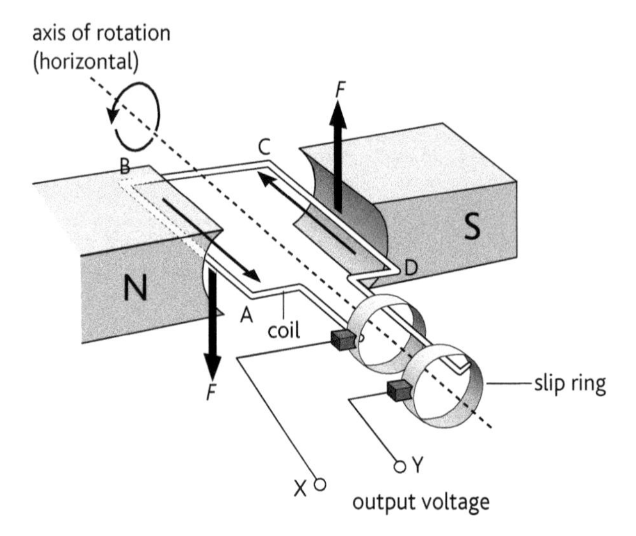

Figure 7.2.5 *A simple a.c. generator*

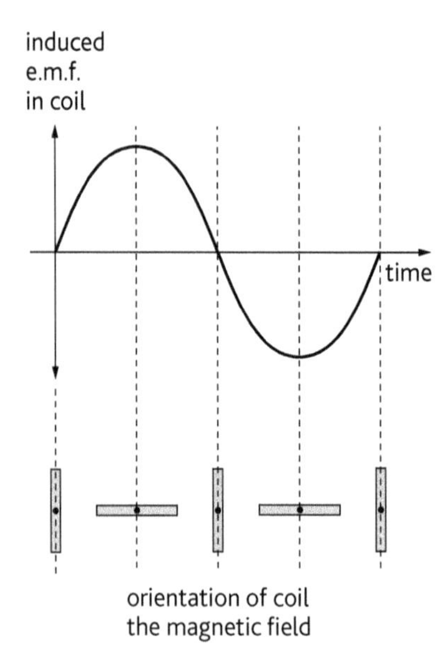

Figure 7.2.6 *The output voltage from an a.c. generator*

Factors that affect the speed of rotation of a d.c. or a.c. motor:

1. The magnitude of the field between the magnetic poles.
2. The magnitude of the current flowing in the coil.
3. The number of turns in the coil.

Factors that affect the magnitude of the e.m.f. induced in the coil in a d.c. or a.c. motor:

1. The speed of rotation.
2. The number of turns in the coil.
3. The magnitude of the field between the magnetic poles.

Key points

- In a d.c. motor, the current is made to flow in the same direction all the time by using a split-ring commutator.
- In an a.c. motor, there are slip rings.
- When a coil is rotated in a magnetic field an e.m.f. is induced in it.

Exam tip

If you are asked to explain the operation of a d.c. motor:

1. Draw a simple sketch and label the north and south poles of the magnet, the coil and the split-ring commutator.
2. Label the edges of the coil.
3. Show the direction of the current in the coil.
4. Mention Fleming's left hand rule to predict the direction of the force acting on the coil.

7.3 Examples on electromagnetic induction

Learning outcomes

On completion of this section, you should be able to:

- provide explanations for various examples of electromagnetic induction.

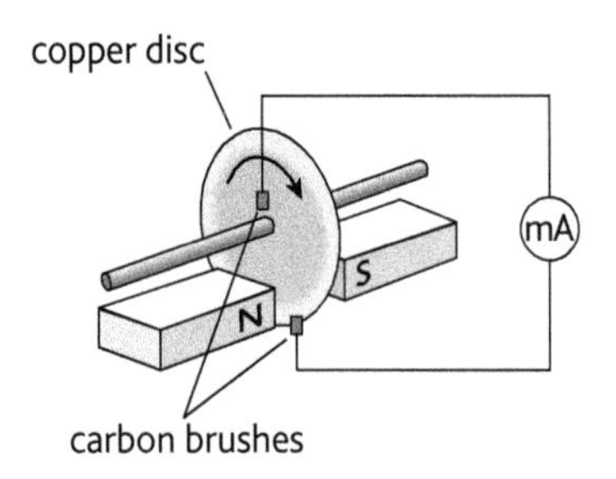

Figure 7.3.1 *Faraday's disc*

Faraday's disc

Consider a copper disc situated between the poles of two magnets (Figure 7.3.1). As the disc rotates, an e.m.f. is induced between the centre of the rotation of the disc and its circumference. As the disc rotates, the electrons inside it experience a force according to Fleming's left hand rule. In the example shown, the electrons experience an upward force and are pushed towards the centre of the disc. The accumulation of electrons at the centre of the disc results in a potential difference being set up between the centre and the circumference of the disc. The flow of electrons towards the centre stops when the electric force acting on the electrons equals the magnetic force acting on them.

Magnet moving in a coil

Consider Figure 7.3.2.

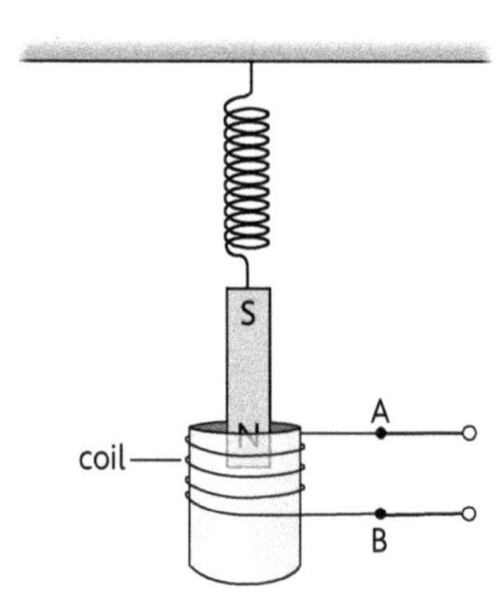

Figure 7.3.2

When the magnet is displaced gently, it begins oscillating with a constant frequency. When a resistor is connected to the terminals AB at time t, the oscillation of the magnet becomes damped. Figure 7.3.3 shows the variation of the amplitude of oscillation of the magnet with time, before and after the resistor was connected to the terminals A and B.

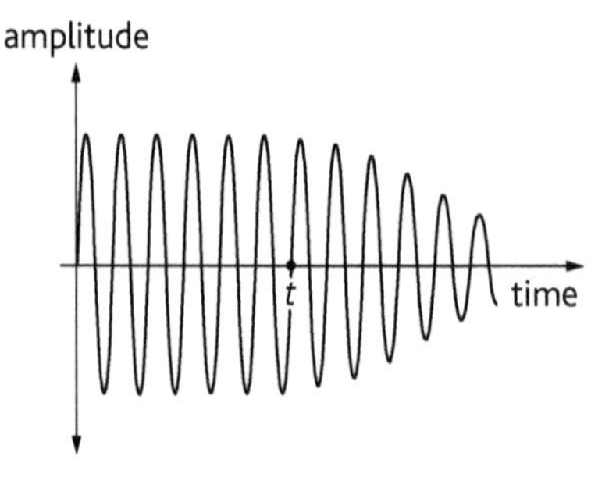

Figure 7.3.3

As the magnet moves towards the coil, an e.m.f. is induced in it. The resistor completes the circuit resulting in a current flowing through it. The current flowing through the resistor causes thermal energy to be dissipated in it. This thermal energy is derived from the movement of the magnet through the coil. Therefore, if thermal energy is being dissipated, the amplitude of oscillation of the magnet will reduce over a period of time. If a resistor of smaller value is used, more energy will be dissipated and the oscillation of the magnet reduces more rapidly.

A current-carrying solenoid placed near to a search coil

Consider the situation where a current-carrying solenoid is placed next to a search coil as shown in Figure 7.3.4.

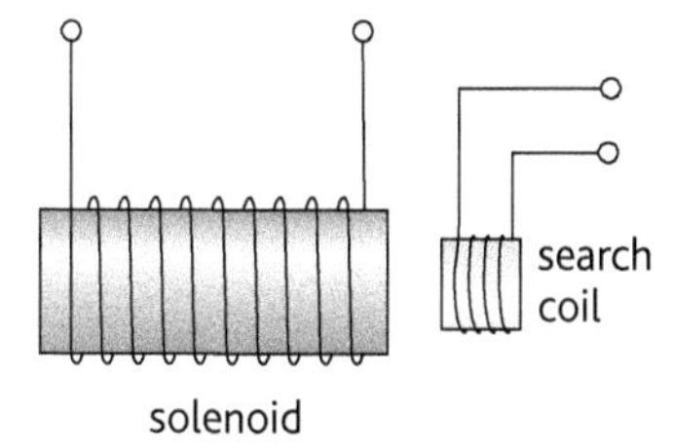

Figure 7.3.4

Figure 7.3.5(a) shows the variation of current flowing through the solenoid with time. The magnetic field strength B produced by the solenoid is directly proportional to the current flowing through it. The variation of the magnetic field strength B with time will be similar to graph (b).

The search coil is situated in the magnetic field produced by the solenoid. An e.m.f. is induced in the search coil according to Faraday's law of electromagnetic induction. The direction of the induced current is determined using Lenz's law.

Induced e.m.f. $E = -\frac{d\phi}{dt}$

Graph (c) shows the variation of the induced e.m.f. in the search coil with time.

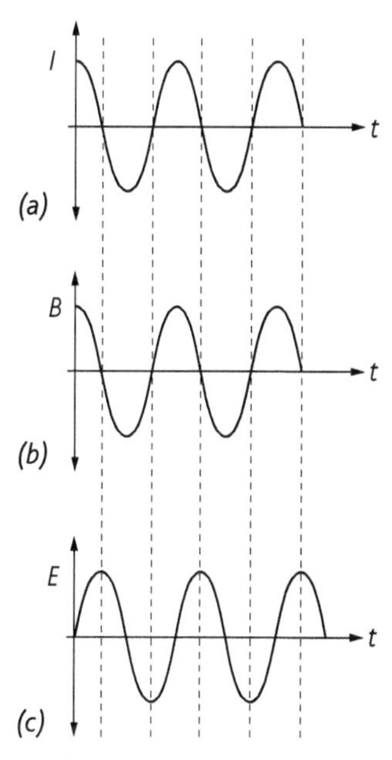

Figure 7.3.5

A washer resting on a solenoid

A light aluminium washer rests on one end of a solenoid as shown in Figure 7.3.6. A large direct current suddenly flows in the solenoid. The washer jumps and immediately falls back.

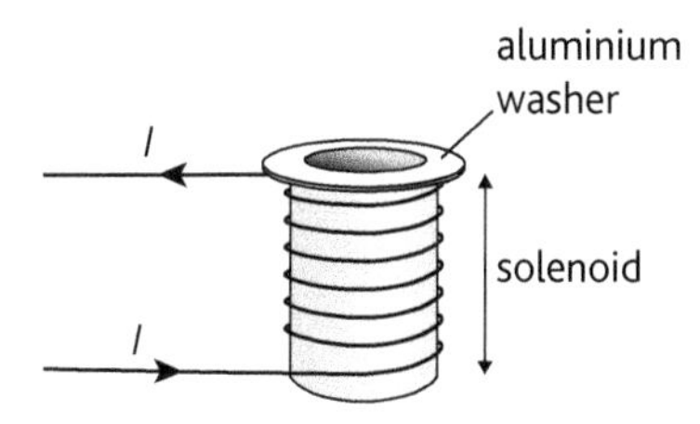

Figure 7.3.6

A magnetic field is produced by a solenoid when a current flows through it. Since the current increases to a large value quickly, the magnetic field changes rapidly. The aluminium washer is situated in the changing magnetic field. An e.m.f. is induced in the washer. Since aluminium is a conductor, a current flows in it because of the induced e.m.f. The induced current produces a magnetic field that opposes the magnetic field produced by the solenoid. This is in agreement with Lenz's law. The repulsive force causes the washer to move upwards. When the current flowing in the solenoid becomes steady, there is no changing flux. No e.m.f. is induced in the washer and it falls.

Exam tip

In the examination you may be given an unfamiliar application of Faraday's and Lenz's laws.

Keep in mind that an e.m.f. is induced in a conductor when magnetic flux is cut. Always remember that an alternating current produces a changing magnetic field.

Revision questions 3

Answers to questions that require calculation can be found on the accompanying CD.

1 **a** Explain what is meant by the terms *magnetic field* and *magnetic flux*. [2]

b Sketch the magnetic field:

i produced by a current-carrying conductor [3]

ii produced by a solenoid [3]

iii produced by a flat circular coil. [3]

2 The diagram shows the top view of a square of wire of side 1.5 cm. It is in a uniform magnetic field of flux density 9.5 mT formed between magnetic north and south poles. The current in the wire is 2.0 A.

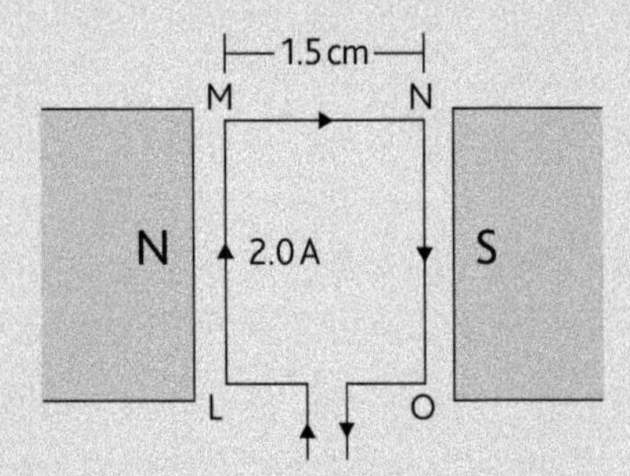

a Determine the sizes and directions of the electromagnetic forces that act on the sides LM and NO of the square of wire. [6]

b Why do no electromagnetic forces act on the sides MN and OL of the square? [2]

3 An aluminium rod of mass 60 g is placed across two parallel horizontal copper tubes that are connected to a low voltage supply. The aluminium rod lies across the centre of and perpendicular to the uniform magnetic field of a permanent magnet as shown in the diagram.

The magnetic field acts over a region measuring 8.0 cm × 6.0 cm. The magnetic flux density of the field between the poles is 0.25 T.

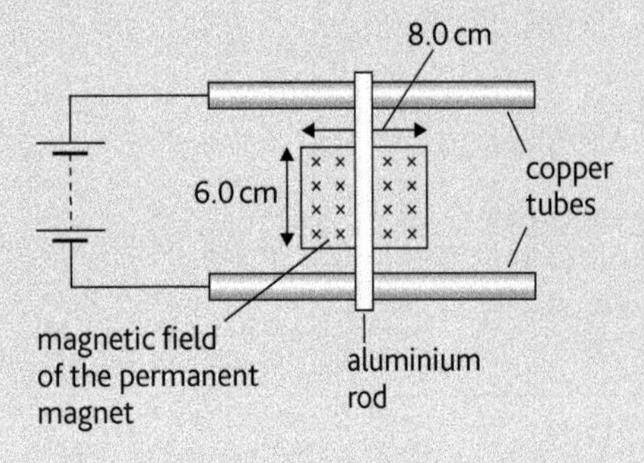

Calculate the initial acceleration of the rod, assuming that it slides without rolling, when the current in the rod is 5.0 A. [5]

4 A strip of aluminium foil is hung over a wooden peg as shown below.

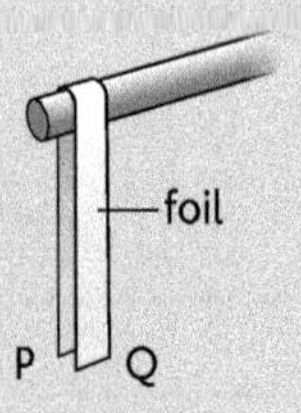

A d.c. power supply is connected between P and Q so that a current flows through the aluminium foil.

a State what is observed with the aluminium foil. [1]

b Explain the observation. [4]

5 Two long straight wires P and Q are parallel to each other. The current in wire P is 4.0 A and that in Q is 6.0 A. The separation of the wires is 2.2 cm. Calculate the force per unit length on wire Q due the current in wire P. [4]

6 **a** Calculate the resistance per metre of a copper wire of diameter 0.06 mm and resistivity $1.7 \times 10^{-8}\,\Omega\,\text{m}$. [2]

b The copper wire is used to construct a solenoid. It is made by winding one layer of close-packed turns on a cardboard tube of length 18 cm and diameter 25 mm. The solenoid is connected in series with a battery of e.m.f. 9.0 V and negligible internal resistance.

Calculate:

i the resistance of the wire used to make the solenoid [4]

ii the magnetic flux density at the centre of the solenoid. [3]

7 A particle has a mass of m and a charge of $+q$.

a State the magnitude and direction of the forces on this particle when it is at rest in:

i a gravitational field [2]

ii an electric field of field strength E [2]

iii a magnetic field of flux density B. [1]

b State the magnitude and direction of the force on this particle when it is moving with a velocity v in a direction normal to:

i a gravitational field [2]

ii an electric field of field strength E [2]

iii a magnetic field of flux density B. [2]

8 a Define the tesla. [3]

b Positive ions are travelling through a vacuum in a narrow beam. The ions enter a region of uniform magnetic field of flux density *B* and are deflected in a semicircular arc as shown below.

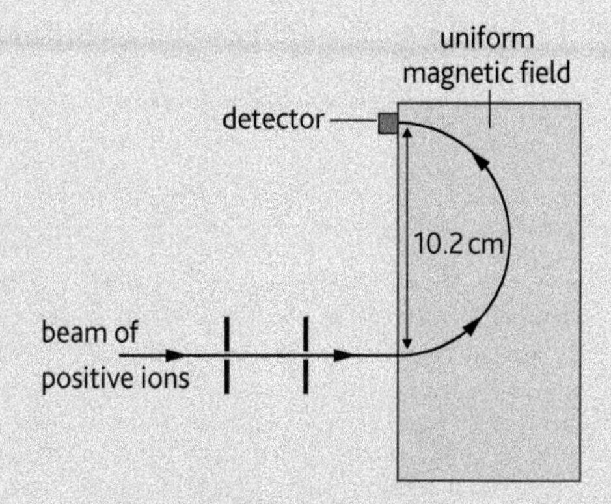

The ions, travelling with a speed of $1.62 \times 10^5\,\text{m s}^{-1}$, are detected at a fixed detector when the diameter of the arc in the magnetic field is 10.2 cm.

i State the direction of the magnetic field. [1]

ii The positive ions have a mass of 2.99×10^{-26} kg and a charge of $+3.2 \times 10^{-19}$ C. Calculate the magnetic flux density of the magnetic field. [3]

iii If the magnetic field strength is increased, sketch a diagram similar to the one above to show the path taken by the ions. [1]

9 An electron is travelling with a velocity of $6.5 \times 10^6\,\text{m s}^{-1}$ at right angles to a magnetic field of magnetic flux density 5.0 mT. Calculate:

a the force acting on the electron [3]

b the centripetal acceleration of the electron [2]

c the radius of the circular path described by the electron. [2]

10 In the velocity selector of a mass spectrometer, positive ions travelling with a speed of $1.2 \times 10^5\,\text{m s}^{-1}$ pass undeflected through a magnetic field of 120 mT which is perpendicular to an electric field.

a Draw a labelled diagram to show the relationship between the two fields. [2]

b Write down an expression for the force exerted by the magnetic field. [1]

c Write down an expression for the force exerted by the electric field. [1]

d Hence calculate the electric field strength of the electric field. [2]

11 An electron travelling with a speed of $3.0 \times 10^7\,\text{m s}^{-1}$ enters a magnetic field of uniform magnetic flux density 7.5 mT, in a direction at right angles to the field.

a Sketch the path of the electron in the magnetic field and show the direction of the field. [2]

b Explain why the path travelled by the electron is circular. [2]

c Explain why the speed of the electron is unchanged. [2]

d Calculate the force acting on the electron. [2]

e Calculate the radius of the path of the electron in the field. [3]

f Calculate the electric field strength required to provide an equal force to that provided by the magnetic field. [2]

g Explain how it is possible to select electrons of a particular speed by the use of electric and magnetic fields. [3]

12 A particle has mass of 9.1×10^{-31} kg and a charge of -1.60×10^{-19} C. It is travelling with a velocity of $4.5 \times 10^7\,\text{m s}^{-1}$ in a uniform magnetic field of flux density 0.30 T. Calculate the radius of curvature of its circular path. [4]

13 A proton of mass 1.67×10^{-27} kg travelling with a speed *v* enters a uniform magnetic field of flux density 2.2 mT. The proton is travelling at right angles to the magnetic field. The proton travels along a circular path while in the magnetic field. The radius of curvature of the path of the proton is 4.2 cm.

a Explain why the proton travels along a circular path when travelling inside the magnetic field. [2]

b Explain why the speed of the proton does not change while travelling inside the magnetic field. [2]

c Calculate the speed *v*. [5]

14 a Define *magnetic flux density* and its SI unit. [3]

b State:

i Faraday's law [2]

ii Lenz's law. [2]

c A straight wire AB is moved at right angles to a magnetic field. State three factors that affect the magnitude of the induced e.m.f. between A and B. [3]

15 a Explain what is meant by electromagnetic induction. [2]

b Describe an experiment to demonstrate electromagnetic induction. [4]

16 A solenoid is designed to produce a magnetic field of flux density 25 mT at its centre. The radius and length of the solenoid are 1.50 cm and 50.0 cm respectively. The current flowing through the solenoid is 10.0 A.
Calculate:

a the minimum number of turns per unit length that must be used [3]

b the total length of wire required for this design. [3]

17 A small copper disc of diameter 50 mm rotates on an axis at 12.0 revolutions per second in a magnetic field of flux density 1.4×10^{-2} T. The magnetic field acts at right angles to the plane of rotation.

a Explain why an e.m.f. is generated between the axle and the rim of the disc when it rotates. [3]

b Calculate:

i the magnetic flux cut every revolution [3]

ii the potential difference maintained between the rim and the axle of the disc. [3]

18 A coil of cross-sectional area $3.5 \times 10^{-4}\,m^2$ and 90 turns is placed in a uniform magnetic field.

a The plane of the coil is at right angles to the magnetic field. Calculate the magnetic flux density if the flux linkage for the coil is 2.0×10^{-4} Wb. [3]

b The coil is now placed in magnetic field of flux density 0.38 T. The angle between the axis of the coil and the magnetic field is 30°. Calculate the flux linkage for the coil. [3]

19 A circular coil of radius 1.5 cm has 1800 turns. The coil is placed at right angles to a magnetic field of flux density 75 mT. The direction of the magnetic field is reversed in a time of 25 ms. Calculate the average magnitude of the induced e.m.f. across the ends of the coil. [5]

20 A straight wire of length 15 cm is travelling at a constant speed of $2.5\,m\,s^{-1}$ in a uniform magnetic field of flux density 45 mT. Calculate the e.m.f. across the ends of the wire. [3]

21 An aircraft has a wing span of 50 m. It is flying horizontally at $600\,km\,h^{-1}$ in a region where the vertical component of the Earth's magnetic field is 4.5×10^{-5} T. Calculate the potential difference induced between one wing tip and the other. [3]

22 A large solenoid is 50 cm long and has 80 turns. Calculate the magnetic flux density inside the solenoid when a current of 2.0 A flows in it. [3]

23 A thin copper ring encloses an area of $1.8 \times 10^{-3}\,m^{-2}$. The plane of the ring is normal to a uniform magnetic field. The magnetic field strength increases at a constant rate of $4.0 \times 10^{-2}\,T\,s^{-1}$. Calculate the e.m.f. induced in the ring. [3]

24 A metal framed window is 1.2 m high and 0.8 m wide. It pivots about a vertical edge and faces due south.

a Calculate the magnetic flux through the closed window. (Horizontal component of the Earth's magnetic field = 18 μT. Vertical component of the Earth's magnetic field = 45 μT.) [2]

b The window is opened through an angle of 90° in a time of 0.90 s. Calculate the average e.m.f. induced in the window frame. [3]

25 A circular coil of diameter 140 mm has 600 turns. It is placed so that its plane is perpendicular to a horizontal magnetic field of uniform flux density 50 mT.

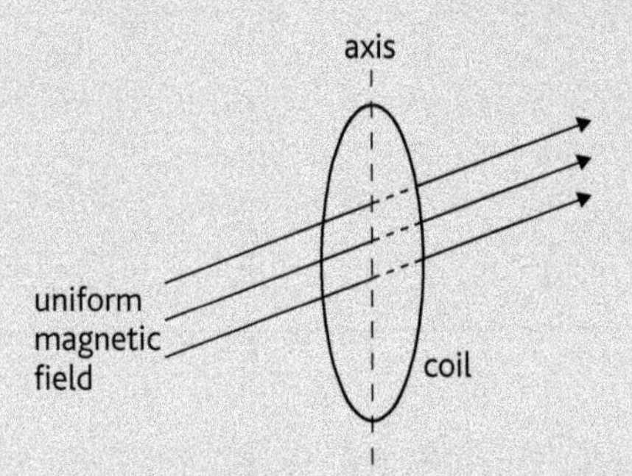

a Calculate the magnetic flux passing through the coil when in this position. [2]

b The coil is rotated through 90° about a vertical axis in a time of 90 ms.
Calculate:

i the change in magnetic flux linkage produced by this rotation [2]

ii the average e.m.f. induced in the coil when it is rotated. [2]

26 With the aid of a clearly labelled diagram, explain what is meant by the *Hall effect*. [8]

27 A rectangular slice of a semiconductor is 2.2 mm thick and carries a current of 120 mA. A magnetic field of flux density 0.6 T acts on the semiconductor as shown in the diagram. A potential difference of 7.65 mV is produced across the semiconductor due to the Hall effect. The main charge carriers in the semiconductor are electrons ($e = -1.6 \times 10^{-19}$ C). Calculate the number of charge carriers per unit volume. Explain your calculation. [8]

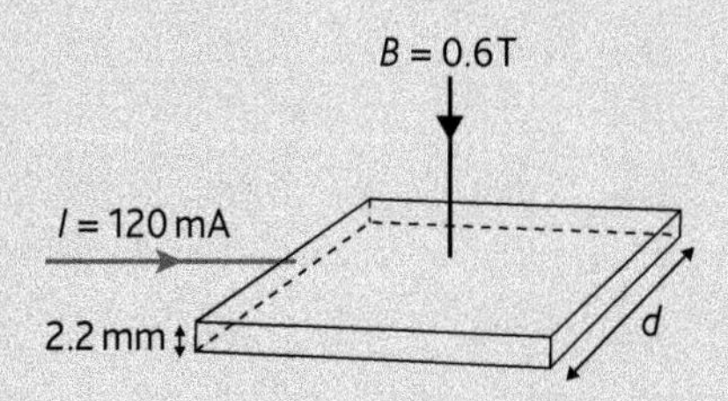

28 a Draw a simple diagram of a simple d.c. motor. [4]

b Explain the operation of a simple d.c. motor. [5]

29 a Use the laws of electromagnetic induction to explain the operation of a simple a.c. generator. [5]

b Sketch the variation of the induced e.m.f. in the a.c. generator with time. [3]

Module 1 Practice exam questions

Answers to the multiple-choice questions and to selected structured questions can be found on the accompanying CD.

Multiple-choice questions

1 A generator produces 120 W of power at a potential of 12 kV. The power is transmitted through overhead cables of total resistance 6 Ω. What is the power loss in the cables?

a 0.6 mW **b** 600 W **c** 6 W **d** 0.2 mW

2 The diagram below shows the relationship between a direct current I in a conductor and the potential difference V across it. When $V < 1.5$ V, the current flowing through the conductor is negligible.

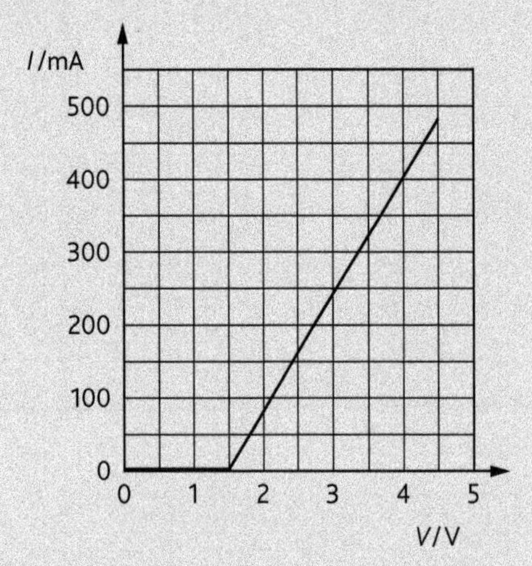

Which of the following statements about the conductor is correct?

a It obeys Ohm's law when $V > 1.5$ V and when $V = 4$ V, its resistance is 10 Ω.

b It obeys Ohm's law when $V > 1.5$ V, but its resistance is constant.

c It does not obey Ohm's law, but when $V > 1.5$ V its resistance is 10 Ω.

d It does not obey Ohm's law, but when $V = 4$ V its resistance is 10 Ω.

3 Two horizontal metal plates are separated by 2.0 mm. The lower plate is at a potential of −8 V. What potential should be applied to the upper plate to create an electric field of strength 3000 V m^{-1} *upwards* in the space between the plates?

a −14 V **b** −2 V **c** +2 V **d** +14 V

4 A cell of e.m.f. 4.0 V and negligible internal resistance is connected to a network of resistors shown below.

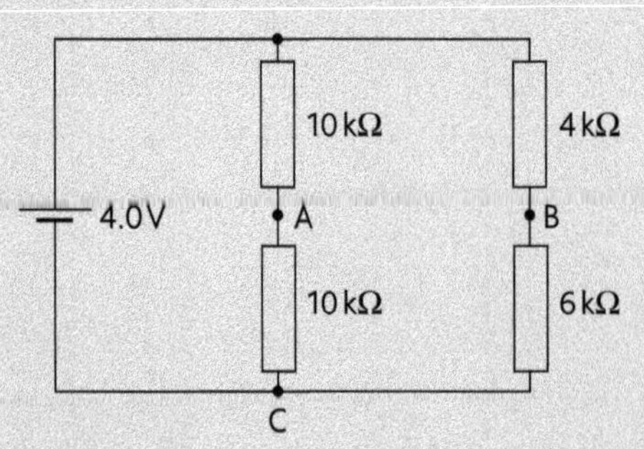

The potential difference between C and A is V_1 and the potential difference between C and B is V_2. What is the value $V_1 - V_2$?

a −1.0 V **b** +1.0 V **c** +0.40 V **d** −0.40 V

5 A wire of length 0.6 m is travelling at a speed of 12 m s^{-1} perpendicular to a magnetic field of flux density 3.0×10^{-5} T. What is the magnitude of the e.m.f. generated between the ends of the wire?

a 2.16×10^{-4} V **b** 1.20×10^{-4} V

c 4.20×10^{-4} V **d** 0 V

6 A flat circular coil has a radius of 2.5 cm and 120 turns. A current of 2.0 A passes through the coil. What is the magnetic field strength at the centre of the coil?

a 2.0 mT **b** 3.0 mT **c** 1.5 mT **d** 6.0 mT

7 Two identical capacitors of capacitance C are connected in series with a potential difference of V. The total energy stored in the capacitors is E_1.

Two identical capacitors of capacitance C are connected in parallel with a potential difference of V. The total energy stored in the capacitors is E_2.

What is the value of $\frac{E_1}{E_2}$?

a ½ **b** ¼ **c** 4 **d** 2

8 Which of the following is correct?

i Kirchhoff's first law is a consequence of the conservation of energy.

ii Kirchhoff's second law is a consequence of the conservation of charge.

iii Kirchhoff's second law states that the algebraic sum of the e.m.f.s around any loop in a circuit is equal to the algebraic sum of the p.d.s around the loop.

a i only **b** i and ii only

c iii only **d** i, ii and iii

9 The diagram below shows two parallel wires A and B in the plane of the paper. A is fixed and B is free to move.

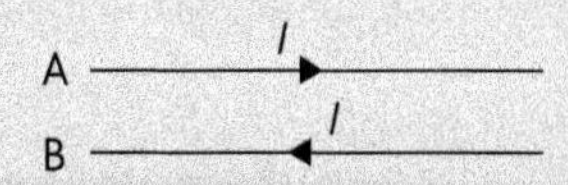

When the same current *I* passes through each wire in opposite directions, which way does B move?

- **a** Upwards out of the paper
- **b** Away from A in the plane of the paper
- **c** Towards A in the plane of the paper
- **d** Downwards out of the paper

10 When an electron, travelling in a vacuum, enters a uniform magnetic field of flux density *B* at right angles to its path, which way is it deflected?

- **a** In the direction of *B* into a parabolic path
- **b** In the direction opposite *B* into a circular path
- **c** In a direction perpendicular to *B* into a circular path
- **d** In the direction of *B* into a parabolic path

Structured questions

11 a Define electric charge. [2]

b Define the SI unit of charge. [2]

c A current of 2.1 A flows through a 75 W light bulb for a period of 480 s.

Determine:

- **i** the charge that flowed through the bulb [2]
- **ii** the energy dissipated in the lamp [2]
- **iii** the potential difference across the lamp [2]
- **iv** the number of electrons that flowed through the lamp. [2]

12 A battery has an e.m.f. of 9 V and internal resistance of 0.6 Ω. The battery is connected across a resistor of resistance 12 Ω. Calculate:

- **a** the current in the circuit [2]
- **b** the potential difference across the 12 Ω resistor [2]
- **c** the power dissipated in the 12 Ω resistor [2]
- **d** the power supplied by the battery [2]
- **e** the fraction of the power is dissipated in the battery. [1]

13 a Explain what is meant by the term *drift velocity*. [2]

b A sample of a conductor has a cross-sectional area *A*. The main charge carriers are electrons and each has a charge of *e*. There are *n* charge carriers per unit volume and the drift velocity of the charge carriers is *v*. Derive an expression for the current *I*, flowing through the conductor. [4]

14 The *I*–*V* characteristic of a thermistor is shown below.

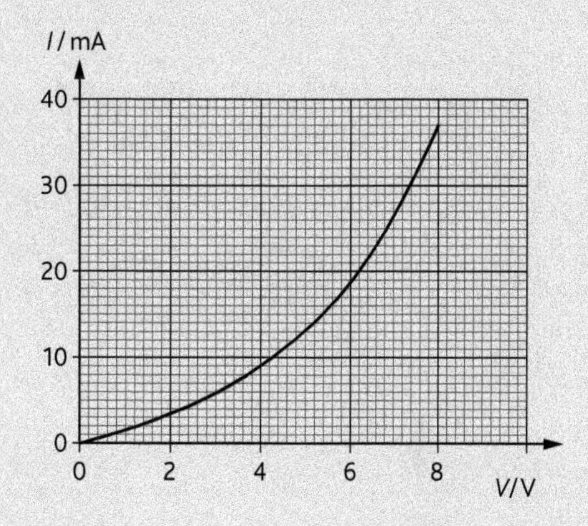

The thermistor is connected as shown below.

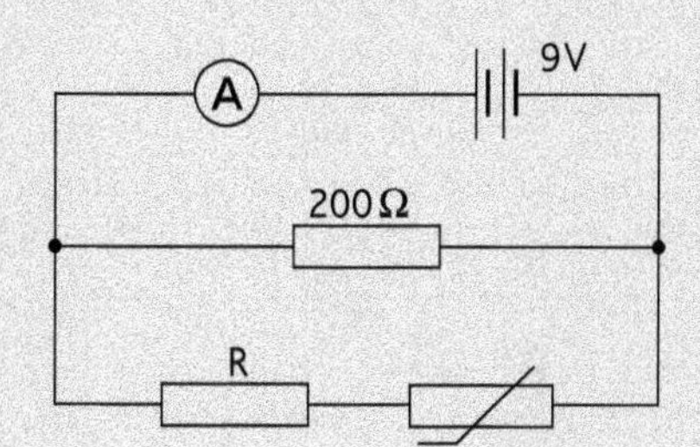

The ammeter reading is 70 mA. Calculate:

- **a** the current flowing through the 200 Ω resistor [2]
- **b** the current flowing through the thermistor [2]
- **c** the potential difference across the thermistor [1]
- **d** the value of *R*. [3]

15 a Explain the difference between electromotive force and terminal potential difference. [4]

b Define the volt. [1]

c State Kirchhoff's first and second laws and state the physical law on which each is based. [6]

d In the circuit below, batteries P and Q have negligible internal resistance. Battery R has an internal resistance of 2 Ω.

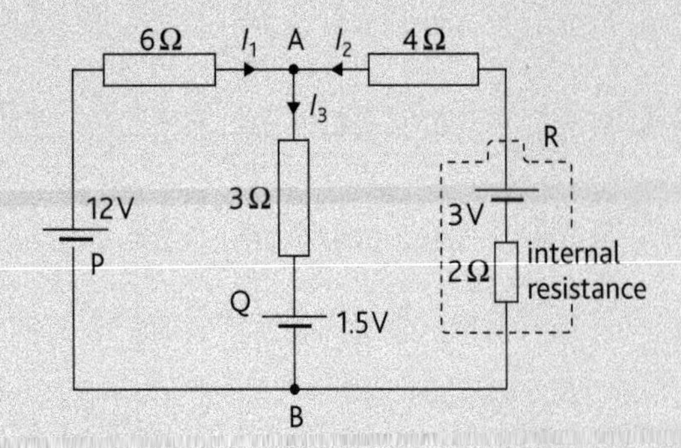

Determine:

i the currents I_1, I_2 and I_3 [8]
ii the potential difference across A and B [2]
iii the terminal potential difference of battery R. [2]

16 a State Coulomb's law for electrostatic charges. [2]

b Define the terms *electric field strength* and *electric potential*. [2]

c State the relationship between electric field strength and electric potential. [1]

d Two point charges A and B have charges 30 μC and −5 μC. A and B are 10 cm apart. Calculate:

i the electric field strength at the midpoint between the two charges [3]
ii the electric potential at the midpoint between the two charges. [3]

17 A 12.0 μF capacitor is charged by a 12.0 V d.c. supply and is then discharged through a 1.2 MΩ resistor.

a Calculate the charge on the capacitor just before being discharged. [2]

b Calculate the time constant for the circuit. [1]

c After 5.0 s, calculate:

i the charge on the capacitor [3]
ii the p.d. across the capacitor [3]
iii the current in the circuit. [3]

18 a Define the terms *magnetic flux density* and the *tesla*. [4]

b Sketch the magnetic flux pattern due to a long straight wire carrying a current *I*. [3]

c Write an equation to determine the magnetic flux density at a distance of *r* from the wire. [2]

d Two straight wires X and Y, each of length *l* are separated a distance of *r*. X carries a current of I_1 and Y carries a current of I_2. I_1 and I_2 are flowing in the same direction.

i Explain why wire X exerts a force on wire Y. [4]
ii State whether the force is attractive or repulsive. [1]
iii Write down an expression for force acting on wire Y. [2]

19 a State:

i Faraday's law of electromagnetic induction [1]
ii Lenz's law. [1]

b Using a bar magnet and coil as an example, explain how Lenz's law can be considered an example of the law of conservation of energy. [3]

c A flat circular coil of radius 1.8 cm consists of 450 turns. The coil is placed at right angles to a uniform magnetic field of flux density 35 mT.

i Calculate the magnetic flux passing through the coil. [2]
ii The field is reduced to zero and it takes 0.15 s to do so. Calculate the e.m.f. induced in the coil. [3]

20 A student wants to construct a solenoid to produce a magnetic field of flux density of 25 mT at its centre. The solenoid has a radius of 1.50 cm and length of 38.0 cm. The solenoid wire carries a current of 11 A.

Calculate:

a the minimum number of turns per unit length that is required for this solenoid [3]

b the total length of wire required to construct the solenoid. [3]

21 Using a diagram, explain the origin of the Hall effect. [7]

On your diagram indicate the direction of the Hall voltage. [1]

22 In a mass spectrometer, negative ions travelling at $2.2 \times 10^5\,\mathrm{m\,s^{-1}}$ pass undeflected through a magnetic field of 0.18 T which is perpendicular to an electric field.

a Draw a labelled diagram to show the orientation of the magnetic and electric fields. [2]

b Write an equation for the force acting on the ions due to the electric field. [1]

c Write an equation for the force acting on the ions due to the magnetic field. [1]

d Calculate the field strength *E* of the electric field. [2]

23 a Define the term capacitance and the farad. [2]

b Derive an expression for two capacitors in series. [3]

c Three initially uncharged capacitors of capacitance 1.5 μF, 0.25 μF and 2.2 μF are connected to a 12 V battery as shown.

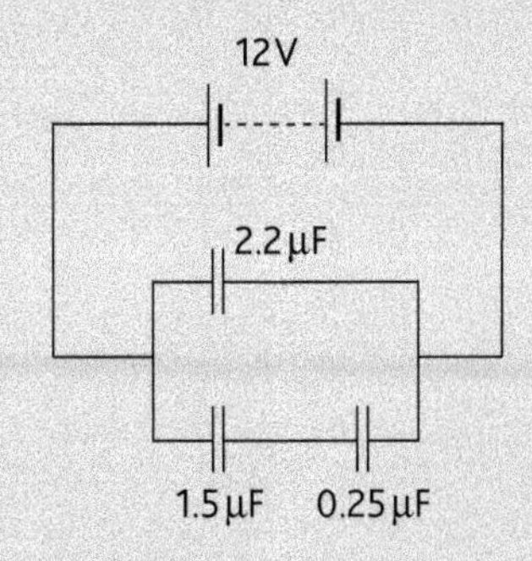

Calculate:

- **i** the capacitance of the 1.5 μF and 0.25 μF in series [2]
- **ii** the combined capacitance in the circuit [1]
- **iii** the total charge supplied by the battery [2]
- **iv** the charge on the 2.2 μF capacitor [1]
- **v** the charge on the 1.5 μF and 0.25 μF capacitors [2]
- **vi** the energy stored in the 1.5 μF capacitor. [2]

24 The diagram below shows a circuit of four similar resistors connected to a battery of e.m.f. 9.0 V and internal resistance of 0.1 Ω.

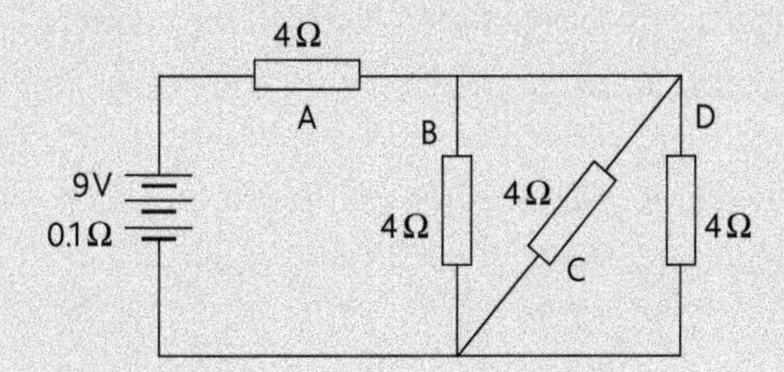

Calculate:

- **a** the equivalent resistance of the external circuit [3]
- **b** the current flowing through the battery [2]
- **c** the potential difference across resistors A, B, C and D [4]
- **d** the current flowing through each of the resistors A, B, C and D. [4]

25 a Define the term magnetic flux density and give its SI unit. [3]

b Sketch a diagram to show the magnetic field around a long straight current-carrying conductor. [2]

c A coil of wire consisting of two loops is suspended from a fixed point as shown.

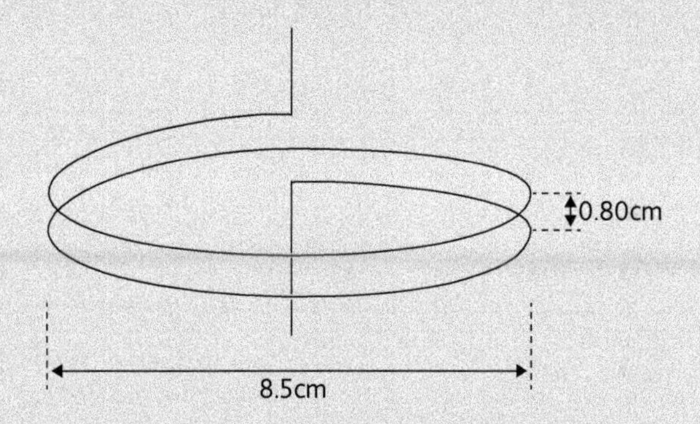

Each loop of the wire has a diameter of 8.5 cm and the separation of the loops is 0.80 cm. The coil is connected into a circuit such that the lower end of the coil is free to move.

- **i** When a current flows throught the coil, the separation of the loops of the coil decreases. Explain why this occurs. [4]
- **ii** Each loop of the coil may be considered as a long straight wire. When the current is flowing through the coil, a mass of 0.35 g is hung from the free end of the coil in order to return the loops of the coil to their original separation of 0.80 cm. Calculate the current flowing through the coil. [4]

26 A uniform electric field exists between two horizontal metal plates. The separation of the plates is 12 mm and a potential difference of 1000 V is applied across the plates. A particle O of charge 4.8×10^{-19} C and mass 5.0×10^{-27} kg starts from rest at the lower plate and is moved vertically to the top plate by the electric field.

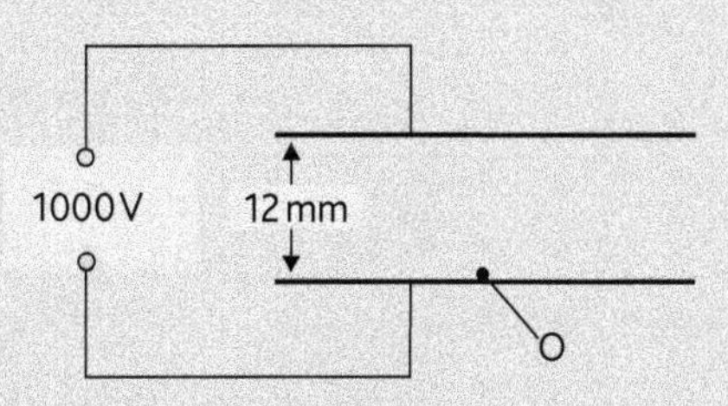

Calculate:

- **a** the electric field strength between the plates [2]
- **b** the work done on O by the electric field [2]
- **c** the gain in gravitational potential energy of O [2]
- **d** the gain in kinetic energy of O [1]
- **e** the speed of O when it reaches the top plate. [2]

8 Alternating currents

8.1 Alternating currents

Learning outcomes

On completion of this section, you should be able to:

- use the terms *frequency*, *peak value* and *root mean square value* when applied to an alternating current or voltage
- use the equation $x = x_0 \sin \omega t$ to represent an alternating current or voltage
- discuss the advantages of using alternating currents and high voltages for the transmission of electrical energy.

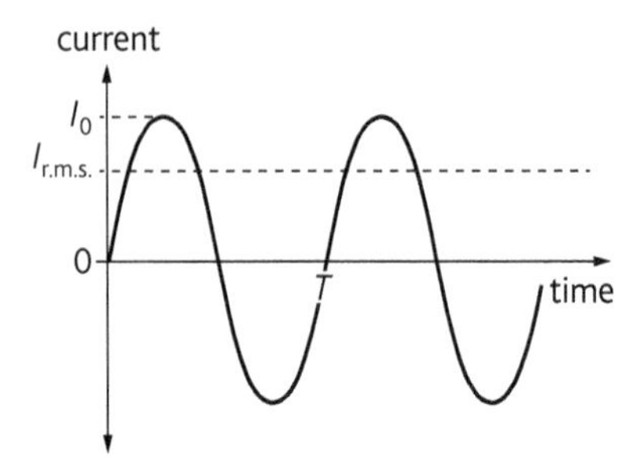

Figure 8.1.1 An alternating current

Alternating currents

Electricity reaching your home is transmitted using an alternating current (a.c.). An alternating current differs from a direct current (d.c.) in that the flow of electrons reverses direction regularly with time. A direct current flows in one direction only. An alternating current can be represented graphically as shown in Figure 8.1.1.

The alternating current shown can be represented mathematically by the following equation.

Equation

$I = I_0 \sin \omega t$

I – the value of the current (at time t)/A
I_0 – the peak value of the current/A
ω – angular frequency/rad s^{-1}
t – time/s

$\omega = 2\pi f$

ω – angular frequency/rad s^{-1}
f – frequency/Hz

The **peak value** of the current I_0 is the maximum value of the current.

The **period** T of the alternating current is defined as the time taken for one complete cycle.

The **frequency** f is the number of complete cycles generated per second.

The frequency and period are related by the following equation:

$$f = \frac{1}{T}$$

Power considerations

The current and the voltage supplied by a power station vary sinusoidally with time. As a result the power supplied also varies sinusoidally with time. This means that there are times when the supplied power is zero and times when it is at a maximum. Figure 8.1.2 shows how the power supplied by the a.c. mains varies with time.

Definition

The **root mean square** value of an alternating current, $I_{r.m.s.}$, is the value of the steady direct current which delivers the same average power as the a.c. to a resistive load.

Equation

$$I_{r.m.s.} = \frac{I_0}{\sqrt{2}}$$

$I_{r.m.s.}$ – root mean square current/A
I_0 – peak value of current/A

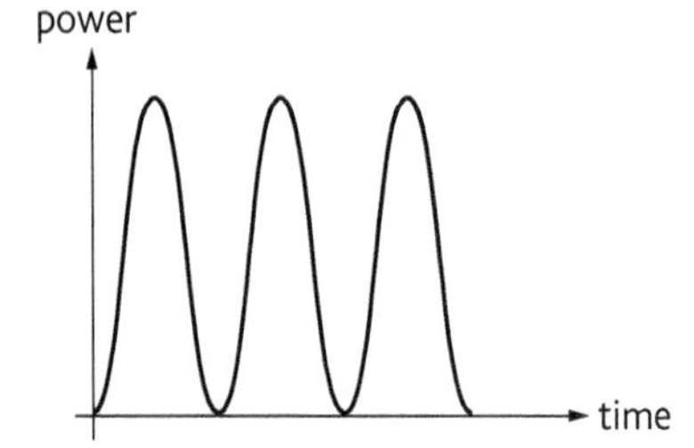

Figure 8.1.2

Example

An alternating power supply is represented by the following equation in volts:

$$V = 230 \sin 314t$$

State:

a the peak value of the power supply

b the r.m.s. value of the power supply

c the frequency of the power supply.

---

a $V_0 = 230\,\text{V}$

b $V_{\text{r.m.s.}} = \dfrac{V_0}{\sqrt{2}} = \dfrac{230}{\sqrt{2}} = 163\,\text{V}$

c Comparing the voltage expression with $V = V_0 \sin \omega t$ we can see that

$$\omega = 2\pi f = 314$$

$$\therefore \quad f = \frac{314}{2\pi} = 50\,\text{Hz}$$

Transmission of electrical energy

Electrical energy is generated in power stations. This energy is then distributed by transmission cables to homes and factories. The power supplied by the power station is given by $P = IV$. The power loss in the transmission cables is given by $P = I^2R$. The power station can either transmit the power at a high voltage and low current or a low voltage and high current. The transmission network is designed to transmit power by using a.c. at very high voltages. That is to say, for a given power, electrical energy is transmitted at high voltages and low currents. The reason for doing this is that the power loss in the transmission cables is proportional to the square of the current flowing through them. By making the current smaller, power losses in the transmission lines are minimised.

Alternating current is used because it can easily be stepped up or stepped down by using transformers (see 8.2). Transformers are very efficient at stepping up and stepping down voltages. They operate using the principle of electromagnetic induction, which is possible because an alternating magnetic field is produced by the alternating current. Transformers do not operate with direct current, however, because there is no changing magnetic field.

Key points

- In an alternating current the flow of electrons reverses direction regularly with time.
- The peak value of an alternating current is the maximum value of the current.
- The root mean square $I_{\text{r.m.s.}}$ value of an alternating current is the value of the steady direct current which delivers the same average power as the a.c. to a resistive load.
- Electrical energy is transmitted using a.c. at high voltages.

Learning outcomes

On completion of this section, you should be able to:

- explain the principle of operation of a simple transformer
- use the relationship $\frac{N_s}{N_p} = \frac{V_s}{V_p} = \frac{I_p}{I_s}$ for an ideal transformer
- discuss the energy losses in a transformer and state how they are minimised.

The simple iron-cored transformer

Transformers are devices used to change the voltage of an alternating power supply. Figure 8.2.1 shows a simple iron-cored transformer.

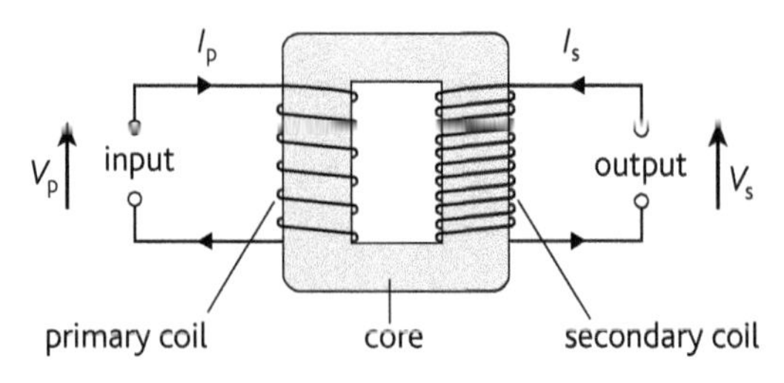

Figure 8.2.1 *A simple iron-cored transformer*

A transformer consists of a primary and a secondary coil wound around a soft iron core. When an alternating supply is connected to the primary coil, the alternating current produces an alternating magnetic field in the soft iron core. Since the secondary coil is situated in this alternating magnetic field, an alternating e.m.f. is induced in it. It should be noted that the two coils are electrically isolated from each other. Energy is transmitted from the primary coil to the secondary coil via the magnetic field in the soft iron core. Transformers do not work with direct currents because they do not produce a varying magnetic field to induce an e.m.f. in the secondary coil.

The primary coil has N_p turns of copper wire and the secondary coil has N_s turns of copper wire. If we apply an alternating voltage V_p on the primary coil we get an alternating voltage V_s on the secondary coil. The currents in the primary and secondary coils are I_p and I_s respectively. For an ideal transformer the input power is equal to the output power.

Equation

For an ideal transformer (no power losses)

Power input = power output

$$I_pV_p = I_sV_s$$

I_p – current in the primary coil/A
V_p – voltage across primary coil/V
I_s – current in the secondary coil/A
V_s – voltage across the secondary coil/V

There is a relationship that relates the voltages across the coils to the number of turns in each coil. The ratio of the voltages is equal to the ratio of the number of turns. This means that if we adjust the turns ratio, we can adjust the output voltage. Therefore a transformer can either step up or step down a voltage depending on the turns ratio.

For a step-up transformer $N_s > N_p$ and the transformer will step up the voltage.

For a step-down transformer, $N_s < N_p$ and the transformer will step down the voltage.

Equation

$$\frac{N_p}{N_s} = \frac{V_p}{V_s}$$

N_p – number of turns of wire on primary coil
V_p – voltage across primary coil/V
N_s – number of turns of wire on secondary coil
V_s – voltage across secondary coil/V

Exam tip

- Either use peak values throughout the calculations or use r.m.s. values throughout the calculations. Do not mix the two.
- Whenever sketching a diagram of a step-up transformer, ensure that there are more turns on the secondary coil than on the primary coil.
- Whenever sketching a diagram of a step-down transformer, ensure that there are fewer turns on the secondary coil than on the primary coil.
- A transformer will step up a voltage by a certain factor, but it will reduce the current by that same factor. Remember the power output cannot be greater than the power input.

Energy losses in a transformer

An ideal transformer has 100% efficiency and therefore has no power losses. In a real transformer, there are power losses. A real transformer can be about 98% efficient. Transformers lose some power because of the following:

- Thermal energy is lost because of the resistance of the windings of the coils ($P = I^2R$). In order to reduce this, the windings are made of thick copper.
- Thermal energy is lost in the core of the transformer. The core warms up as the magnetic flux continually reverses direction. (The core is made of soft iron instead of steel because soft iron magnetises and demagnetises more easily. Soft iron is therefore said to have a high magnetic permeability.)
- Thermal energy is lost in eddy currents. As the soft iron core is situated in a changing magnetic field, e.m.f.s are induced in the soft iron core at right angles to the magnetic field. Since the soft iron core is a conductor, currents begin flowing. These currents are known as **eddy currents**. These currents cause energy to be dissipated ($P = I^2R$) in the iron core. In order to reduce the energy losses due to eddy currents, the soft iron core is made up of numerous thin sheets of metal rather than a single block of iron. e.m.f.s are still induced in the soft iron core, but the induced currents produced cannot flow easily between the sheets of metal.

Example

An ideal transformer is constructed such that the ratio of the number of secondary turns to the number of primary turns is 1 : 15. A 240 V supply is connected to the primary coil and a 4 Ω resistor is connected to the secondary coil. Calculate the current in the primary coil.

$$\frac{V_p}{V_s} = \frac{N_p}{N_s} = \frac{15}{1}$$

$$\frac{240}{V_s} = \frac{15}{1}$$

Voltage across secondary coil $V_s = \frac{240}{15} = 16\,\text{V}$

Current flowing through 4 Ω resistor $= \frac{V_s}{R} = \frac{16}{4} = 4\,\text{A}$

$$I_pV_p = I_sV_s$$

$$I_p = \frac{I_sV_s}{V_p}$$

Therefore, current flowing in primary coil $= \frac{16 \times 4}{240} = 0.27\,\text{A}$

Key points

- A transformer can step up or step down an alternating voltage.
- Transformers only work on a.c.
- Energy is lost in a transformer as a result of eddy currents, resistance of the copper windings and the changing magnetic field in the iron core.

8.3 Semiconductors

Learning outcomes

On completion of this section, you should be able to:

- describe the electrical properties of semiconductors and distinguish between p-type and n-type materials
- explain the formation of the depletion layer at a p-n junction
- discuss the flow of current when the p-n junction diode is forward-biased or reverse-biased
- discuss the *I–V* characteristic of the p-n junction diode
- recall that a junction transistor is basically two p-n junctions.

Definitions

A p-type material is one in which the majority of charge carriers are holes.

An n-type material is one in which the majority of charge carriers are electrons.

Definition

The depletion region is the region on either side of the p-n junction where there are no net charge carriers.

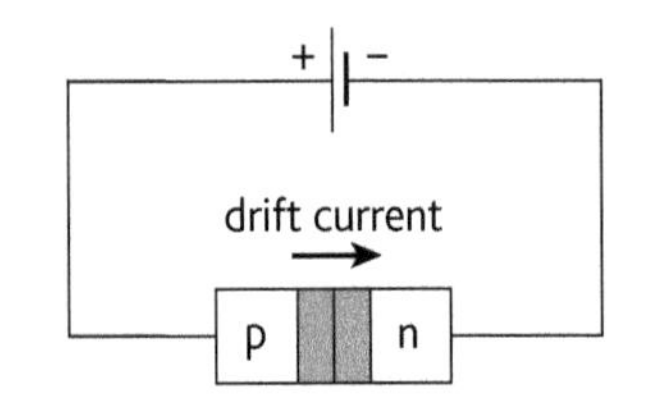

Figure 8.3.2 *Forward-biased junction*

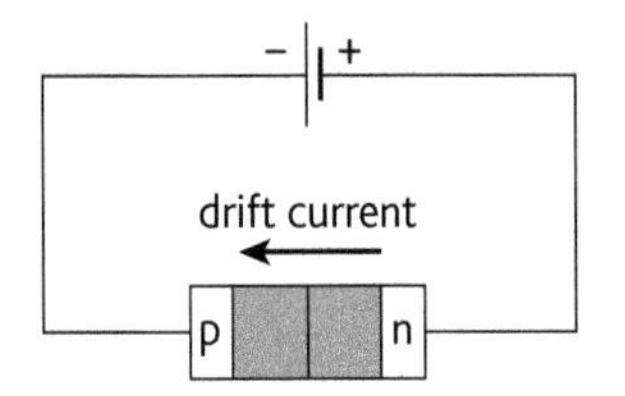

Figure 8.3.3 *Reverse-biased junction*

p-type and n-type materials

Semiconductor materials, as the name implies, have conductivities that make them neither a good insulator nor a good conductor. Pure elements such as silicon and germanium are called **intrinsic semiconductors**. In a silicon atom, there are four electrons in its outermost or valence shell. In its pure state, all the electrons are held tightly in covalent bonds. In a process called **doping**, an impurity (another element) is added to the silicon to enhance its conductive properties. If an element having five electrons in its valence shell (e.g. phosphorus) is added to silicon, an **n-type material** is formed. Four of the valence shell electrons are used for bonding and there is one left over. There is therefore an excess of negative charge carriers (electrons) which allows for conduction. If an element having three electrons in its valence shell (e.g. boron) is added to silicon, a **p-type material** is formed. The three valence shell electrons are used for bonding. This means that one bond is short of one electron. This deficiency is referred to as a **hole**. There is therefore an excess of positive charge carriers (holes) which allows for conduction.

The amount of impurity added to the silicon is approximately 1 part impurity to 10^6 parts pure semiconductor. This means that only 1 in 10^6 atoms produce electron–hole pairs for conduction. The amount added affects the resistivity of the doped semiconductor. The resistivity of silicon is approximately 0.1–60 Ω m.

In a pure metal such as copper free electrons are the main charge carriers. Each copper atom produces at least one conduction electron. Copper therefore has a low resistivity ($\rho = 1.68 \times 10^{-8}\,\Omega\,m$).

The p-n junction

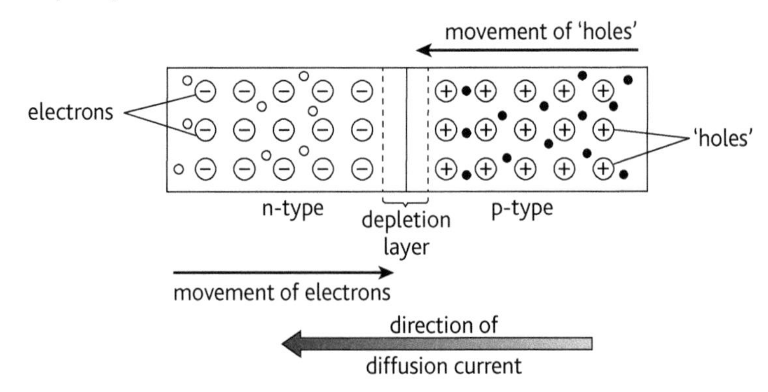

Figure 8.3.1 *The p-n junction*

Consider what happens when a p-type material is placed against an n-type material. The charge carriers in the n-type material (electrons) diffuse across the p-n boundary into the p-type material. The charge carriers in the p-type material (holes) diffuse across the p-n boundary into the n-type material. The n-type material has lost some electrons and acquires a positive potential with respect to the p-type material. This prevents further movement of electrons across the boundary. The p-type material has gained electrons and acquires a negative charge with respect to the n-type material. This prevents further movement of holes across the boundary. After some time, the movement of electrons and holes stops because of the potential difference across the junction. This potential difference is called the contact or **barrier potential**. The region around the boundary that has become depleted of electrons and holes is called the **depletion region** or **depletion layer**. The depletion region is about 1 μm thick.

A **diffusion current** is a current that occurs because of a difference in the concentration of holes and electrons in a p-n junction (Figure 8.3.1).

Forward-biased p-n junction

If a cell is connected across a p-n junction so that the positive terminal is connected to the p-type material and the negative terminal is connected to the n-type material (Figure 8.3.2), the junction is said to be **forward-biased**. 'Holes' in the p-type material are repelled by the positive terminal of the cell. Electrons in the n-type material are repelled by the negative terminal of the cell. When the applied voltage is greater than the barrier potential, electrons are pushed from the n-type to p-type material. At the same time, 'holes' are pushed from the p-type to the n-type material. The depletion region then decreases in width. A **drift current** flows across the p-n junction. This is a current that occurs when a potential difference is applied across a p-n junction.

Reverse-biased p-n junction

If a cell is connected across a p-n junction so that the positive terminal is connected to the n-type material and the negative terminal is connected to the p-type material (Figure 8.3.3), the junction is said to be **reverse-biased**. 'Holes' in the p-type material are attracted to the negative terminal of the cell. Electrons in the n-type material are attracted to the positive terminal of the cell. Electrons and 'holes' move away from the junction and the depletion region increases in width. The barrier potential also increases. Eventually, electrons and 'holes' stop moving and the drift current ceases to flow.

A p-n junction is referred to as a junction diode. The junction diode is used to rectify alternating currents.

The *I–V* characteristic of a semiconductor diode

Figure 8.3.4 shows the *I–V* characteristic of a diode. When the potential difference across the diode is positive, the diode is in the forward-biased configuration. Between 0 V and 0.7 V, no current flows through the diode. When the potential difference increases beyond 0.7 V, the diode begins conducting. A drift current now flows through it.

When the potential difference across the diode is negative, the diode is in the reverse-biased configuration. The current flow is practically zero. The minimal current that does flow is called the leakage current. At a particular voltage, called the breakdown voltage, the diode suddenly conducts a very large current.

The junction transistor

The junction transistor is essentially made up of two p-n junctions. It can be made such that a p-type material is sandwiched between two n-type materials (Figure 8.3.5). This type of transistor is called an n-p-n transistor. It can also be made such that an n-type material is sandwiched between two p-type materials (Figure 8.3.6). This type of transistor is called a p-n-p transistor.

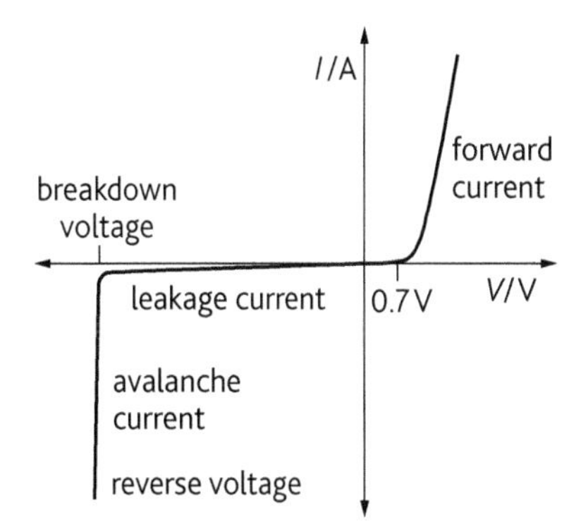

Figure 8.3.4 *The I–V characteristic of a semiconductor diode*

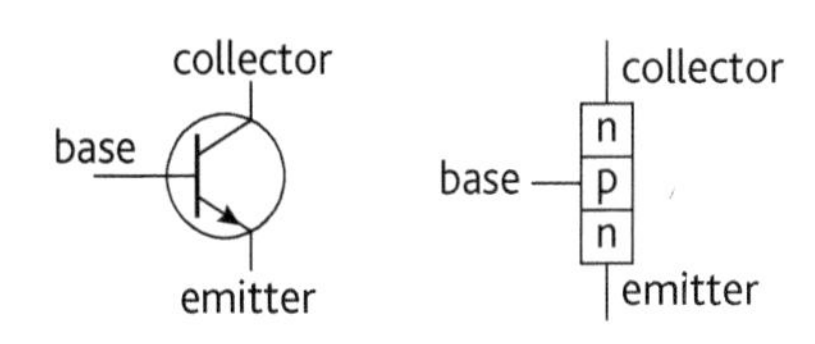

Figure 8.3.5 *An n-p-n transistor*

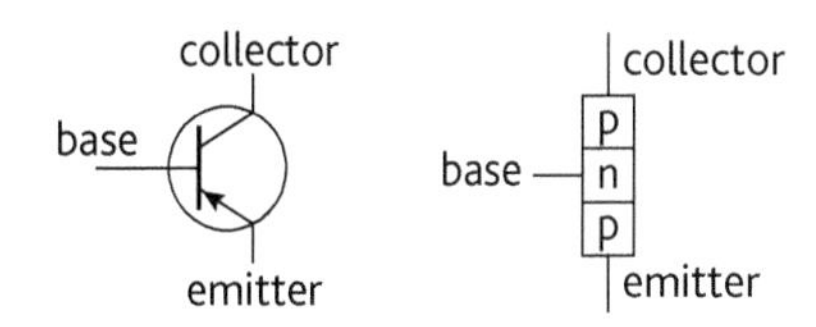

Figure 8.3.6 *A p-n-p transistor*

Key points

- A semiconductor material is neither a good conductor nor a good insulator.
- Doping is the process of adding a controlled amount of an impurity to a semiconductor to enhance its conductive properties.
- A p-type material is one in which the majority of charge carriers are holes.
- An n-type material is one in which the majority of charge carriers are electrons.
- The depletion region is the region on either side of the p-n junction where there are no net charge carriers.
- The junction transistor is essentially two p-n junctions.

Learning outcomes

On completion of this section, you should be able to:

- use a diode for half-wave rectification
- use the bridge rectifier for full-wave rectification
- represent half-wave and full-wave rectification graphically
- discuss the use of a capacitor for smoothing a rectified a.c. wave.

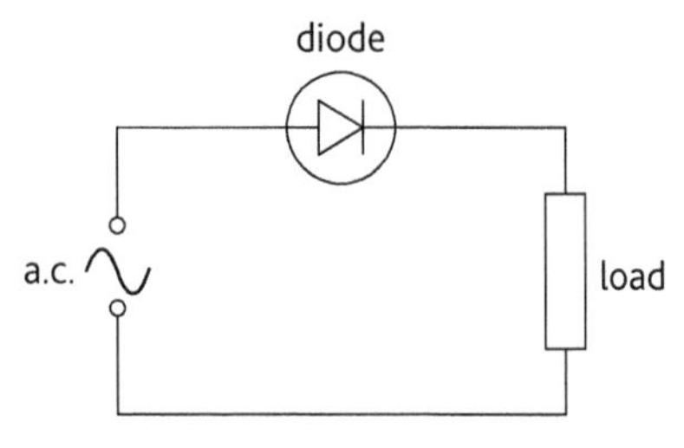

Figure 8.4.1 *Half-wave rectification*

Rectification

Rectification is a process whereby an a.c. voltage is converted into a d.c. voltage.

Half-wave rectification

A single diode can be used to rectify an a.c. voltage. Figure 8.4.1 shows how this is achieved. The input supply to the circuit is an alternating voltage. The graphs in Figure 8.4.2 show the input voltage and the potential difference across the load (output voltage). When the input voltage is positive (positive half-cycle), the diode is forward-biased and conducts an electric current. When the input voltage is negative, the diode is reverse-biased and does not conduct an electric current. The output current is zero in this half-cycle. Notice that the output voltage is always either positive or zero. The input voltage has been converted from an a.c. voltage into a d.c. voltage.

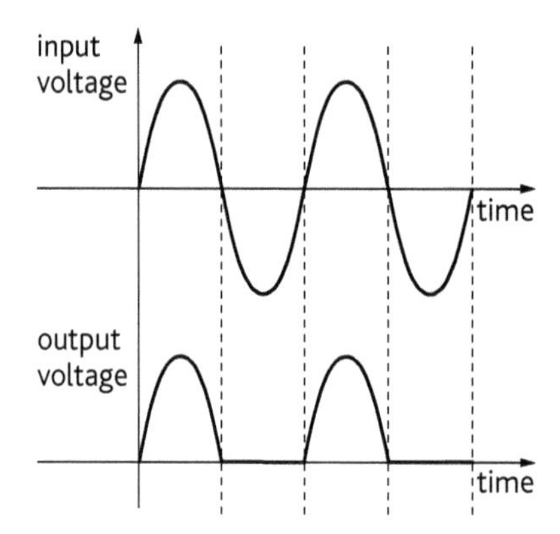

Figure 8.4.2 *Graphical representation of half-wave rectification*

Full-wave rectification

Four diodes can be used to rectify an a.c. voltage. When used in this manner, the process is called full-wave rectification. Figure 8.4.3 shows how this is achieved. The graphs in Figure 8.4.4 show the input voltage and the potential difference across the load (output voltage).

Consider the situation when the input voltage is positive (positive-half cycle). A current flows through D_2, because it is forward-biased. The current then flows through the load. On its return back to the a.c. power supply, the current flows through D_4. Consider the situation when the input voltage is negative (negative-half cycle). A current flows through D_3, because it is forward-biased. The current then flows through the load. On its return back to the a.c. power supply, the current flows through D_1. Notice that full-wave rectification is much more efficient at converting an a.c. voltage into a d.c. voltage than half-wave rectification. (Compare the output voltage graphs for both methods.)

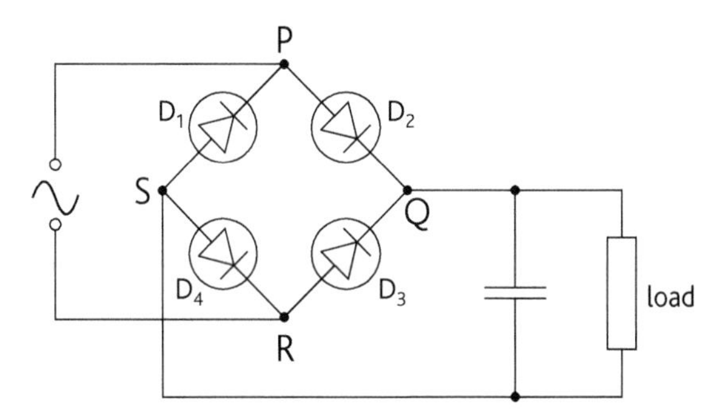

Figure 8.4.3 Full-wave rectification

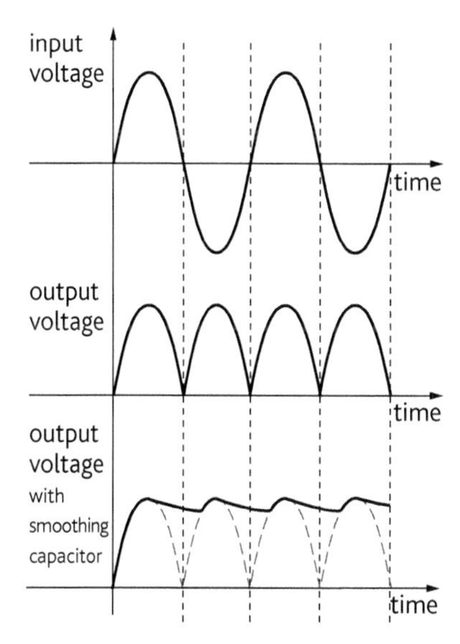

Figure 8.4.4 Graphical representation of full-wave rectification

Function of the capacitor

In Figure 8.4.3 a capacitor is inserted in parallel with the load. The function of this capacitor is to help smooth the output voltage so that it has fewer ripples. When the output voltage increases, the capacitor charges. As soon as the output voltage begins to fall, the capacitor begins to discharge through the load. The time constant for the circuit is $\tau = CR$. When a larger capacitance is placed in parallel with the load, the time constant increases. This means that the capacitor takes a longer time to discharge. The size of the ripples in the output voltage is made even smaller by increasing the size of the capacitor.

Key points

- Rectification is a process whereby an a.c. voltage is converted into a d.c. voltage.
- Half-wave rectification is achieved by using a single diode.
- Full-wave rectification is achieved by using four diodes.
- A capacitor inserted in parallel with the load helps reduce the ripples in the output voltage.
- Increasing the size of the capacitor increases the time constant in the circuit. This reduces the ripples in the output voltage even further.

Revision questions 4

Answers to questions that require calculation can be found on the accompanying CD.

1 a Explain what is meant by the root mean square (r.m.s.) value of an alternating voltage. [2]

b An alternating voltage V is represented by the equation $V = 120\sin(120\pi t)$, where V is measured in volts and t is in seconds.

Determine:

i the peak voltage [1]

ii the r.m.s. voltage [1]

iii the frequency. [1]

c The alternating voltage is applied across a resistor such that the mean power output from the resistor is 1.2 kW. Calculate the resistance of the resistor. [2]

2 An alternating voltage supply has a peak value of 20 V and frequency of 60 Hz.

Determine:

a the period [2]

b the root mean square voltage [2]

c the mean power dissipated in a resistor when the alternating supply is connected in series with a resistor of resistance 2.0 Ω. [2]

3 Explain what is meant by the following terms:

a A semiconductor [1]

b Doping [1]

c A p-type semiconductor [1]

d An n-type semiconductor [1]

e Depletion region [1]

4 a Describe the structure and principle of operation of an iron-cored transformer. [4]

b The peak power input to a transformer is 60 W. The transformer has a turns ratio of $N_s/N_p = 20$ and the sinusoidal input voltage has a value of $8\,V_{r.m.s.}$ Assuming that the transformer is ideal, calculate:

i the r.m.s. value for the output voltage [2]

ii the mean power input [1]

iii the r.m.s. value of the input current [2]

iv the r.m.s. value of the output current. [1]

5 a Explain:

i why the supply to the primary coil must be alternating current and not direct current [2]

ii why for a constant power input, the output current must decrease if the input voltage increases. [2]

b The graph below shows the variation with time t of the current I_p in the primary coil of a transformer.

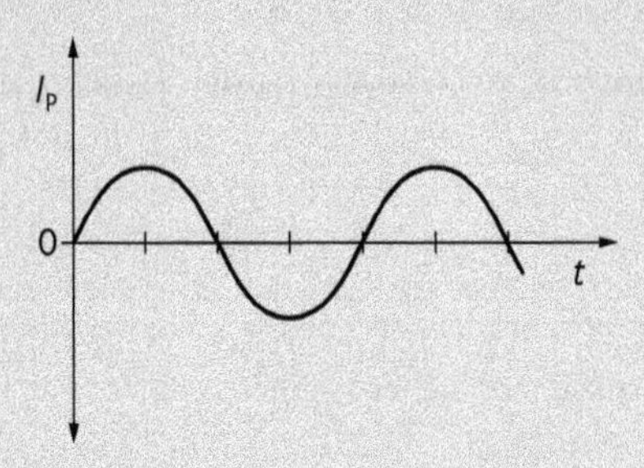

i Sketch a graph to show the variation with time of the magnetic flux in the core of the transformer. [2]

ii Sketch a graph to show the variation with time of the e.m.f. induced in the secondary coil. [3]

6 An ideal transformer has 4000 turns on its primary coil. It is used to convert a mains supply of $220\,V_{r.m.s.}$ to an alternating voltage having a peak value of 6.0 V. The output from the transformer is rectified using a full-wave rectifier and connected to a resistor R.

a Calculate the number of turns on the secondary coil of the transformer. [3]

b Sketch a diagram of a full-wave rectifier. [4]

c Sketch a graph to show the variation with time t of the voltage across the resistor. [2]

d Add a circuit component X to your diagram in **b** to show how the voltage across the resistor can be smoothed. [1]

e Explain how component X works in the circuit. [2]

f On your graph in **c** show what happens to the voltage across the capacitor when component X is added. [2]

7 For an ideal transformer:

a explain why the coils are wound on a core made of iron [1]

b explain why thermal energy is generated in the core. [2]

8 a Explain why it is necessary to use high voltages for the efficient transmission of electrical energy. [3]

b Explain why it is advantageous to use alternating current when transmitting electrical energy. [2]

9 In a real transformer, the core on which the primary and secondary coils are wound is laminated. This reduces energy losses due to currents induced in the core. Explain:

a how these currents arise in the core [3]

b why laminating the core reduces energy losses due to the currents. [2]

10 A charger for a car battery operates using a 120 $V_{r.m.s.}$ a.c. supply. The charger provides 10 A d.c. at 14.0 V. The battery charger is 90% efficient. Calculate the current drawn from the a.c. supply. [4]

11 An alternating supply is connected in series with a resistor R. The variation with time t of the current I in the resistor is given by the expression $I = 8.2 \sin(377t)$, where I is measured in amps and t is measured in seconds.

a Calculate the frequency of the supply. [2]

b Calculate the r.m.s. current. [2]

c Given that the mean power dissipated in the resistor cannot exceed 350 W, calculate the minimum value of R. [2]

12 A sinusoidal alternating voltage supply is connected to a bridge rectifier consisting of four diodes. The output of the rectifier is connected to a resistor R and a capacitor C.

a Sketch a diagram of the circuit described above. [3]

b State the function of the capacitor in the circuit. [1]

c The variation with time t of the potential difference V across the resistor R is shown below.

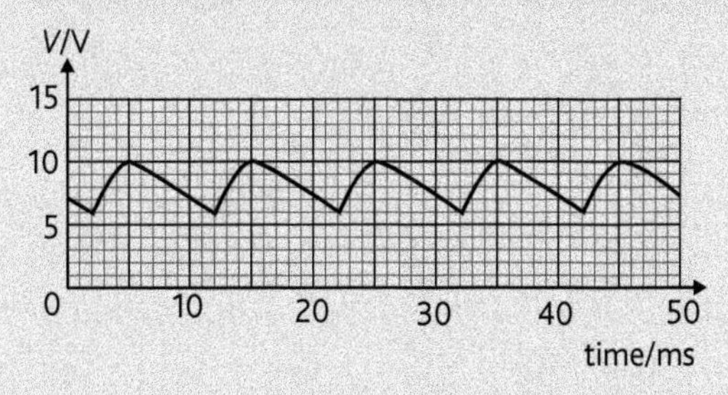

Using the graph:

i determine the peak voltage [1]

ii determine the root mean square voltage [1]

iii calculate the frequency of the alternating voltage. [2]

The capacitor C has a capacitance of 10.0 μF. For a single discharge of the capacitor through the resistor R, use the diagram to:

iv determine the change in potential difference [1]

v determine the change in charge on each plate of the capacitor [2]

vi calculate the average current in the resistor R [2]

vii estimate the value of R. [2]

9 Analogue electronics

9.1 Transducers

Learning outcomes

On completion of this section, you should be able to:

- explain the use of the light-dependent resistor (LDR), the thermistor and the microphone as input devices
- describe the operation of the light-emitting diode (LED), the buzzer and the relay as output devices.

Input devices

The LDR, thermistor and microphone are used as input devices in electronic circuits.

The light-dependent resistor (LDR)

A light-dependent resistor (LDR) is a device whose resistance decreases when exposed to sunlight. When placed in an enclosure (complete darkness), its resistance is approximately 10 MΩ. When exposed to bright sunlight, its resistance decreases to approximately 100 Ω. The symbol for an LDR is shown in Figure 9.1.1.

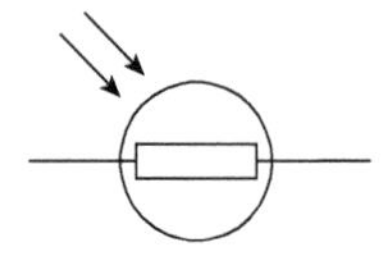

Figure 9.1.1 *The symbol for an LDR*

The thermistor

A thermistor is a device whose resistance varies with temperature. A thermistor having a negative temperature coefficient is one whose resistance decreases with increasing temperature. The symbol for a thermistor is shown in Figure 9.1.2.

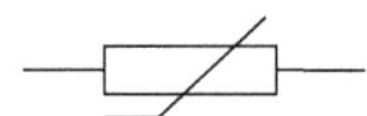

Figure 9.1.2 *The symbol for a thermistor*

The microphone

A microphone is a device that converts sound waves into electrical signals. The circuit symbol for a microphone is shown in Figure 9.1.3.

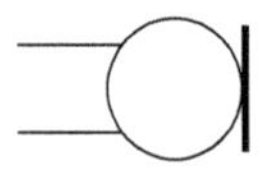

Figure 9.1.3 *The symbol for a microphone*

Output devices

The LED, buzzer and relay are used as output devices in electronic circuits.

The light-emitting diode (LED)

A light-emitting diode (LED) is a device that emits light when a current flows through it. Since it is a diode, an electric current can only flow through it in one direction. The LED only begins to conduct when the anode is approximately 1.8 V more positive than the cathode. When the diode begins to conduct, the potential difference across it remains at 1.8 V, whatever the magnitude of the current flowing through it. The symbol for an LED is shown in Figure 9.1.4. A protective resistor is usually connected in series with an LED. The resistor limits the current flowing through the LED and prevents damage to it.

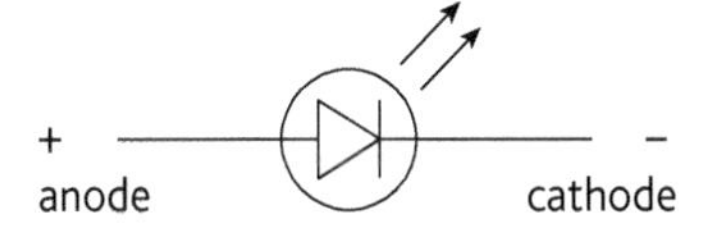

Figure 9.1.4 *The symbol for an LED*

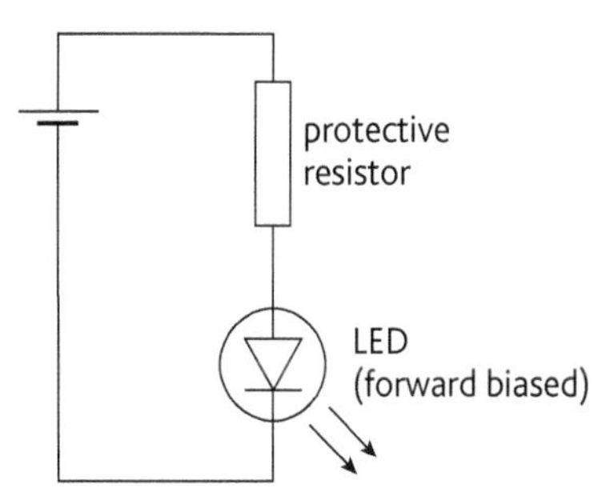

Figure 9.1.5 *LED with a protective resistor*

The buzzer

The buzzer is a device that produces sound when a power supply is connected to it. It consists of an electric oscillator circuit which is connected to a solid-state sounder. The symbol for the sounder is shown in Figure 9.1.6.

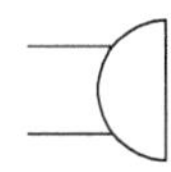

Figure 9.1.6 *The symbol for a buzzer*

The relay

A relay is a device that consists of one or more switch contacts which are controlled by an electromagnet. When a current flows through the electromagnet, the relay becomes energised and can either close or open the switch contacts. Relays can be 'normally open' (NO) or 'normally closed' (NC). For a normally open relay, the contacts close when the relay is energised. For a normally closed relay, the contacts open when the relay is energised. The symbol for a relay is shown in Figure 9.1.7.

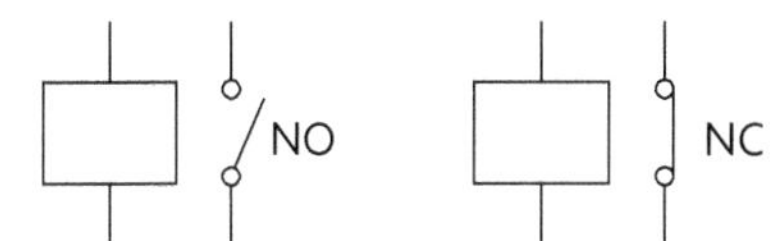

Figure 9.1.7 *The symbol for a relay*

Key points

- The LDR, thermistor and microphone are used as input devices in electronic circuits.
- The LED, buzzer and relay are used as output devices in electronic circuits.

9.2 Operational amplifiers

Learning outcomes

On completion of this section, you should be able to:

- describe the properties of the ideal operational amplifier
- compare the properties of a real op-amp with an ideal op-amp
- use an op-amp as a comparator
- understand the term *saturation*.

Properties of an ideal operational amplifier

An operational amplifier (op-amp) is an integrated circuit (IC). It has two inputs and one output. Sometimes there are more than just operational amplifiers in a single integrated circuit. There will also be connections for supplying power to the operational amplifier. Figure 9.2.1 shows an operational amplifier and the symbol used to represent it.

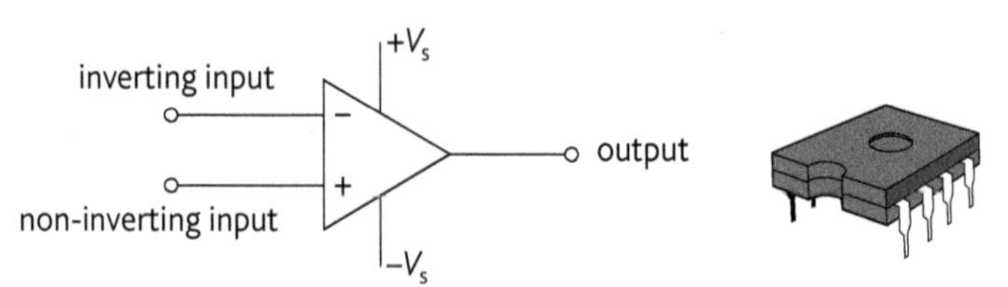

Figure 9.2.1 *An operational amplifier*

The properties of an ideal operational amplifier are as follows.

- It has an infinite input impedance.
- Essentially we are saying that no current enters it through either of its inputs.
- It has an infinite **open-loop gain**.
- This is the gain without positive or negative feedback (see 9.3). If there were only a slight difference between the input signals, the output would saturate (limited by the power supply voltage).
- It has zero output impedance.
- There will be no voltage drop across the output circuit of the operational amplifier.
- It has infinite bandwidth. (The bandwidth of an amplifier is the range of frequencies for which the gain is constant.)
- It amplifies all frequencies in an input signal by the same amount.
- It has an infinite slew rate. (The slew rate of an amplifier is the rate of change in the output voltage caused by a step change in the input voltage.)
- If a sudden change were made to the input voltage, the output would change instantly without any delay.

A real operational amplifier behaves as follows.

- The input impedance is approximately $10^{12}\,\Omega$.
- The open-loop gain is approximately 10^5 for d.c. signals.
- The output impedance is approximately $100\,\Omega$.
- The bandwidth is limited.
- The slew rate is limited.

Figure 9.2.2 shows how power is connected to an operational amplifier. Note that all voltages are measured with respect to a common potential (earth).

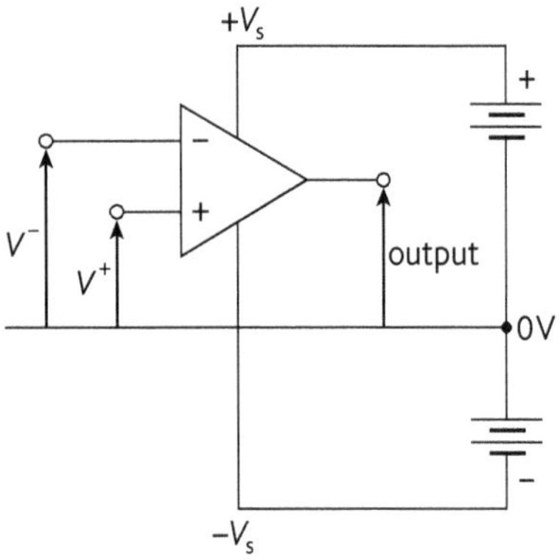

Figure 9.2.2 *Wiring an operational amplifier*

Using an operational amplifier as a comparator

An operational amplifier can be used to compare two voltages. In Figure 9.2.2, the operational amplifier is said to be in open-loop mode. This is because there is no feedback connection from the output of the operational amplifier to either of the inputs.

In order to determine the output voltage, you simply multiply the difference in voltage between the inverting and non-inverting terminals by the open-loop gain.

- If $V^+ > V^-$ the output voltage will be positive. (Positive saturation $+V_s$)
- If $V^+ < V^-$ the output voltage will be negative. (Negative saturation $-V_s$)
- If $V^+ = V^-$ the output voltage will be zero.

When the operational amplifier is set up as shown in Figure 9.2.3, it is being used as a comparator. The gain of the operational amplifier is typically 10^5.

Suppose $V^+ = 1.5\,\text{V}$ and $V^- = 1.8\,\text{V}$ and the supply voltage to the operational amplifier is $+15\,\text{V}$ and $-15\,\text{V}$.

Then $V_{out} = A_0(V^+ - V^-) = 10^5(1.5 - 1.8) = -30\,000\,\text{V}$

The maximum output voltage of the operational amplifier can either be +15 V or −15 V. Therefore, the output voltage will be −15 V and not −30 000 V. The output is said to be **saturated**.

Normally, when an operational amplifier is being used as a comparator, potential divider circuits are used to provide the voltages at its inputs. One of the inputs is fixed at a particular voltage (the reference voltage). The voltage at the other terminal is then compared with the reference voltage.

We can use electrical components whose resistance varies with some physical property to design some practical circuits. For example, suppose you were working for an aircraft manufacturer. You were told that the temperature of the wings of an aircraft should never drop below a particular value or else icing of the wings would occur. You are asked to design a simple circuit to light an LED or sound a buzzer if the temperature drops below this value. A possible circuit is illustrated in Figure 9.2.3.

In this circuit, the voltage at the inverting terminal is fixed by the potential divider. A thermistor is used as the sensing device. The resistance of the thermistor varies with temperature. The thermistor is attached to the wing of the aircraft. The output of the circuit is connected to a buzzer or LED. When the temperature of the thermistor varies, the voltage at the non-inverting terminal varies as well. If the voltage at the non-inverting terminal is greater than the voltage at the inverting terminal, the op-amp will saturate to +15 V. This will cause the buzzer to sound or the LED to light indicating to the pilot that there is a potential problem.

Another use of a comparator is in the production of a square-wave voltage from an alternating input voltage. Figure 9.2.4 shows how a sine wave can be converted into a square wave. As soon as $V^- > 0$, the output saturates to −9 V. When $V^- < 0$, the output saturates to +9 V.

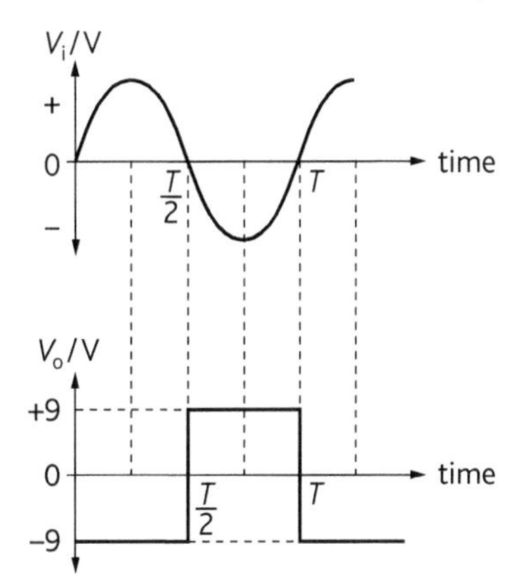

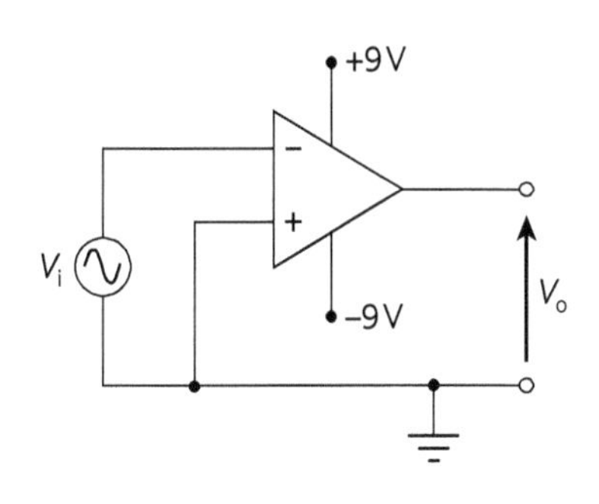

Figure 9.2.4

Equation

The output voltage of an operational amplifier is given by

$V_{out} = A_0(V^+ - V^-)$

V_{out} – output voltage/V
A_0 – open-loop gain
V^+ – non-inverting input/V
V^- – inverting input/V

Exam tip

Always note the supply voltage of the operational amplifier, as this cannot be exceeded. The output will always saturate to this value when the operational amplifier is in open-loop mode.

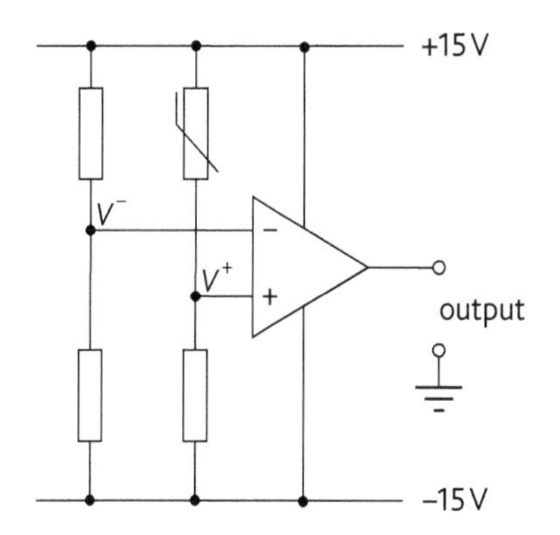

Figure 9.2.3 *Using an op-amp as a comparator.*

Key points

- An ideal operational amplifier has an infinite input impedance, an infinite open-loop gain, zero output impedance, infinite bandwidth and infinite slew rate.
- It has two inputs: inverting and non-inverting.
- An operational amplifier can be used as a comparator in open-loop mode.
- When used as a comparator, the output voltage cannot exceed the voltage of the power supply.

9.3 Inverting and non-inverting amplifiers

Learning outcomes

On completion of this section, you should be able to:

- represent an inverting and non-inverting op-amp with a single input
- use the concept of the virtual earth in the inverting amplifier
- derive an expression for the gain of an inverting and non-inverting amplifier
- explain the meaning of gain and bandwidth of an amplifier
- explain the gain–frequency curve for a typical op-amp
- determine bandwidth from a gain–frequency curve.

Feedback

Feedback is the process of taking a fraction of the output signal and adding it to the input signal being fed into an amplifier. There are two kinds of feedback. There is **negative feedback** and **positive feedback**.

When negative feedback is used, there is a reduction in gain of the amplifier. There are numerous advantages to using negative feedback. These include:

- increased bandwidth
- less distortion
- greater operating stability.

When positive feedback is used, there is increased instability and a reduction in bandwidth.

The gain of an inverting amplifier

The **gain** of an amplifier is the ratio of the output voltage to the input voltage.

Figure 9.3.1 shows an inverting amplifier with negative feedback.

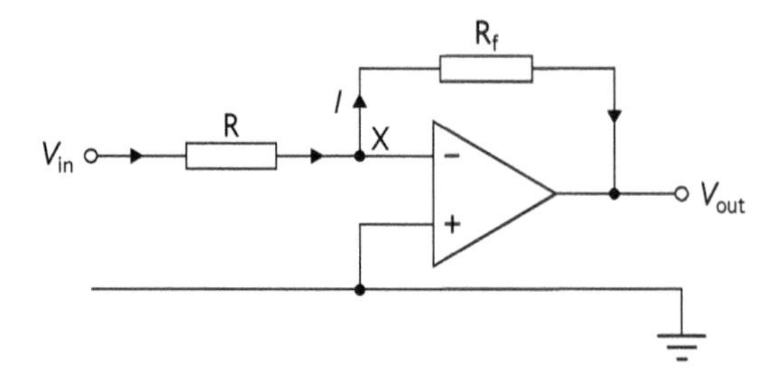

Figure 9.3.1 *An inverting amplifier*

Assumptions:

- Since the open-loop gain of the operational amplifier is high, the slightest difference in potential between the inverting and non-inverting terminals will cause the output voltage to be saturated. In order for the output of the operational amplifier not to be saturated, $V^+ = V^-$.
- No current enters the terminals of the operational amplifier since it has a high input impedance.

Since the potential at the non-inverting terminal is at earth potential (0 V), the potential at the inverting terminal has to be 0 V. This has to be true in order for the output of the operational amplifier not to be saturated. The point X (inverting terminal) is called a **virtual earth**.

Potential difference across R $= V_{in} - 0$

Potential difference across R_f $= 0 - V_{out}$

Since no current enters the operational amplifier:

$$\text{Current in R} = \text{current in } R_f$$

$$\frac{V_{in} - 0}{R} = \frac{0 - V_{out}}{R_f}$$

$$\frac{V_{in}}{R} = \frac{-V_{out}}{R_f}$$

$$-V_{out}R = V_{in}R_f$$

Voltage gain: $$\frac{V_{out}}{V_{in}} = -\frac{R_f}{R}$$

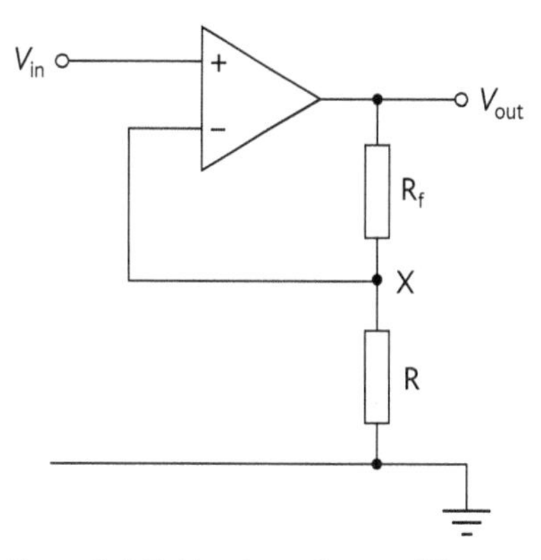

Figure 9.3.2 Non-inverting amplifier with negative feedback

The gain of a non-inverting amplifier

Figure 9.3.2 shows a non-inverting amplifier with negative feedback.

Assumptions:

- Since the open-loop gain of the operational amplifier is high, the slightest difference in potential between the inverting and non-inverting terminals will cause the output voltage to be saturated. In order for the output of the operational amplifier not to be saturated, $V^+ = V^-$.
- No current enters the terminals of the operational amplifier since it has a high input impedance.

Since the potential at the non-inverting terminal is V_{in}, the potential at the inverting terminal has to be V_{in} also. This has to be true in order for the output of the operational amplifier not to be saturated.

The potential at X is V_{in}. We can write the potential at X in terms of the output voltage by thinking of the resistors as making up a potential divider circuit.

$$V_{in} = \left(\frac{R}{R + R_f}\right)V_{out}$$

$$\frac{V_{out}}{V_{in}} = \left(\frac{R + R_f}{R}\right)$$

Voltage gain: $$\frac{V_{out}}{V_{in}} = 1 + \frac{R_f}{R}$$

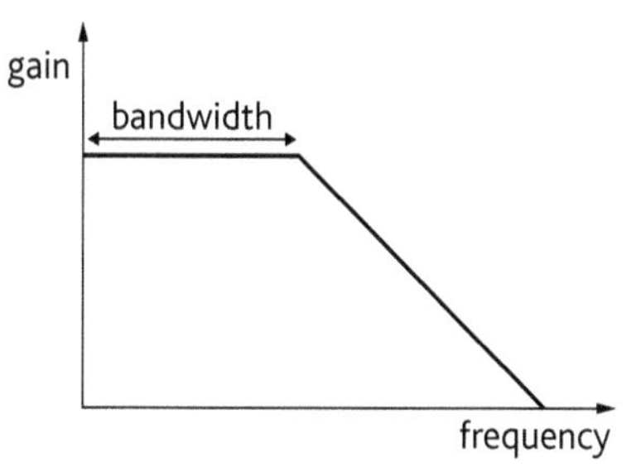

Figure 9.3.3 The frequency response curve of an amplifier

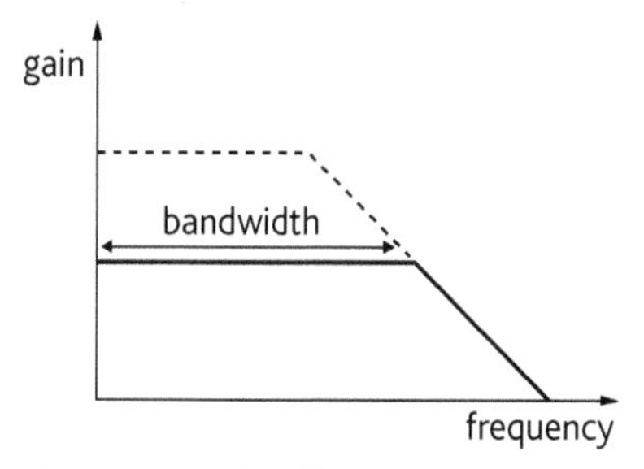

Figure 9.3.4 The effect of negative feedback on gain and bandwidth

The frequency response of an op-amp

Figure 9.3.3 shows the frequency response of an op-amp. The graph is obtained by varying the frequency of the input signal and measuring the gain of the op-amp at each input frequency. The **bandwidth** of an amplifier is the range of frequencies for which the gain of an amplifier remains constant. At low frequencies, the gain of the amplifier is high and constant. At higher frequencies, the gain decreases. The frequency at which the gain starts to decrease (i.e. the point at which there is a sudden change in the slope) is called the cut-off or break frequency.

A logarithmic scale is typically used on the gain and frequency axes. A logarithmic scale allows for a large range of values to be shown on a graph by compressing the larger values and expanding the smaller values. Gain is plotted on a logarithmic scale if it is to be expressed as decibels. The graph shows how bandwidth is determined from a frequency response curve. Figure 9.3.4 shows the effect on gain and bandwidth when negative feedback is used. As the gain is reduced, the bandwidth increases.

Key points

- Feedback is the process of taking a fraction of the output signal and adding it to the input signal being fed into an amplifier.
- The voltage gain of an inverting amplifier is $-\frac{R_f}{R}$.
- The voltage gain of a non-inverting amplifier is $1 + \frac{R_f}{R}$.
- When negative feedback is used, the gain of the amplifier decreases and the bandwidth increases.

9.4 The summing amplifier and the voltage follower

Learning outcomes

On completion of this section, you should be able to:

- describe the use of the inverting amplifier as a summing amplifier
- solve problems relating to summing amplifier circuits
- describe the use of the op-amp as a voltage follower.

The summing amplifier

A summing amplifier combines several inputs into one output signal. In the music industry, it is often required that several signals (guitar, piano and voice etc.) can be combined by using a summing amplifier.

The voltage gain of a summing amplifier

In order for the output not to be saturated, the potential at the inverting and non-inverting terminals of the op-amp must be the same. Since the potential at the non-inverting terminal is 0 V, the potential at the point X must also be at zero potential (virtual earth).

Assume that each input source is positive and currents are flowing as shown in Figure 9.4.1.

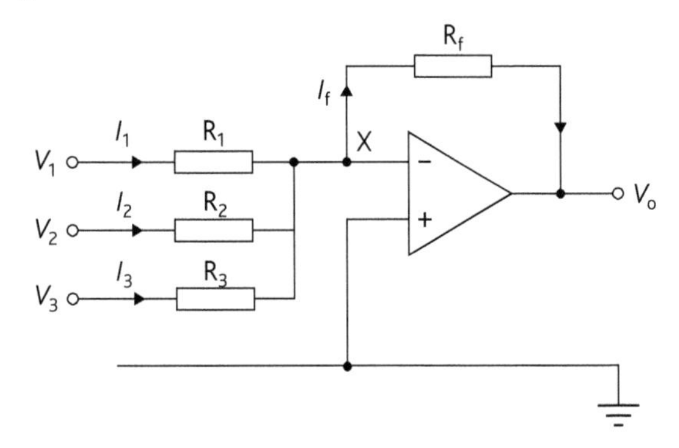

Figure 9.4.1 *A summing amplifier*

Using Kirchhoff's first law, we can write,

$$I_1 + I_2 + I_3 = I_f$$

$$\frac{V_1 - 0}{R_1} + \frac{V_2 - 0}{R_2} + \frac{V_3 - 0}{R_3} = \frac{0 - V_0}{R_f}$$

$$V_0 = -\left(\frac{R_f}{R_1}V_1 + \frac{R_f}{R_2}V_2 + \frac{R_f}{R_3}V_3\right)$$

If all the resistors have the same value, then

$$V_0 = -(V_1 + V_2 + V_3)$$

Example

An ideal operational amplifier is used in its inverting mode in the mixer circuit shown in Figure 9.4.2.

Calculate the output voltage.

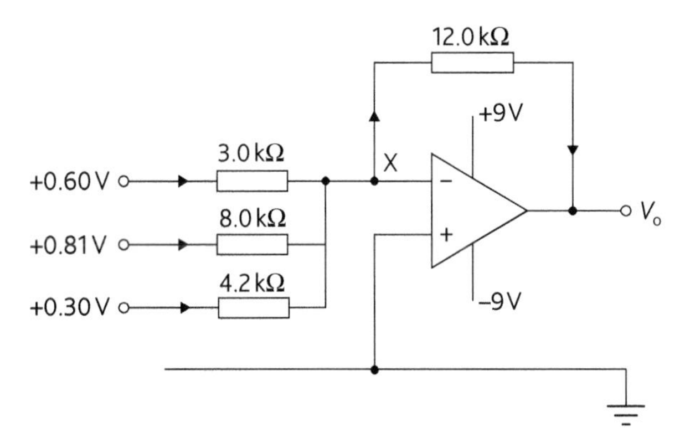

Figure 9.4.2

$$V_0 = -\left(\frac{R_f}{R_1}V_1 + \frac{R_f}{R_2}V_2 + \frac{R_f}{R_3}V_3\right)$$

$$V_0 = -\left(\frac{12}{3}(0.60) + \frac{12}{8}(0.81) + \frac{12}{4.2}(0.30)\right)$$

$$= -4.47\,\text{V}$$

Using a summing amplifier to perform digital to analogue conversion

In the world of electronics, analogue signals have to be converted into digital signals by a device called an analogue to digital converter (ADC). Digital signals can also be converted into an analogue signal.

A summing amplifier can convert a digital signal into an analogue signal. A simple 3-bit digital to analogue converter (DAC) is constructed using a summing amplifier as shown in Figure 9.4.3. A, B and C are the digital inputs. The values of the resistances are scaled to represent the weights of the different bits. The resistor R corresponds to the most significant bit (MSB). The resistor 4R corresponds to the least significant bit (LSB). Since there are three inputs, there are $2^3 = 8$ possible logic combinations. Each combination will correspond to a unique output voltage.

Suppose a logic 1 at an input is represented by a +5 V and that a logic 0 at an input is represented by 0 V. Suppose the output voltage of the DAC is V_{out}.

V_{out} is determined by $V_{out} = -\left(V_1 + \frac{1}{2}V_2 + \frac{1}{4}V_3\right)$

Therefore, if the input digital signal is 010, $V_{out} = -\left(0 + \frac{1}{2}(5) + \frac{1}{4}(0)\right)$

$$= -2.5\,\text{V}$$

Table 9.4.1 illustrates all the possible combinations of input and output.

Table 9.4.1

A	B	C	V_{out}/V
0	0	0	0.00
0	0	1	–1.25
0	1	0	–2.50
0	1	1	–3.75
1	0	0	–5.00
1	0	1	–6.25
1	1	0	–7.50
1	1	1	–8.75

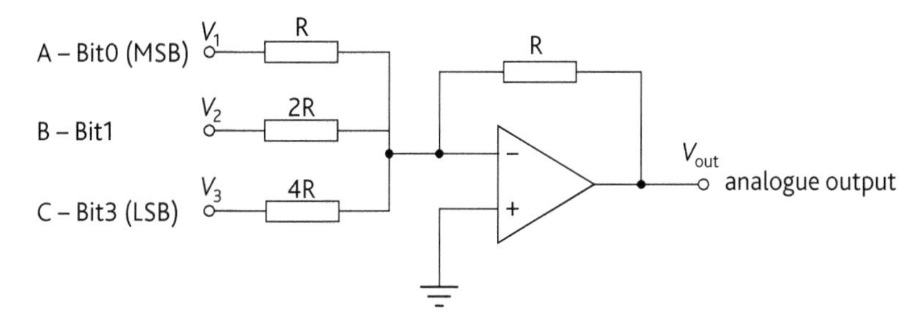

Figure 9.4.3 *A simple digital to analogue converter*

The voltage follower

The circuit in Figure 9.4.4 represents a voltage follower. The voltage gain of this circuit is 1. The output voltage is equal to the input voltage.

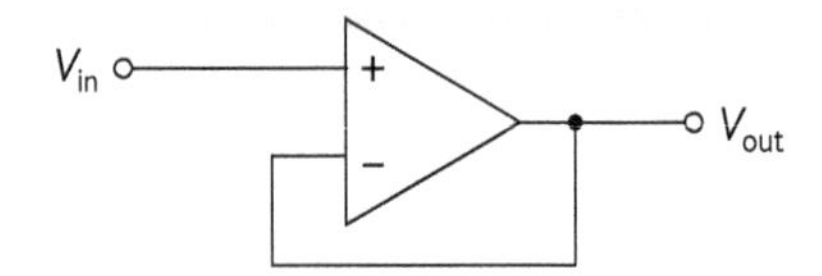

Figure 9.4.4 A voltage follower (buffer)

The voltage follower circuit is used to connect high impedance sources to low impedance loads. It is often referred to as a buffer.

Cascade amplifiers

Several amplifier circuits can be connected together in such a way that the output of one acts as the input of another. The amplifiers are said to be **cascaded**.

Suppose three amplifier circuits have voltage gains A_1, A_2 and A_3 respectively. When cascaded, the overall voltage gain is given by $A = A_1 \times A_2 \times A_3$.

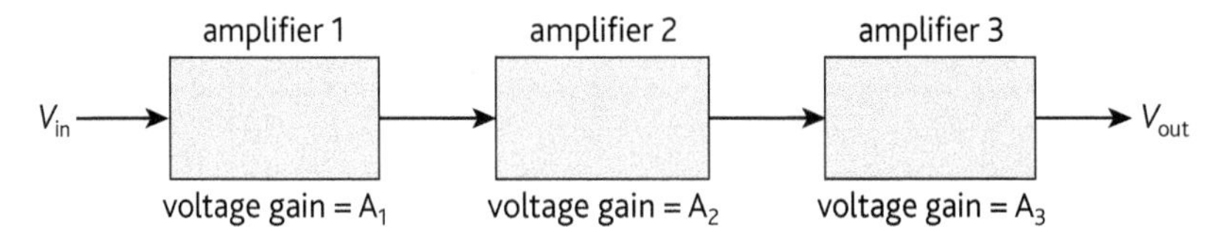

Figure 9.4.5 Cascaded amplifiers

Example

Figure 9.4.6 shows a cascade amplifier circuit. Calculate the overall voltage gain of the circuit.

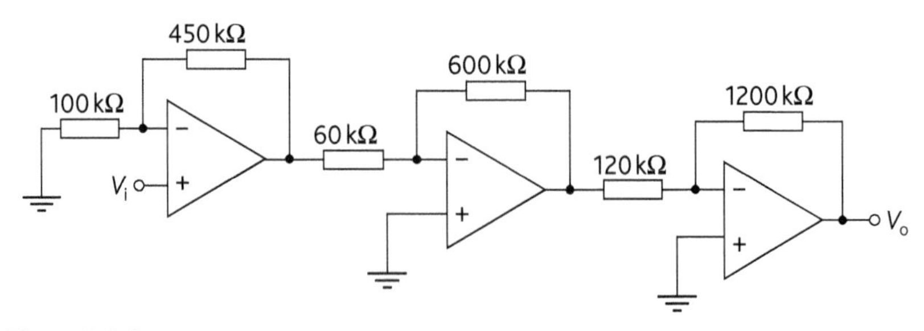

Figure 9.4.6

The first amplifier is non-inverting.

The voltage gain $A_1 = 1 + \frac{R_f}{R} = 1 + \frac{450}{100} = 5.5$

The second amplifier is inverting.

The voltage gain $A_2 = -\frac{R_f}{R} = -\frac{600}{60} = -10$

The third amplifier is inverting.

The voltage gain $A_2 = -\frac{R_f}{R} = -\frac{1200}{120} = -10$

Overall voltage gain $= A_1 \times A_2 \times A_3 = 5.5 \times -10 \times -10 = 550$

The transfer characteristic of an operational amplifier

Figure 9.4.7 shows a circuit that can be used to measure the open-loop gain of an operational amplifier. The power supply connected to the operational amplifier is $+V_s$ and $-V_s$. A signal generator is connected to the input of the operational amplifier. An oscilloscope is used to measure the input voltage V_i. An oscilloscope is used to measure the output voltage V_0. The frequency of the signal generator is set to a desired value. The input voltage V_i is then adjusted to a very low value (μV) and the corresponding output voltage V_0 is measured. Since the open-loop gain of the operational amplifier is very large (10^5), the output saturates when $V^+ > V^-$ or $V^+ < V^-$.

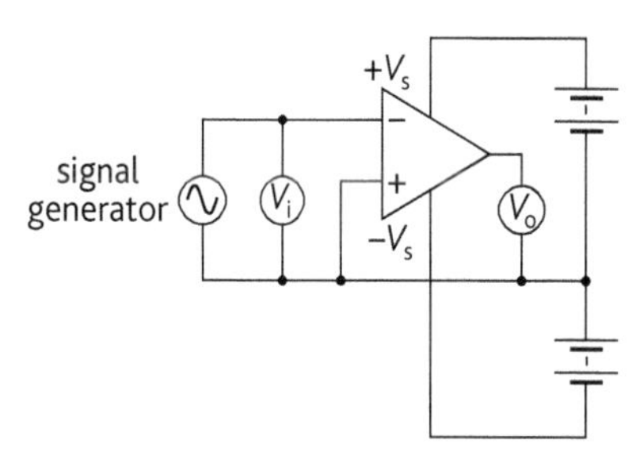

Figure 9.4.7 Circuit used to measure the open-loop gain of an operational amplifier

A plot of the output voltage V_0 against input voltage V_i is known as the transfer characteristic of the operational amplifier. Figure 9.4.8 shows the transfer characteristic of an operational amplifier. The open-loop gain of the operational amplifier is found by determining the gradient of the linear region of the graph.

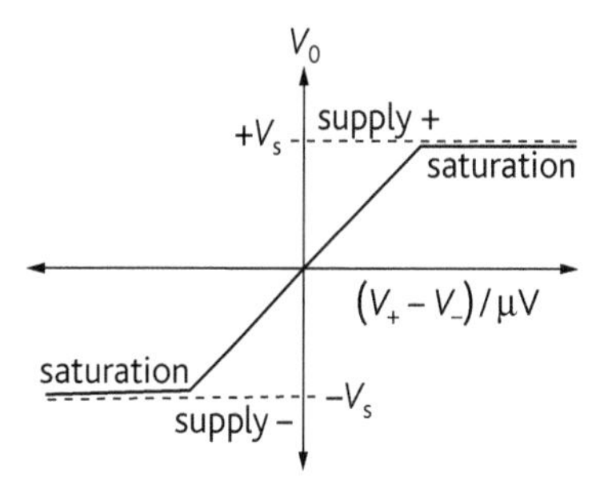

Figure 9.4.8 The transfer characteristic of an operational amplifier

The circuit in Figure 9.4.7 can be used to plot the frequency response of an operational amplifier. The frequency of the input signal is varied and the corresponding gain (V_0/V_i) of the operational amplifier is measured. Figure 9.4.9 shows the variation of the open-loop gain with frequency of the input signal.

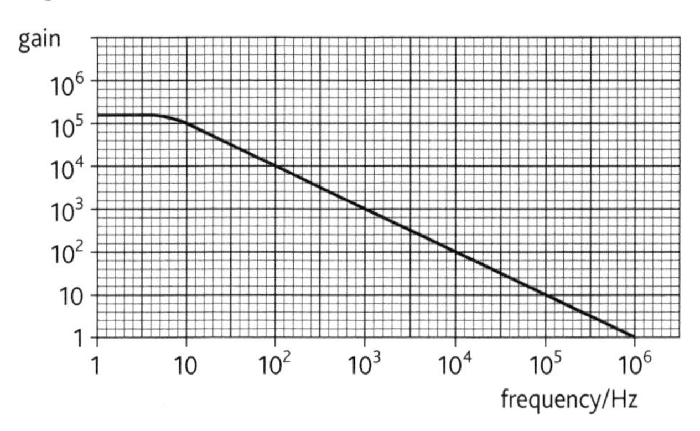

Figure 9.4.9

Example

A student is asked to design a circuit to combine two signals V_1 and V_2 to form signal V_0 according to the relationship $V_0 = -3V_1 - 4V_2$. The minimum input resistance for both signals should be no less than 100 kΩ. Design a circuit to achieve this goal.

This design requires the use of a summing amplifier with two inputs.

The output signal of a summing amplifier with two inputs is given by

$$V_0 = -\left(\frac{R_f}{R_1}V_1 + \frac{R_f}{R_2}V_2\right),$$

where R_f is the feedback resistance and R_1 and R_2 represent the input resistances for the two inputs.

Comparing this equation with $V_0 = -3V_1 - 4V_2$ we get

$$\frac{R_f}{R_1} = 3 \text{ and } \frac{R_f}{R_2} = 4$$

We also know that R_1 and R_2 must be greater than 100 kΩ.

The next step is to choose values of resistors that satisfy the conditions above.

One possible solution is R_f = 1200 kΩ, R_1 = 400 kΩ and R_2 = 300 kΩ.

Key points

- A summing amplifier combines several inputs into one output signal.
- The voltage follower circuit or buffer is used to connect high impedance sources to low impedance loads.
- The gain of a cascade amplifier is equal to the product of the gains of each amplifier in the circuit.

9.5 Analysing op-amp circuits

Learning outcomes

On completion of this section, you should be able to:

- analyse simple amplifier circuits
- analyse the response of amplifier circuits to input signals, using timing diagrams
- understand the term *clipping* when applied to amplifiers.

Example

Figure 9.5.1 shows the variation with frequency f of the voltage gain G, without feedback, of an ideal operational amplifier (op-amp). The op-amp is used in the amplifier circuit shown in Figure 9.5.2.

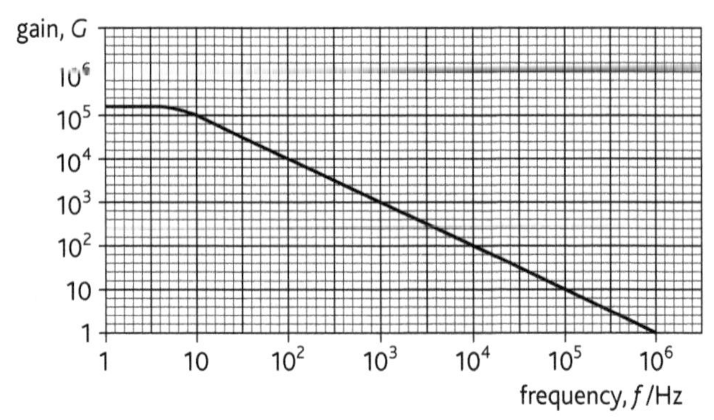

Figure 9.5.1

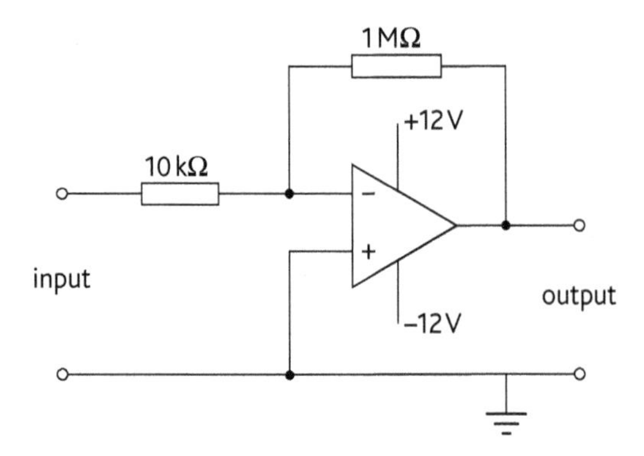

Figure 9.5.2

Calculate:

a the gain A, of the amplifier

b the bandwidth of the amplifier

c the peak output voltage for an input signal of peak value 0.1 V and frequency:

i 100 Hz

ii 1×10^5 Hz.

d A signal generator is attached to the input of the amplifier. The waveform produced by the signal generator is shown in Figure 9.5.3. The frequency of the signal is 1×10^4 Hz. Sketch the waveform of the output signal.

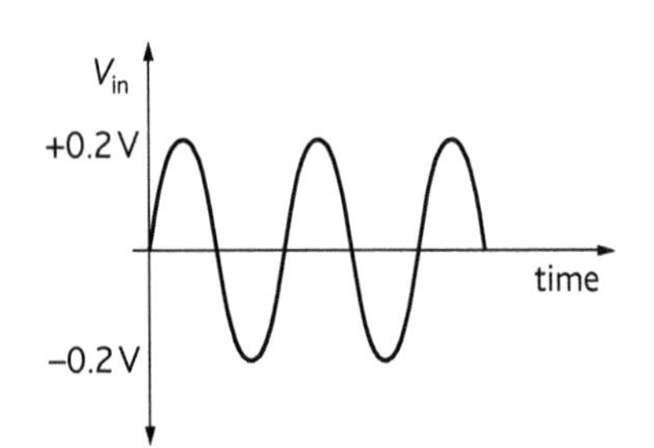

Figure 9.5.3

a Gain $A = -\dfrac{R_f}{R_i} = -\dfrac{1 \times 10^6}{10 \times 10^3} = -10^2 = -100$

b A gain of 10^2 corresponds to a bandwidth of 10^4 Hz (according to the frequency response curve).

c i From the graph, $G = 100$ when $f = 100$ Hz.
(Remember with feedback, the gain is reduced. The gain with feedback is 100.)

Peak output voltage = $G \times$ peak input voltage = $100 \times 0.1 = 10$ V

ii From the graph, $G = 10$ when $f = 1 \times 10^5$ Hz

Peak output voltage = $G \times$ peak input voltage = $10 \times 0.1 = 1.0$ V

d The peak output voltage will be $10^2 \times 0.2 = 20$ V. The output will saturate to +12 V. The output signal is 'clipped'.

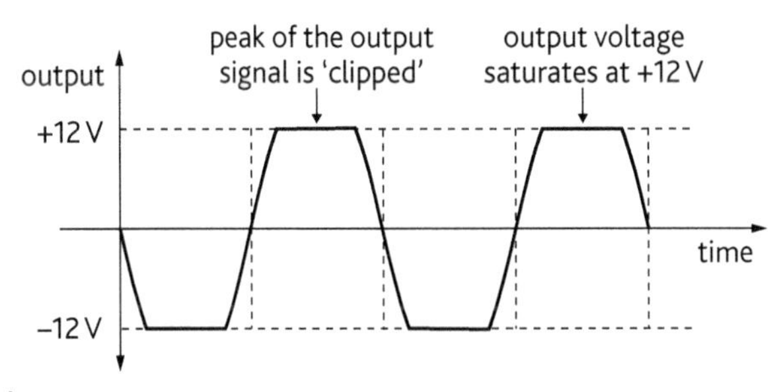

Figure 9.5.4

Example

A particular plant requires that the ambient temperature be between 18 °C and 21 °C for the plant to survive. A student designs a circuit that can be used to monitor the temperature of the room where the plant grows (Figure 9.5.5).

The LEDs L_1 and L_2 emit light when the output from the appropriate op-amp is positive and high. The thermistor has a negative temperature coefficient. At a temperature of 18 °C the potential difference across R is 4.5 V.

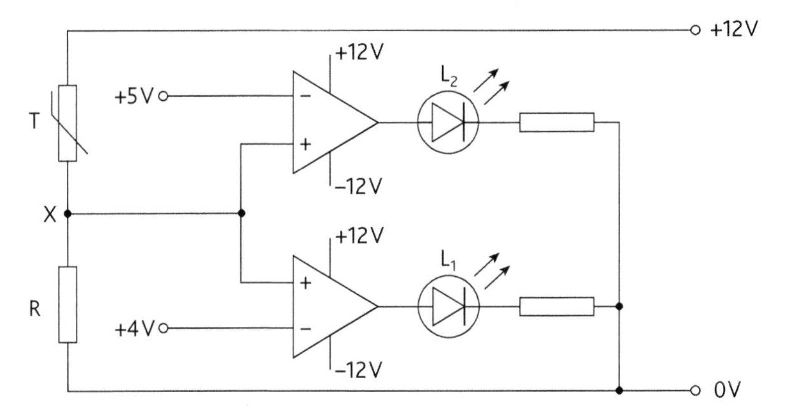

Figure 9.5.5

The potential at the non-inverting inputs of both op-amps is 4.5 V. The output of the op-amp at the top saturates to −12 V, because $V^+ < V^-$. Since L_2 is reverse-biased, L_2 will not light. The output of the op-amp at the bottom saturates to +12 V, because $V^+ > V^-$. Since L_1 is forward-biased, L_1 will light.

L_1 lights up at the lower temperature limit of 18 °C.

Consider what happens when the temperature of the room increases.

When the temperature of the room increases, the temperature of the thermistor increases and its resistance decreases. The potential difference across the thermistor decreases. At the same time, the potential difference across the resistor R increases beyond +4.5 V. When the potential at X increases beyond +5.0 V, the output of the op-amp at the top saturates to +12 V. Since L_2 is now forward-biased, L_2 will light.

L_2 can therefore be used as an indicator to determine when the temperature of the room exceeds 21 °C.

10 Digital electronics

10.1 Logic gates

Learning outcomes

On completion of this section, you should be able to:

- describe the function of the following logic gates: NOT, AND, NAND, OR, NOR, EXOR, EXNOR
- use truth tables to represent the function of no more than two inputs.

Digital systems operate using logic. Logic involves the use of binary variables. A binary variable can have only two possible values. In digital circuits, the binary variable is a voltage. Typically, a 5 V power supply is used in digital circuits.

- A low voltage of 0 V represents logic 0.
- A high voltage of +5 V represents logic 1.

Digital circuits can be:

- **combinational** – the output of the circuit is determined by the current inputs
- **sequential** – the output of the circuit is determined by the current inputs and previous outputs.

Sequential digital circuits are classified as:

- **synchronous** – a clock input is used to drive all the circuit operations
- **asynchronous** – the circuit operations are driven by changes in the input signals.

Integrated circuits (ICs) can be built to perform logic functions. **Logic gates** are the building blocks of digital circuits. The logic function performed by a particular logic gate is represented by a **truth table**.

NOT gate

Table 10.1.1 *Truth table for a NOT gate*

A	B
1	0
0	1

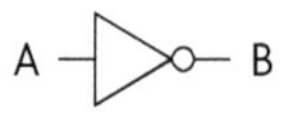

A B

Figure 10.1.1 *NOT gate*

The NOT gate inverts the logic state of the input. It is sometimes called an inverter.

AND gate

Table 10.1.2 *Truth table for an AND gate*

A	B	C
0	0	0
0	1	0
1	0	0
1	1	1

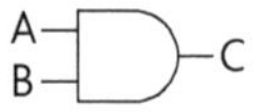

A B C

Figure 10.1.2 *AND gate*

NAND gate

Table 10.1.3 Truth table for a NAND gate

A	B	C
0	0	1
0	1	1
1	0	1
1	1	0

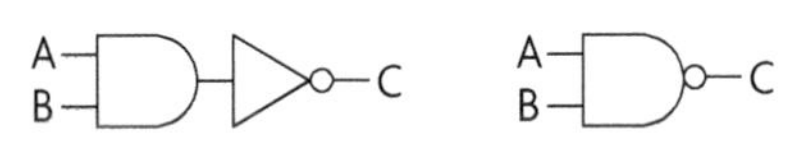

Figure 10.1.3 *NAND gate*

OR gate

Table 10.1.4 Truth table for an OR gate

A	B	C
0	0	0
0	1	1
1	0	1
1	1	1

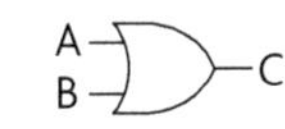

Figure 10.1.4 *OR gate*

NOR gate

Table 10.1.5 Truth table for a NOR gate

A	B	C
0	0	1
0	1	0
1	0	0
1	1	0

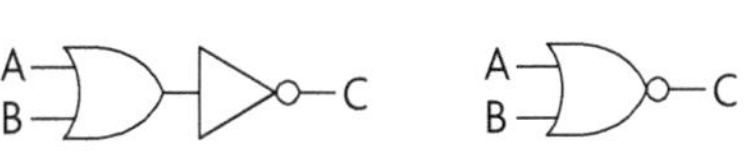

Figure 10.1.5 *NOR gate*

EXOR gate

Table 10.1.6 Truth table for an EXOR gate

A	B	C
0	0	0
0	1	1
1	0	1
1	1	0

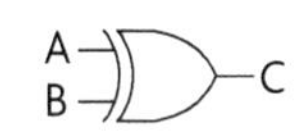

Figure 10.1.6 *EXOR gate*

EXNOR gate

Table 10.1.7 Truth table for an EXNOR gate

A	B	C
0	0	1
0	1	0
1	0	0
1	1	1

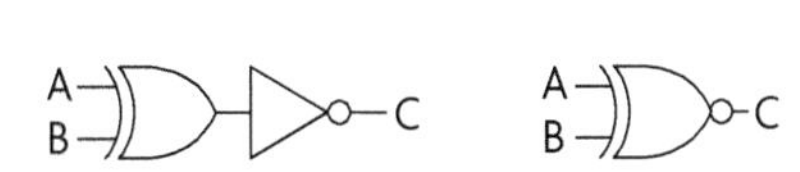

Figure 10.1.7 *EXNOR gate*

Key points

- Digital electronics involves the use of binary logic.
- Binary logic has only two possible values: logic 0 and logic 1.
- Digital circuits can be combinational or sequential.
- Digital sequential circuits can be synchronous or asynchronous.
- Logic gates are the building blocks of digital circuits.
- The logic function of a logic gate is represented using a truth table.

10.2 Equivalent logic gates

Learning outcomes

On completion of this section, you should be able to:

- construct AND gates from NOR gates
- construct OR gates from NAND gates
- redesign a circuit to contain only NOR gates or NAND gates
- reduce circuits to the minimum gate count.

Equivalent logic gates

NAND and NOR gates are called 'universal gates'. This is because each can be used to form all possible logic gates. Figure 10.2.1 illustrates how all other logic gates can be made by using only NAND gates or only NOR gates.

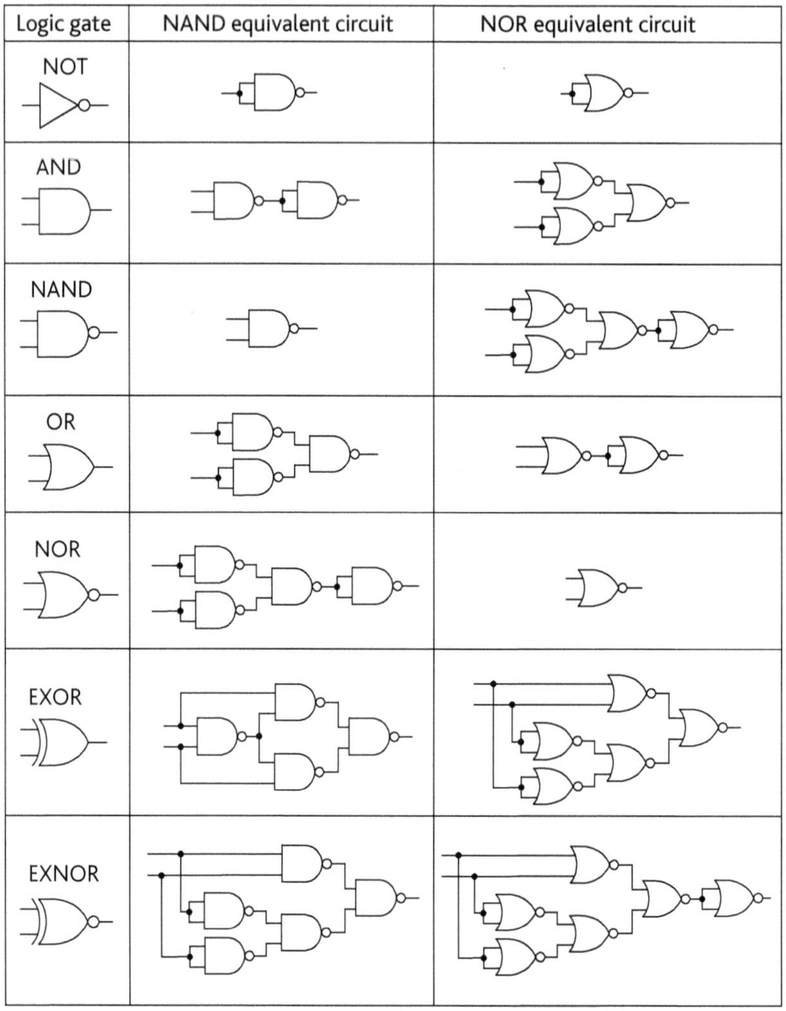

Logic gate	NAND equivalent circuit	NOR equivalent circuit
NOT		
AND		
NAND		
OR		
NOR		
EXOR		
EXNOR		

Figure 10.2.1 *Equivalent logic gates*

Redesigning a circuit to use only NOR gates

Example

Consider the logic circuit in Figure 10.2.2. Draw the equivalent circuit using only NOR gates and minimise the number of gates in the circuit.

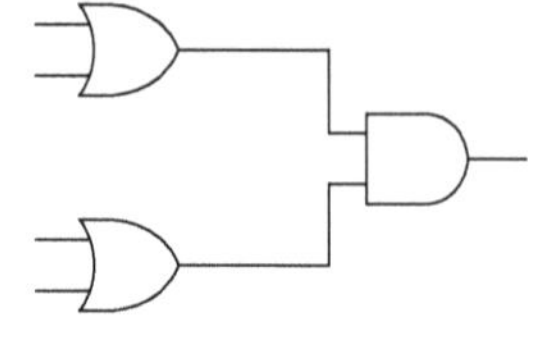

Figure 10.2.2

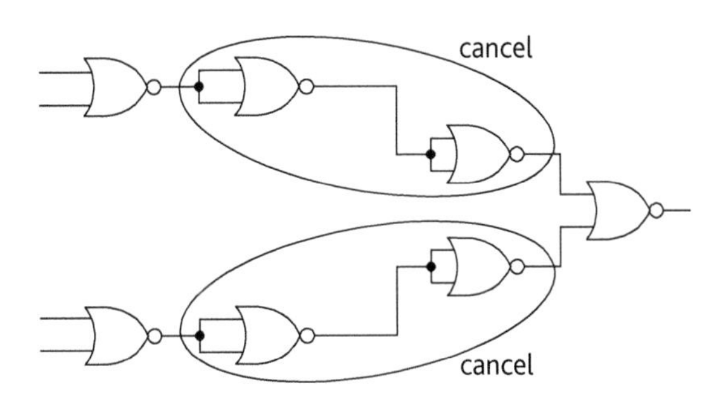

Figure 10.2.3 Replace all logic gates with NOR gates only

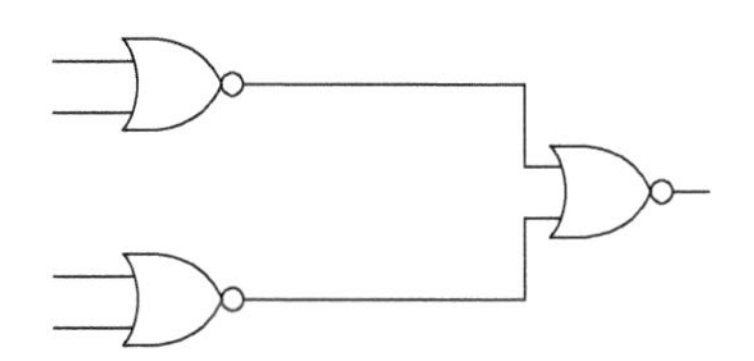

Figure 10.2.4 Minimise the number of gates

Redesigning a circuit to use only NAND gates

Example

You are provided with a quad-NAND chip.

a Replace all the logic gates in Figure 10.2.5 with NAND gates.

b Minimise the number of gates so that the logic circuit could be made by using a single quad-NAND chip.

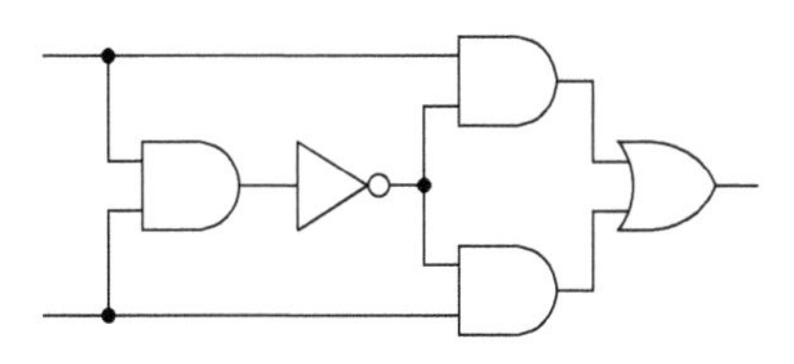

Figure 10.2.5

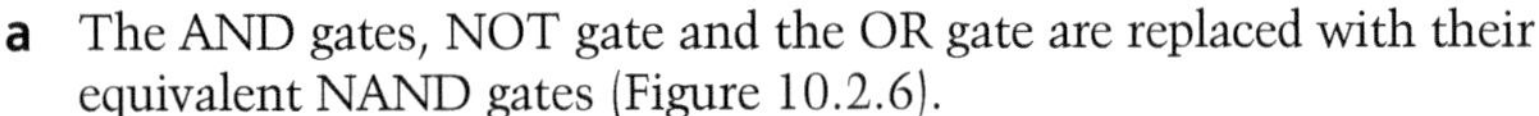

a The AND gates, NOT gate and the OR gate are replaced with their equivalent NAND gates (Figure 10.2.6).

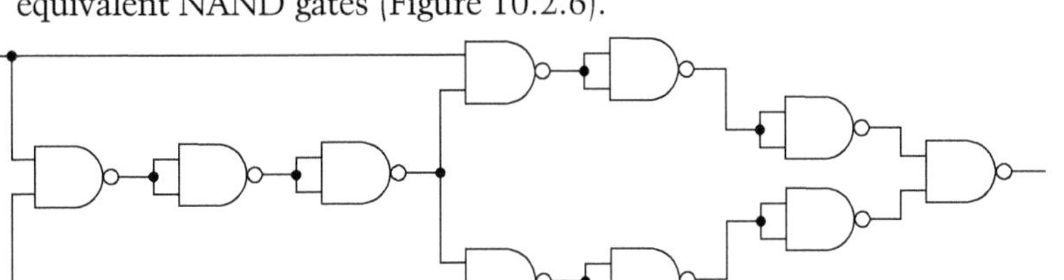

Figure 10.2.6 Replace all logic gates with NAND gates only

b See Figure 10.2.7.

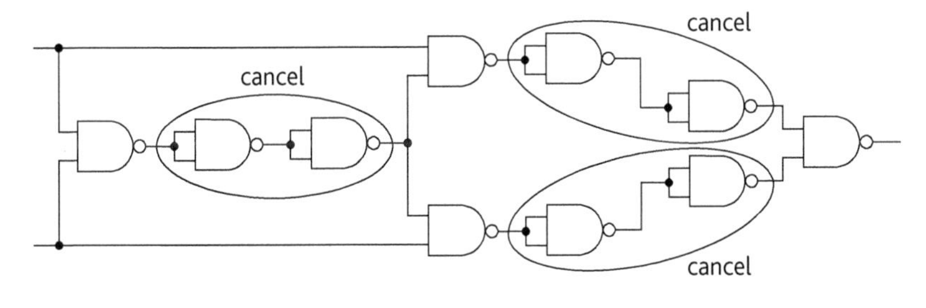

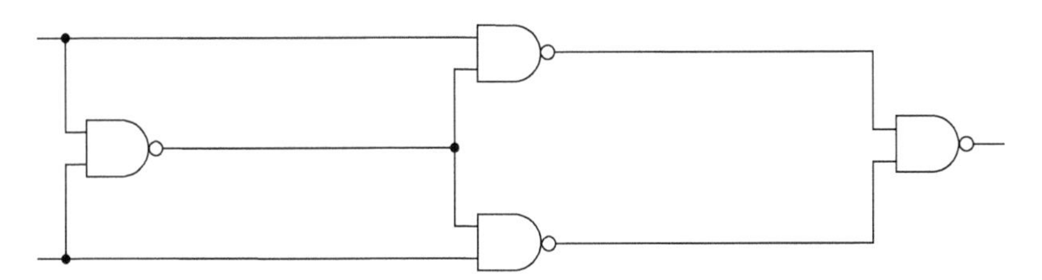

Figure 10.2.7 Minimise the number of gates

Key points

- All logic gates can be made by using only NAND gates or only NOR gates.
- In order to minimise the number of logic gates used in a circuit, any two sequential gates that perform inverting functions are removed.

10.3 Logic gates and timing diagrams

Learning outcomes

On completion of this section, you should be able to:

- understand how to use timing diagrams.

Timing diagrams

Timing diagrams are very useful in analysing digital circuits. They illustrate the logic states of various inputs and outputs over a period of time.

Example

Consider the logic circuit in Figure 10.3.1. The inputs are I_1 and I_2 and the output is X.

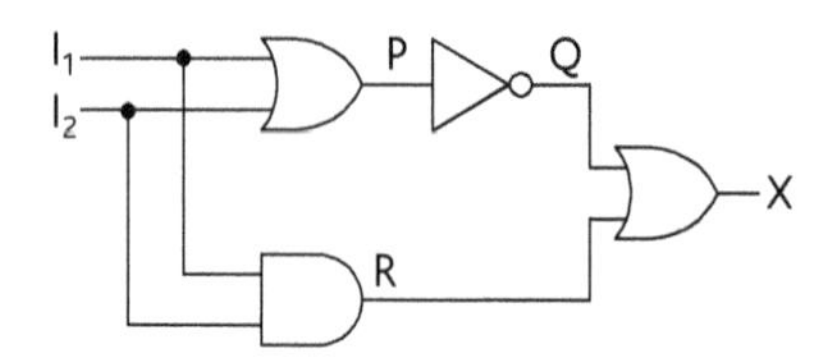

Figure 10.3.1

The truth table is illustrated below:

I_1	I_2	P	Q	R	X
0	0	0	1	0	1
0	1	1	0	0	0
1	0	1	0	0	0
1	1	1	0	1	1

The inputs I_1 and I_2 are shown in the timing diagram in Figure 10.3.2. In order to draw the timing diagram for the output X, the inputs I_1 and I_2 must be considered.

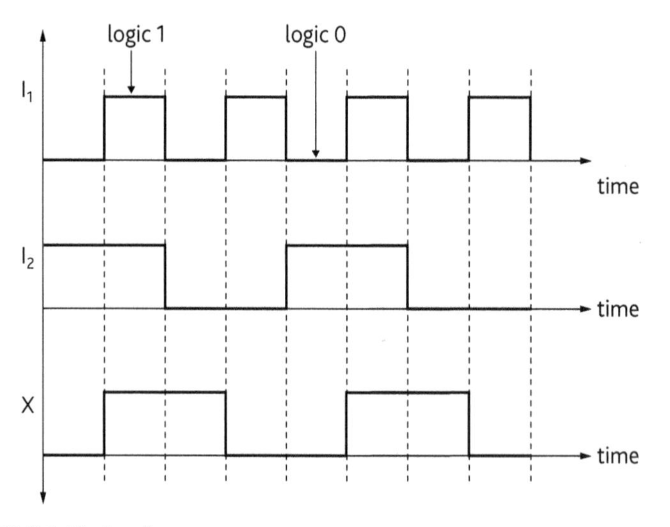

Figure 10.3.2 *Timing diagram*

Example

Consider the logic circuit in Figure 10.3.3. The inputs are I_1 and I_2 and the output is X.

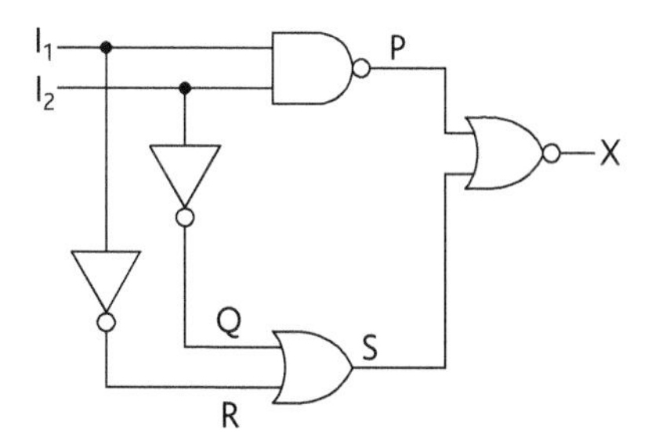

Figure 10.3.3

I_1	I_2	P	Q	R	S	X
0	0	1	1	1	1	0
0	1	1	0	1	1	0
1	0	1	1	0	1	0
1	1	0	0	0	0	1

The inputs I_1 and I_2 are shown in the timing diagram in Figure 10.3.4. In order to draw the timing diagram for the output X, the inputs I_1 and I_2 must be considered.

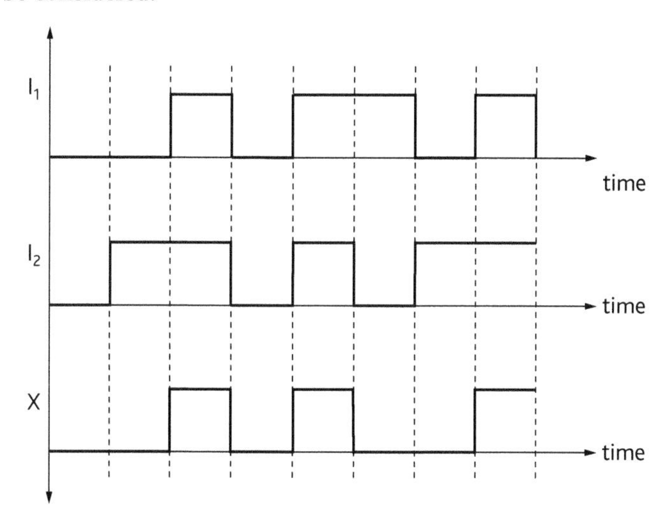

Figure 10.3.4 *Timing diagram*

Key point

- Timing diagrams are useful in analysing logic circuits.

10.4 Applications of logic gates

Learning outcomes

On completion of this section, you should be able to:

- discuss the application of digital systems in homes and industry
- describe applications of logic gates to real world problems.

Application of digital systems in homes and industry

Digital systems are found in all aspects of modern society. We live in the digital age. Engineers have used digital systems to solve various problems and enhance our quality of life.

Applications of digital systems in the home

- Television – flat screens, LEDs, LCDs, plasma
- Music – CD players, MP3 players
- Video – HD TV, DVD players
- Computers – laptops, desktops, tablets
- Refrigerators
- Microwave ovens
- Washing machines
- Home security systems

Applications of digital systems in industry

- Telecommunications – PBXs, digital phones, cellular technology, cellular phones
- Data communications – switches, routers, wireless access points
- Control systems – industrial plants
- Cable TV
- The internet
- Car manufacturing

Using logic gates to solve real world problems

Example

The circuit in Figure 10.4.1 represents a combination lock system that opens a door. In order for the lock to open, the solenoid coil has to be energised. This can only happen when c is a logic 1, which can only happen when a and b are also logic 1. This can only happen when:

- A is logic 1
- B is logic 0
- C is logic 1
- D is logic 0.

Any other combination will cause c to be logic 0 and d to be logic 1, which will sound a buzzer.

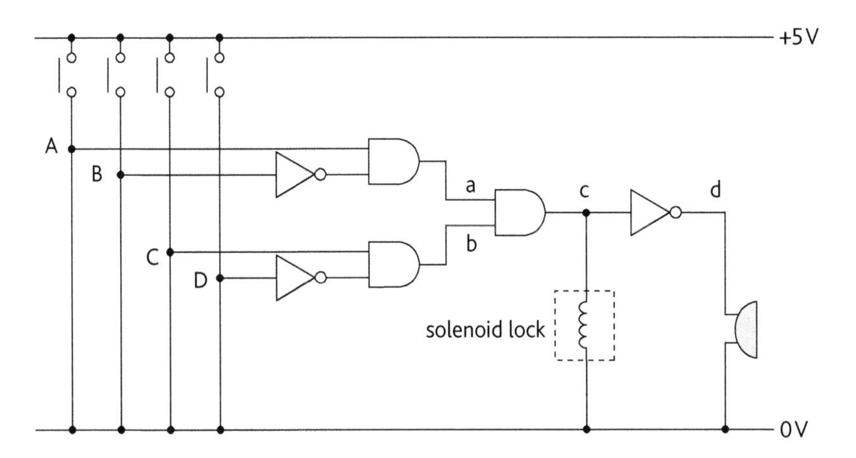

Figure 10.4.1

Example

The circuit in Figure 10.4.2 represents a simple alarm system in a car. A push-button switch is attached to each of the four doors of a car. When the doors of the car are closed, D_1, D_2, D_3 and D_4 are all at logic 0. When the alarm is armed, a logic 1 is sent to one of the inputs of the AND gate. When any door is opened, a logic 1 is sent to the other input of the AND gate. The output of the AND gate is a logic 1 and the buzzer sounds.

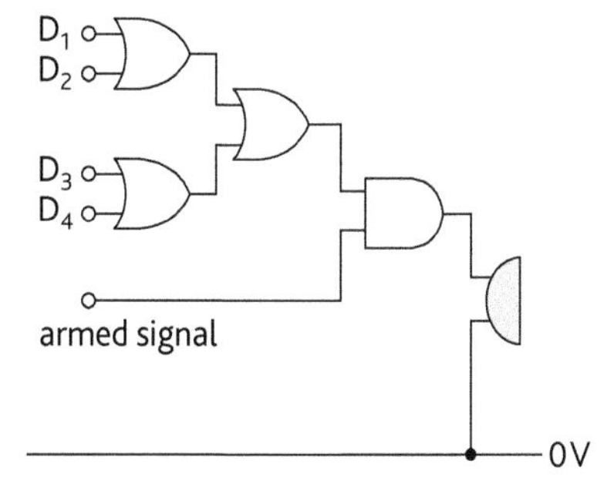

Figure 10.4.2

Example

The circuit in Figure 10.4.3 represents a system that monitors the temperature of water in a tank and the water level. A thermistor is inserted in the tank to measure the temperature and the water level is measured by a float switch.. When the temperature of the water falls below a pre-set value and the water level is higher than the level of the float switch, the output of the AND gate becomes a logic 1. The normally open relay is closed and the water heater is switched on. If the level of water in the tank was not taken into account, the heater could be switched on when there was no water in the tank and the heating element would overheat and burn out.

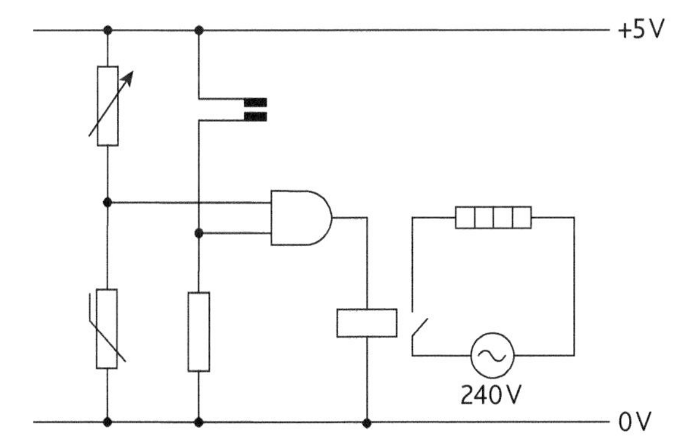

Figure 10.4.3

Key points

- Applications of digital systems can be found in homes and industry.
- Logic gates can be used as control circuits in real world problems.

10.5 Sequential circuits

Learning outcomes

On completion of this section, you should be able to:

- explain the operation of a flip-flop consisting of two NAND gates or two NOR gates
- describe the operation of a triggered bistable
- combine triggered bistables (T flip-flops) to make a 3-bit binary counter
- draw a circuit to show the construction of a half-adder and explain its operation
- use two half-adders and an OR to construct a full-adder.

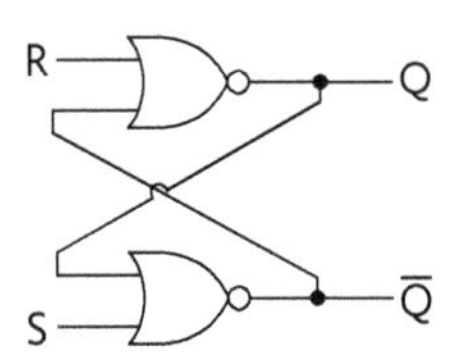

Figure 10.5.1 *SR flip-flop using NOR gates*

Table 10.5.1 *Truth table for an SR flip-flop*

S	R	Q	$\overline{Q}$
0	0	latch	
0	1	0	1
1	0	1	0
1	1	invalid	

Table 10.5.2

Sequence	S	R	Q	$\overline{Q}$
1	0	1	0	1
2	0	0	0	1
3	1	0	1	0
4	0	0	1	0
5	0	1	0	1
6	0	0	0	1

Sequential circuits

In a combinational circuit, the output is at all times dependent on the combination of the inputs. In sequential circuits, the output is dependent on the present inputs and the previous output. A memory element is required to store the previous output. The memory elements used in sequential circuits are **flip-flops** and are able to store one bit of information. A flip-flop or latch is a circuit that has two stable states. It is said to be a **bistable** element. Flip-flops are the basic building blocks of sequential circuits. There are different types of flip-flops (SR, JK, D and T flip-flops).

The SR flip-flop (set–reset flip-flop)

An SR flip-flop is constructed from a pair of cross-coupled NOR gates (Figure 10.5.1).

The inputs are called S (set) and R (reset). The outputs are Q and $\overline{Q}$. $\overline{Q}$ is the complement of Q. (i.e. When Q = 1, $\overline{Q}$ = 0 and when Q = 0, $\overline{Q}$ = 1).

The circuit operates as follows:

- When both inputs are low, the output Q does not change and remains latched in its last state.
- When the R input is set to low and the S input is set to high, the Q output is set to logic 1.
- When the R input is set to high and the S input is set to low, the Q output of the latch is reset to logic 0.
- When the R and S inputs are set to high, the output is unpredictable. This state is an invalid state.

Table 10.5.1 shows that there are only two stable output states for the SR flip-flop.

Example

Consider a sequence of changes that take place at the inputs of an SR flip-flop. Suppose S = 0 and R = 1 initially. The output Q will be set to 0.

- In sequence 2, S = 0 and R = 0, Q will hold its last output, which was 0.
- In sequence 3, S = 1 and R = 0, Q will be set to 1.
- In sequence 4, S = 0 and R = 0, Q will hold its last output, which was 1.
- In sequence 5, S = 0 and R = 1, Q will be reset to 0.
- In sequence 6, S = 0 and R = 0, Q will hold its last output, which was 0.

An SR flip-flop can also be constructed using cross-couple NAND gates as shown in Figure 10.5.2.

Figure 10.5.3 shows the circuit symbol for an SR flip-flop.

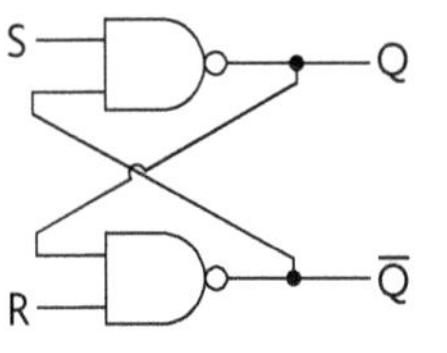

Figure 10.5.2 *SR flip-flop using NAND gates*

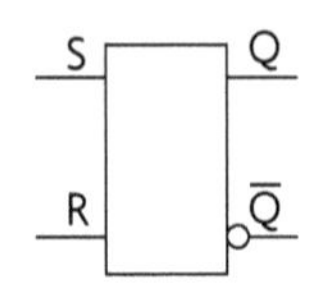

Figure 10.5.3 *Circuit symbol for an SR flip-flop*

The T flip-flop (toggle flip-flop)

The circuit symbol for a T flip-flop (toggle flip-flop) is shown in Figure 10.5.4. If the T input is high, the T flip-flop changes state (toggles) when the clock input changes.

If the T input is low, the T flip-flop holds its previous value.

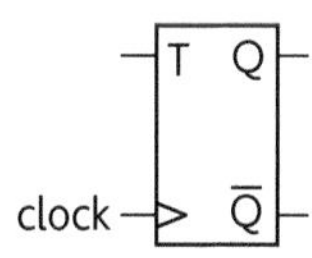

Figure 10.5.4 *Circuit symbol for a T flip-flop*

Table 10.5.3

T	Q (current state)	Q_{next} (next state)	
0	0	0	hold state
0	1	1	hold state
1	0	1	toggle
1	1	0	toggle

The 3-bit binary counter

When several flip-flops are connected together, they form a **register**. A register is used for storing and shifting data from an external source. The data is in the form of 1s and 0s. A **counter** is a register that is able to count the number of clock pulses arriving at its clock input. A counter can be positive edge triggered (i.e. the input changes from low to high) or negative edge triggered (i.e. the input changes from high to low). An n-bit binary counter consists of n flip-flops and has 2^n distinct states. A 3-bit binary counter consists of three flip-flops and has eight states (i.e. $2^3 = 8$).

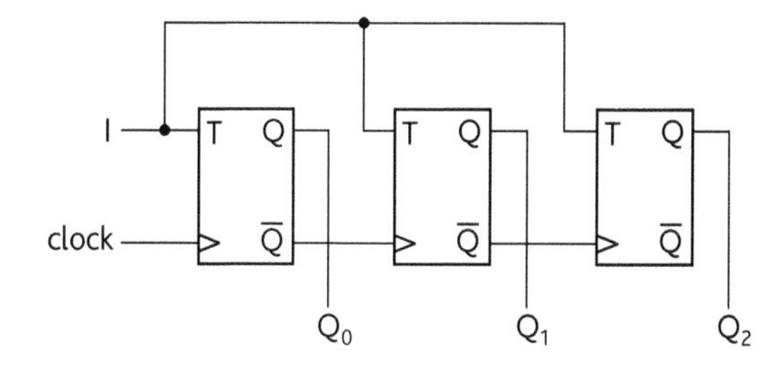

Figure 10.5.5 *A 3-bit binary counter*

Figure 10.5.5 shows the circuit diagram for a 3-bit binary counter consisting of T flip-flops. The output changes when the input detects a negative edged signal (i.e. the input changes from logic 1 to logic 0). The outputs of the counter are Q_0, Q_1 and Q_2. Assuming the outputs are initially all at logic 0, Figure 10.5.6 shows the timing diagram for the operation of the counter.

Table 10.5.4 shows the sequential truth table for the 3-bit binary counter.

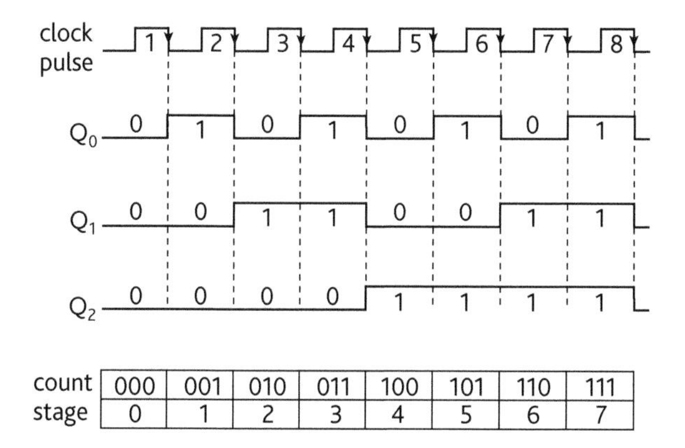

Figure 10.5.6 *The timing diagram for a 3-bit binary counter*

Table 10.5.4 *Sequential truth table for a 3-bit binary counter*

Clock pulse	Q_0	Q_1	Q_2
0	0	0	0
1	1	0	0
2	0	1	0
3	1	1	0
4	0	0	1
5	1	0	1
6	0	1	1
7	1	1	1

Adding binary numbers

When adding 1-bit binary numbers, a sum and a carry bit are produced. Figure 10.5.7 shows how 1-bit binary numbers are added.

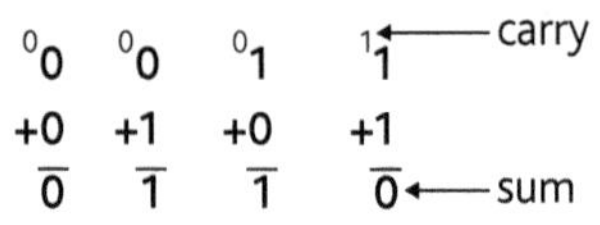

0	0	0	1	← carry
0	0	1	1	
+0	+1	+0	+1	
0	1	1	0	← sum

Figure 10.5.7 *Adding 1-bit binary numbers*

The method can be extended to any number of n-bit binary numbers. Figure 10.5.8 gives an example of adding two 3-bit numbers.

```
 101+
 011
1000
```

Figure 10.5.8 *Adding two 3-bit numbers*

The half-adder

A **half-adder** is an arithmetic circuit used to add two 1-bit numbers (Figure 10.5.9).

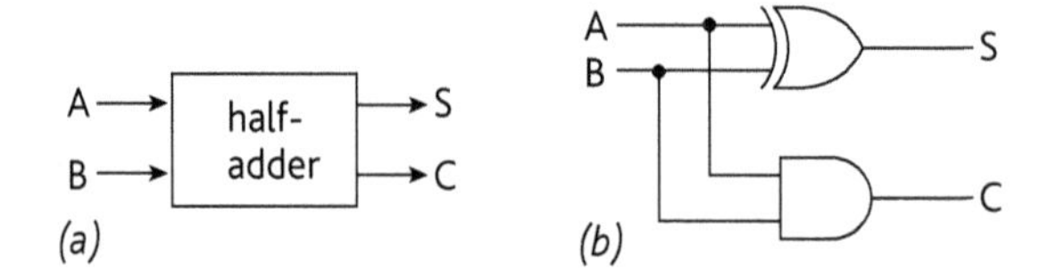

Figure 10.5.9 *A half-adder: (a) typical diagram, (b) logic diagram*

The circuit has two inputs A and B, which represents the two bits to be added. The circuit has two outputs. One output produces the sum (S) and the other produces the carry (C). Table 10.5.5 shows the truth table for a half-adder.

Table 10.5.5 *Truth table for a half-adder*

A	B	S (sum)	C (carry)
0	0	0	0
0	1	1	0
1	0	1	0
1	1	0	1

The full-adder

A **full-adder** circuit adds three one-bit numbers (A, B and Carry in). A full-adder circuit is made by using two half-adders and an OR gate (Figure 10.5.10). A full-adder is typically a component in a cascade of adders. The carry input for the full-adder is from the carry output from the adder above it in the cascade. Table 10.5.6 shows the truth table for a full-adder.

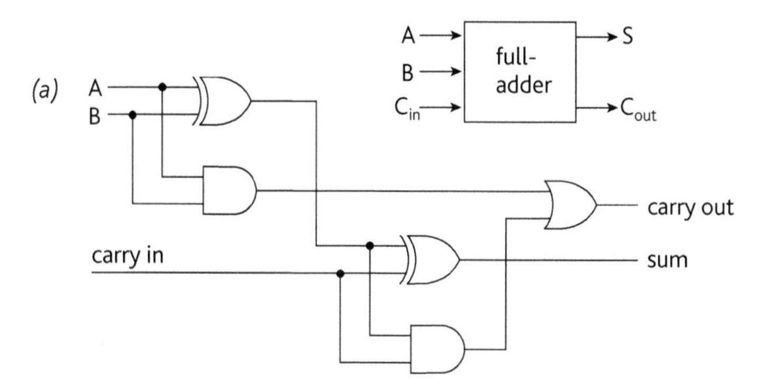

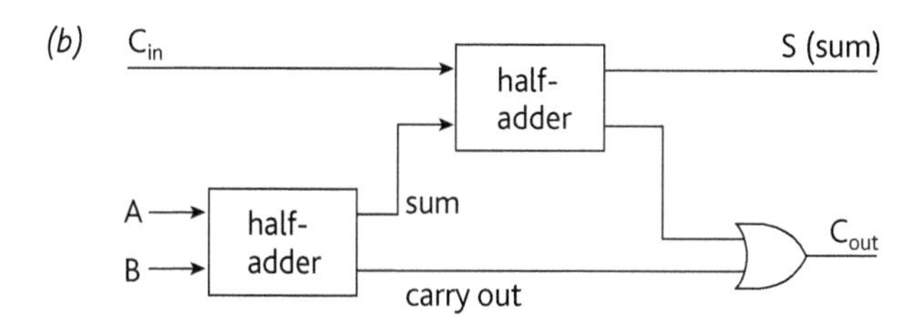

Figure 10.5.10 *A full-adder: (a) typical diagram, (b) logic diagram*

Table 10.5.6 *Truth table for a full-adder*

A	B	C_{in}	S (sum)	C_{out}
0	0	0	0	0
0	0	1	1	0
0	1	0	1	0
0	1	1	0	1
1	0	0	1	0
1	0	1	0	1
1	1	0	0	1
1	1	1	1	1

Key points

- A flip-flop is a bistable device, meaning that is has two stable states.
- A flip-flop is able to store one bit of information.
- A T flip-flop changes state when the clock input changes.
- A 3-bit binary counter can be constructed by using three T flip-flops.
- A half-adder is an arithmetic circuit used to add two 1-bit numbers.
- A full-adder circuit adds three 1-bit numbers.

Revision questions 5

Answers to questions that require calculation can be found on the accompanying CD.

1 a State three characteristics of an ideal operational amplifier. [3]

b An amplifier circuit for a microphone is shown below.

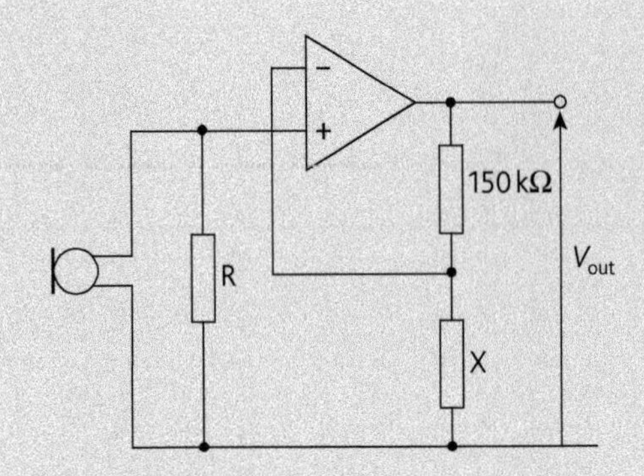

i State the type of feedback used in this amplifier. [1]

ii State two advantages of using this type of feedback. [2]

iii State the type of amplifier being used. [1]

iv The output potential difference V_{out} is 3.8 V for a potential difference across the resistor R of 50 mV. Calculate:

1 the gain of the amplifier [2]

2 the resistance of the resistor X. [2]

2 a The diagram below shows an inverting amplifier.

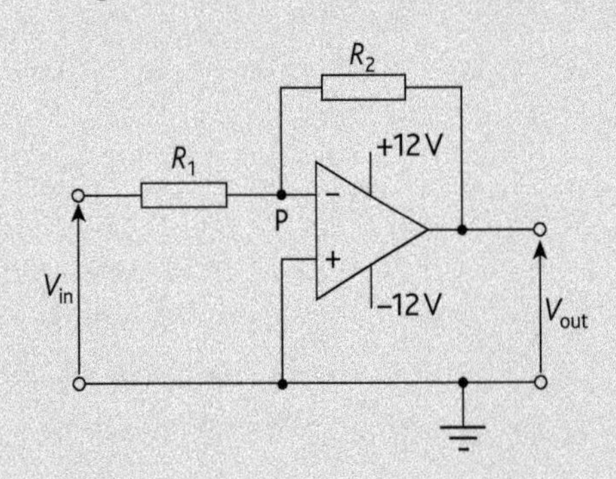

i Explain why the point P is referred to as a virtual earth. [3]

ii Derive an expression for the gain of the amplifier. [4]

b In a particular circuit $R_1 = 10\,k\Omega$ and $R_2 = 100\,k\Omega$.

i Calculate the gain of the amplifier. [1]

ii Determine the output voltage when $V_{in} = +1.1\,V$ [1]

iii Determine the output voltage when $V_{in} = +2.0\,V$ [2]

c A sinusoidal voltage $V_{in} = 2.0 \sin \omega t$ is applied to the input of the amplifier.

i Sketch a graph to show the variation of the input voltage with time. [2]

ii Sketch a graph to show the variation of the output voltage V_{out} with time. [3]

3 Three voltage signals V_1, V_2 and V_3 are to be combined to produce an output:

$$V_o = -2V_1 - 3V_2 - V_3$$

Design a circuit using the following information: [5]

- The input resistance of the inputs must be greater than 10 kΩ.
- All the resistor values must be less than 500 kΩ.

4 Two circuits P and Q are cascaded to produce an output V_o.

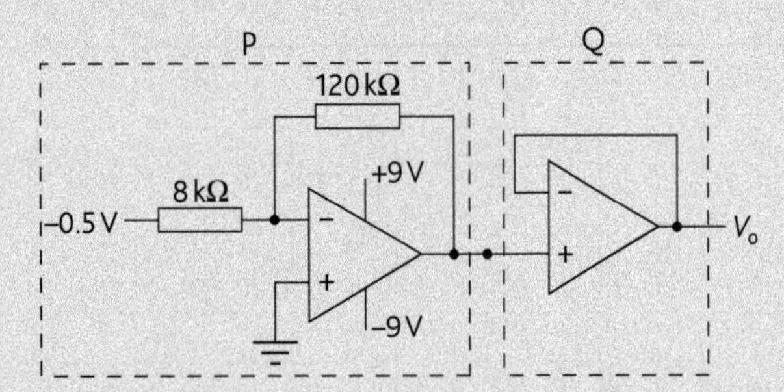

a Calculate V_o. [3]

b State the name of the circuit Q and state its gain. [2]

c Suggest a practical use for circuit Q. [1]

d Determine the minimum voltage which causes V_o to saturate to −9 V. [2]

5 The figure below shows a cascade amplifier circuit.

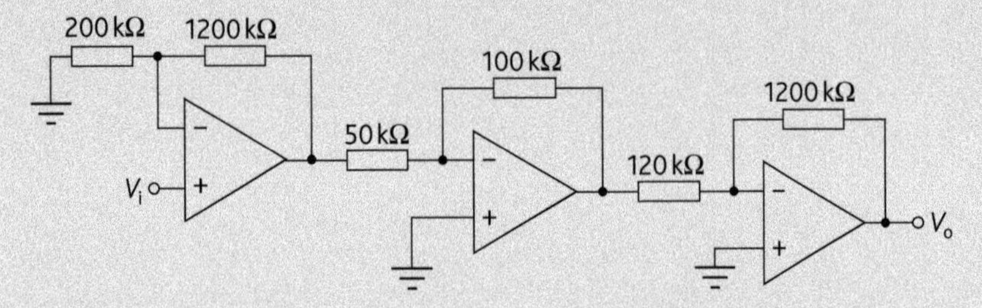

a Calculate the overall gain of the circuit. [4]

b Calculate the output voltage V_o, when $V_i = 20\,mV$. [2]

6 The diagram below shows an amplifier circuit.

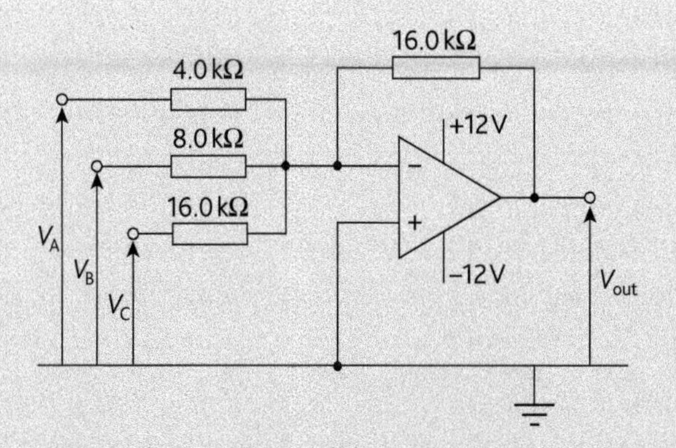

a State the type of amplifier circuit shown. [1]

b For each of the inputs A, B and C, the amplifier may be considered as a single input amplifier. When the amplifier is not saturated, the output potential is V_{out}. Write down an expression for V_{out} in the form:

$V_{out} = -(G_1V_A + G_2V_B + G_3V_C)$, where G_1, G_2 and G_3 are constants. [3]

c This amplifier circuit uses negative feedback. Explain what is meant by the term *negative feedback*. [2]

d State two effects of negative feedback on an amplifier. [2]

e The input potentials V_A, V_B and V_C are either 0 V or +1.0 V. Copy and complete the following table. [3]

V_A/V	V_B/V	V_C/V	V_{out}/V
0	0	1	
0	1	0	
0	1	1	
1	0	0	
1	0	1	
1	1	0	
1	1	1	

f Explain how the circuit can be used as a digital to analogue converter. [3]

7 a Use diagrams to illustrate how the following logic gates can be constructed using NAND gates only.

i NOT gate [1]

ii AND gate [1]

iii OR gate [2]

b i Draw a truth table for the logic circuit shown below. [4]

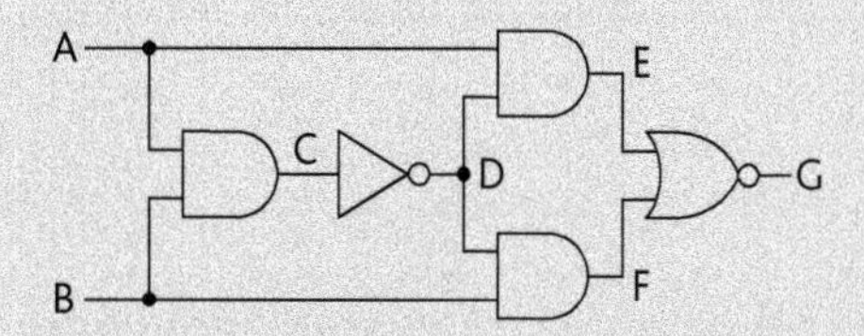

ii Replace all the circuit components with NAND gates. Hence minimise the number of gates. [5]

8 a Explain what is meant by a bistable device. [2]

b State the function of bistable devices in digital circuits. [1]

c Draw the circuit for an SR flip-flop. [1]

d Using a truth table, explain how an SR flip-flop can be used as a latch. [5]

9 a Draw a diagram of a 3-bit binary counter using T flip-flops. [3]

b Use a truth table to explain the operation of a 3-bit binary counter. [5]

10 a Add the following binary numbers: 101 and 011 [3]

b Draw a logic diagram of a half-adder. [2]

c Construct a truth table for a half-adder. [3]

d Draw a diagram to show how a full-adder can be constructed from half-adders and give an example to explain its operation. [4]

11 a Construct a truth table for the following circuit. [4]

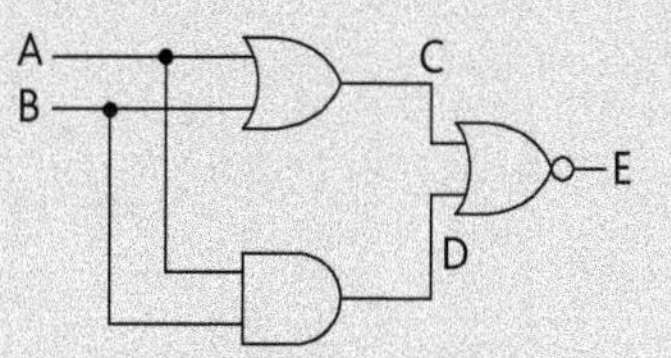

b You are asked to design a logic circuit to control an LED inside a sports car. The sports car has two doors. Each door has a switch which is open when the doors are closed. When a door is opened, the switch closes and the LED lights. When either door is opened, the LED lights. When both doors are closed, the LED does not light.

Construct a truth table for the logic and draw a logic circuit to meet the requirements. [5]

12 The diagram below shows a digital circuit with inputs A and B.

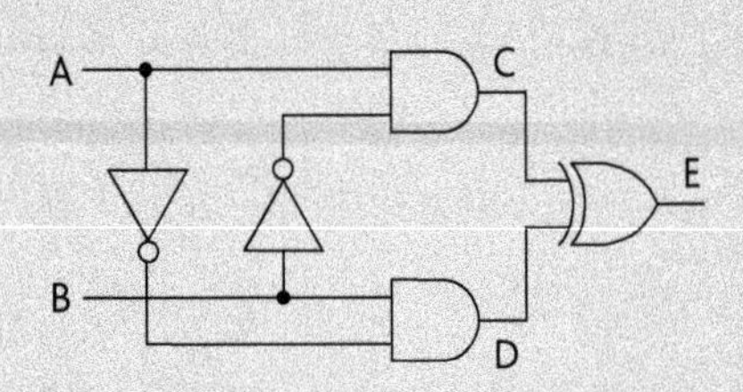

a Construct a truth table for the digital circuit. [4]

b The timing diagram below illustrates the variation of the inputs A and B. Copy and complete the diagram to show the variation of the output E. [3]

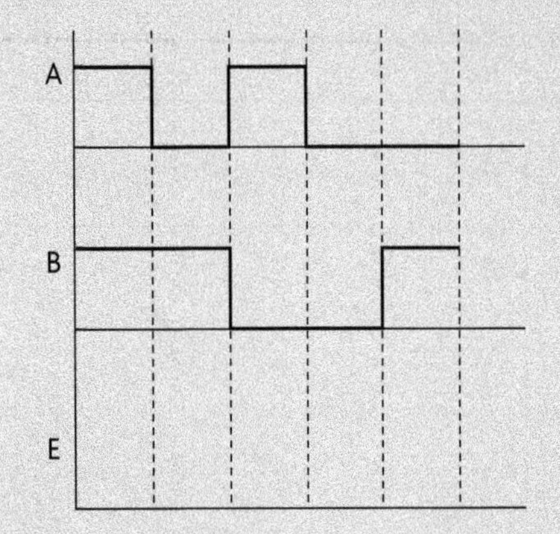

13 For the circuit below, replace all the logic gates with NOR gates. Minimise the number of logic gates in the circuit. [5]

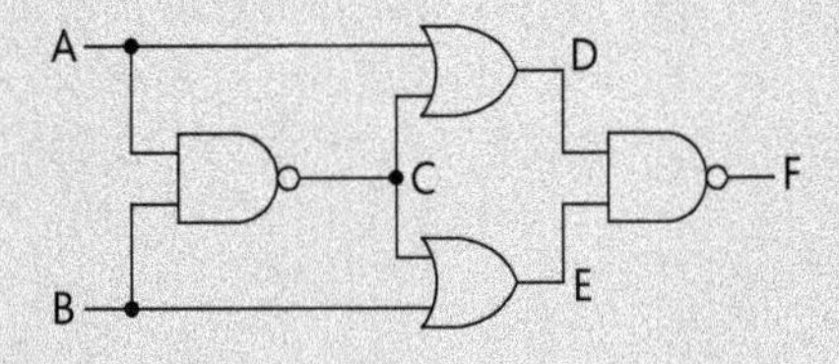

14 Construct a truth table for the following logic circuit. [5]

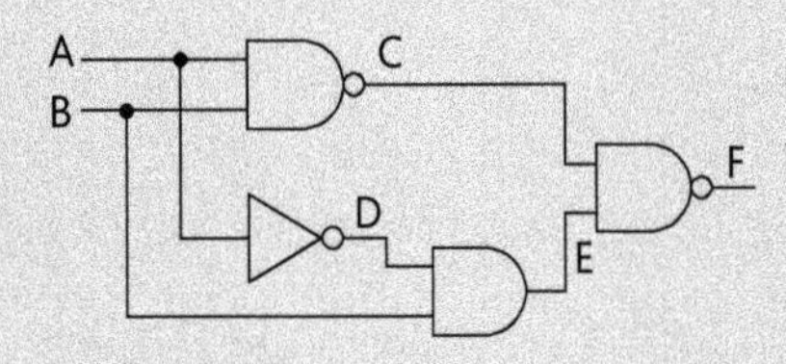

15 The diagram below shows an amplifier circuit.

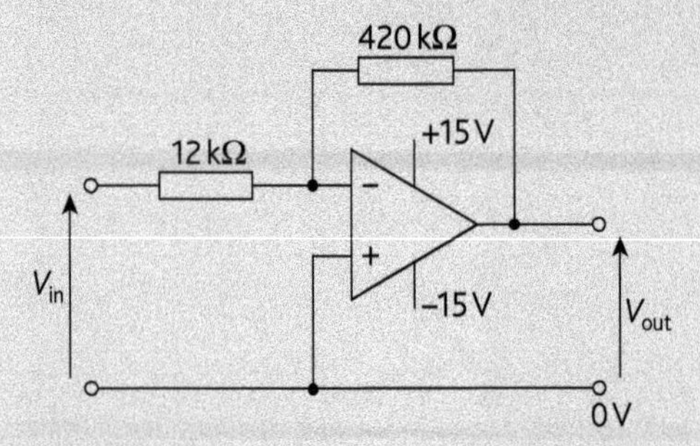

a State the type of circuit. [1]

b Derive an expression for the voltage gain of the circuit. [5]

c Explain the concept of a virtual earth and explain why the inverting terminal is at this potential. [3]

d Calculate the voltage gain of the amplifier. [2]

e Explain what is meant by saturation of an amplifier. [2]

f Calculate the maximum input voltage it can amplify before saturation of the output occurs. [2]

16 The diagram below shows an inverting amplifier.

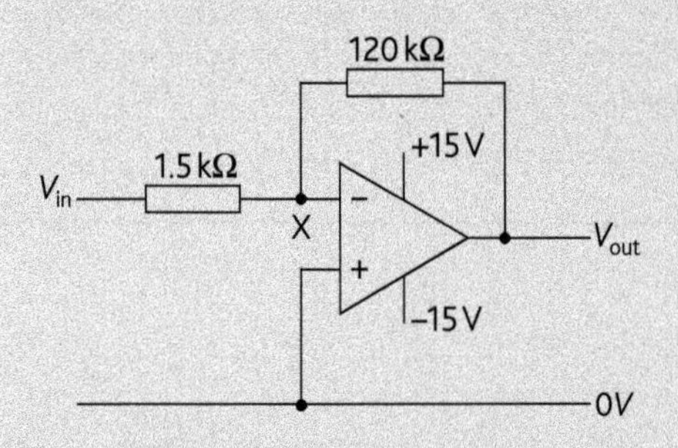

The input voltage V_{in} is 35 mV.

a State the voltage at the point X. [1]

b Calculate the current in the 1.5 kΩ resistor. [2]

c State and explain the current in the 120 kΩ resistor. [2]

d Calculate the output voltage V_{out}. [2]

e Calculate the maximum input voltage before saturation occurs. [2]

17 a Explain what is meant by negative feedback when applied to an amplifier. [2]

b State two advantages of using negative feedback. [2]

c Explain what is meant by the bandwidth of an amplifier. [2]

18 a State four properties of an ideal operational amplifier. [4]

b Sketch a graph of a typical frequency response curve for an operational amplifier. [3]

c Label the bandwidth and gain of the amplifier. [1]

d Illustrate what happens to the gain and bandwidth of the amplifier when negative feedback is used. [2]

19 The diagram below illustrates a summing amplifier.

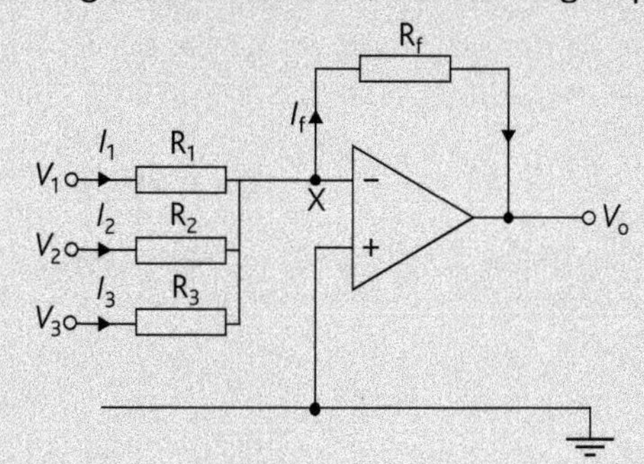

a Derive an expression for the voltage gain of the amplifier. [6]

b State one use of this type of amplifier. [1]

c Explain how a summing amplifier can be used as a digital to analogue converter. [6]

20 a Draw a diagram to show how a single operational amplifier can be used to generate a square wave voltage from a sine wave input voltage. [3]

b Explain the operation of the circuit. [3]

21 The following diagram shows and operational amplifier being used as a summing amplifier.

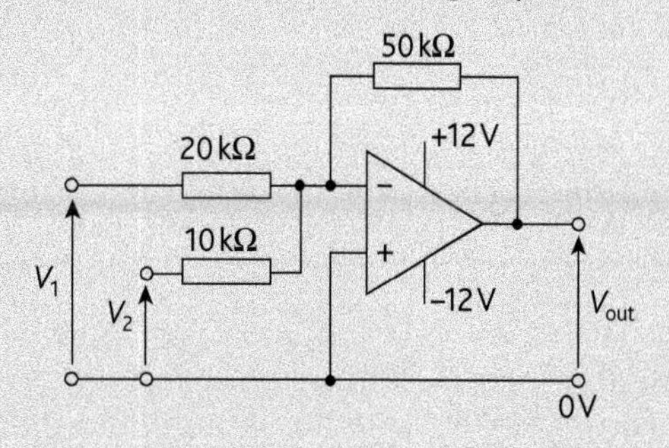

a State the type of feedback being used in the circuit. [1]

b Calculate the output voltage $V_{out,}$ when the input voltages are $V_1 = -2.0$ V and $V_2 = +3.0$ V. [3]

c Calculate the new output voltage if both input voltages are set to −4.0 V. [2]

Module 2 Practice exam questions

Multiple-choice questions

Answers to the multiple-choice questions and to selected structured questions can be found on the accompanying CD.

1 An alternating current of r.m.s. magnitude 4 A and a steady current of I flowing through identical resistors dissipate heat at equal rates. What is the value of I?

a $\frac{4}{\sqrt{2}}$ A b 4 A c 2 A d $\sqrt{2}$ A

2 A transformer is 100% efficient. The ratio of the secondary turns to the primary turns is 1 : 20. A 220 V a.c. supply is connected to the primary coil and a 5.0 Ω resistor is connected across the secondary coil. What is the current in the primary coil?

a 0.11 A b 2.2 A c 0.22 A d 2.0 A

3 Which of the following circuits illustrates how an a.c. supply can be converted into a d.c. supply with the correct polarity?

a

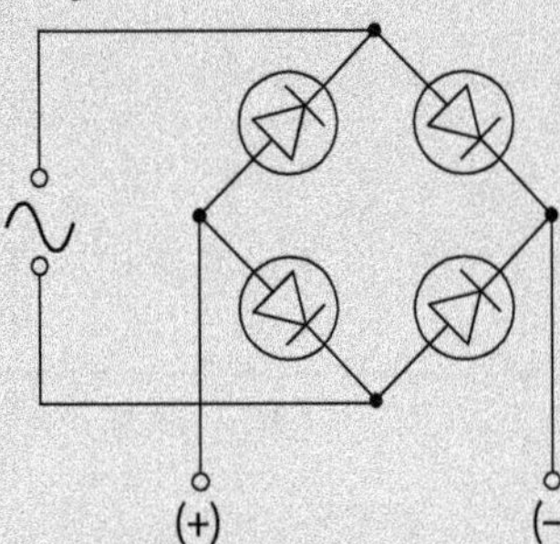

b

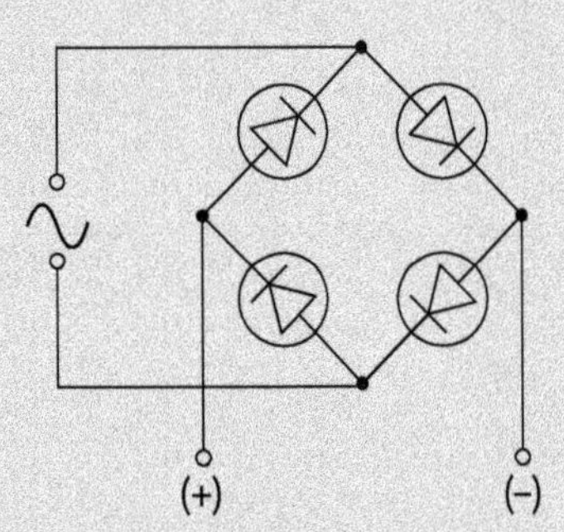

c

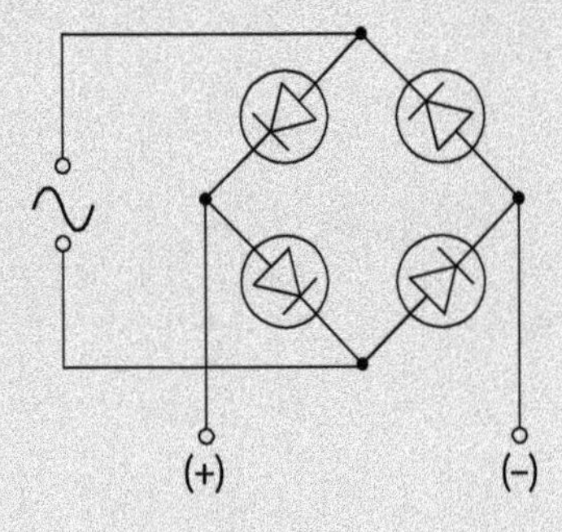

d

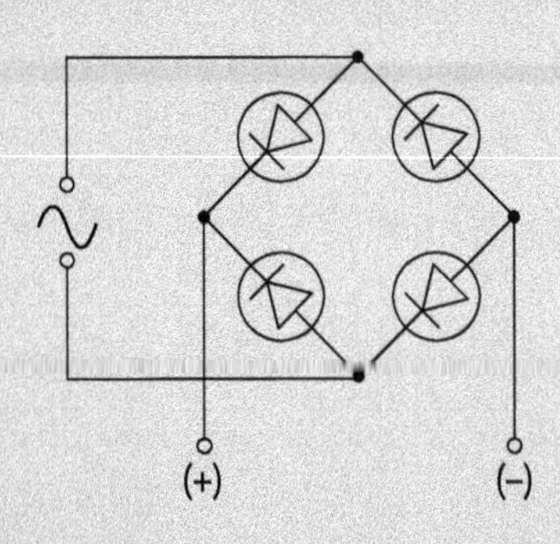

4 Which of the following is true about an ideal operational amplifier?

i It has an infinite open loop gain.
ii It has a high output impedance.
iii It has an infinite input impedance.

a i only b i and iii only
c i and ii only d i, ii and iii

5 An alternating voltage of 0.15 V peak value is connected to the input of the amplifier circuit below.

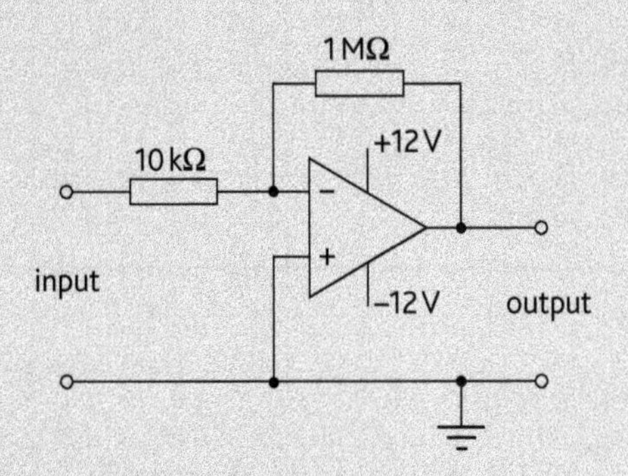

What is the peak output voltage?

a +15 V b +12 V c −15 V d −12 V

6 Which of the following is **not** an advantage of using negative feedback?

a Increased bandwidth
b Less distortion
c Reduced bandwidth
d Greater operating stability

7 An voltage of 2.0 V is applied to the input of a non-inverting amplifier shown below.

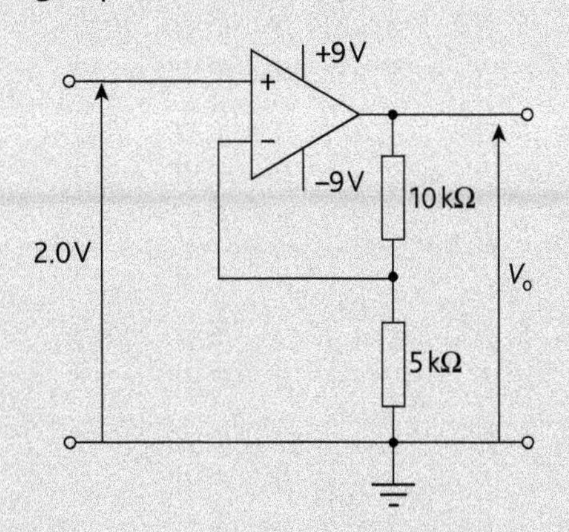

What is the output voltage V_o?

a 4.0 V **b** 6.0 V **c** 9.0 V **d** 2.0 V

8 The following combination of NOR gates is equivalent to what type of logic gate?

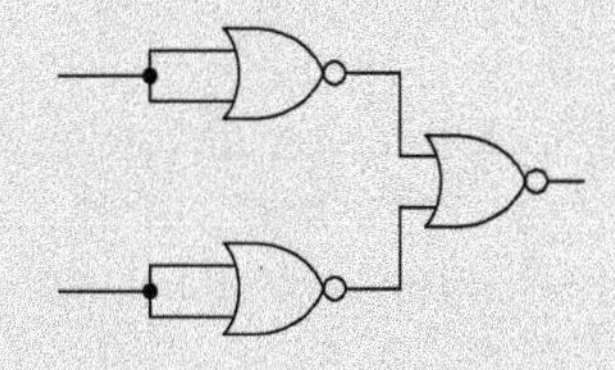

a AND **b** OR **c** NOT **d** XOR

9 The following diagram shows a half-adder. If A = 1 and B = 1, what is the value of the sum and the carry?

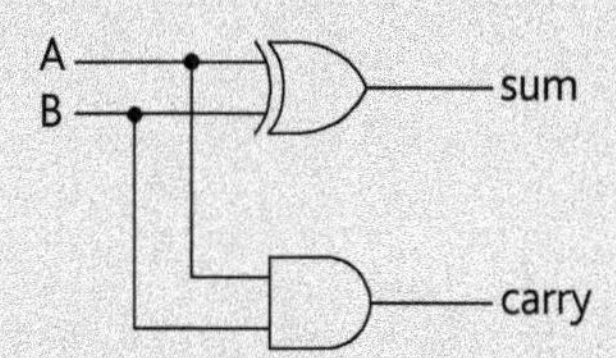

a Sum = 1, Carry = 0
b Sum = 0, Carry = 0
c Sum = 1, Carry = 1
d Sum = 0, Carry = 1

10 Which of the following is true about a flip-flop?

i It is a monostable device.
ii It can be made using cross-coupled NAND gates.
iii It can store one 'bit' of memory.

a i only **b** i and ii only
c ii and iii only **d** i, ii and iii

Structured questions

11 a i Draw a diagram of a transformer that can be used to convert a 120 V a.c. supply into a 12 V a.c. supply. [2]

ii State and explain three design features of the transformer that makes it very efficient. [6]

b A step-up transformer is connected to a 240 $V_{r.m.s.}$ a.c. supply. The output of the transformer is 12 $kV_{r.m.s}$ and is connected to a lighted street sign.

i Explain the distinction between r.m.s. value and peak value of an alternating voltage. [2]

ii Calculate the peak value of the primary voltage. [2]

iii Calculate the ratio $\frac{N_p}{N_s}$ for the transformer. [2]

iv Given that the current flowing through the street sign is 12 mA, calculate the power input to the transformer. [2]

v Calculate the current flowing in the primary coil of the transformer. [2]

c Explain why electrical power is transmitted using a.c. at high voltages. [5]

12 a Explain what is meant by an intrinsic semiconductor and the term *doping*. [2]

b Distinguish between n-type and p-type materials. [2]

c A p-type material and an n-type material are placed against each other. A depletion layer forms.

i Using a diagram, explain what is meant by the depletion layer and explain how it forms. [3]

ii State an approximate value for thickness of the depletion layer. [1]

d State and explain what happens to the depletion region when a p-n junction is:

i forward-biased [2]

ii reverse-biased. [2]

e Distinguish between diffusion current and drift current. [2]

f State one use of a p-n junction diode. [1]

g Sketch an *I–V* characteristic of a p-n junction diode. [2]

13 The figure below shows a non-inverting amplifier.

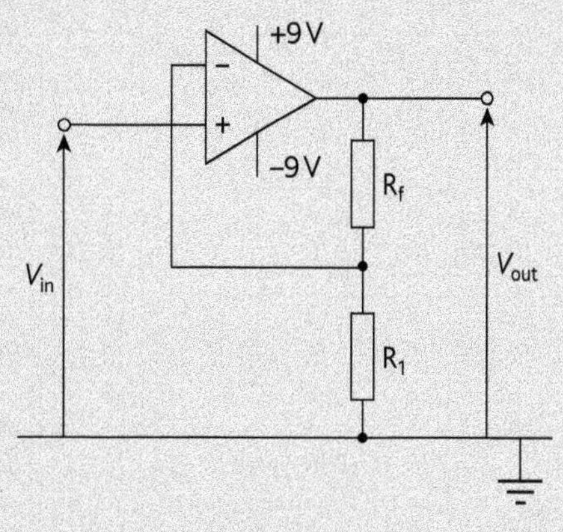

a Derive an expression for the closed-loop gain *A* of the non-inverting amplifier. [5]

b The resistors R_f and R_1 are 500 kΩ and 50 kΩ respectively in a practical circuit.

- **i** Determine the gain A of the amplifier. [2]
- **ii** Determine the output voltage when V_{in} = +100 mV. [2]
- **iii** Determine the maximum value of V_{in} such that the V_{out} is not saturated. [2]
- **iv** Suppose $V_{in} = 0.9 \sin(2\pi(20)t)$. Sketch a graph to show:
  - **1** the variation of V_{in} with time [2]
  - **2** the variation of V_{out} with time. [3]

c You are asked to design to a circuit to combine two signals v_1 and v_2. v_1 is the output signal from a microphone and v_2 is a signal from an electric guitar. The two signals v_1 and v_2 are added as follows: $V_o = 3v_1 - 2v_2$, where V_o is the output of the circuit. The input resistance of both signal inputs must be greater than 100 kΩ. Design a circuit to satisfy these conditions. [4]

14 a An SR flip-flop is called a bistable device.

- **i** Draw a diagram of an SR flip-flop using only NAND gates. [2]
- **ii** Explain what is meant by the term *bistable*. [1]
- **iii** What is the function of flip-flops in digital circuits? [2]

b The inputs to an SR flip-flop are I_1 and I_2. The outputs are Q and $\overline{Q}$. Using the sequential truth table below, explain how the SR flip-flop operates as an electronic latch. [5]

Sequence number	I_1	I_2	Q	$\overline{Q}$
1	0	1	1	0
2	0	0		
3	1	0		
4	0	0		
5	0	1		
6	1	0		

c The timing diagram below shows two inputs P and Q of a logic gate. The output of the logic gate is R.

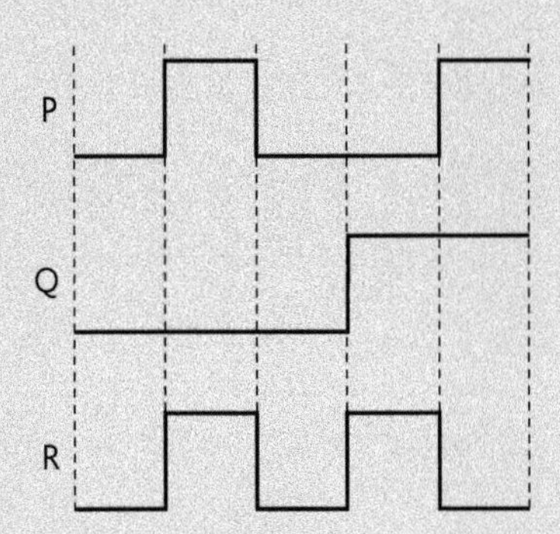

- **i** Complete a truth table with two inputs and one output for the logic gate. [3]
- **ii** Draw a diagram of the logic gate represented by the timing diagrams. [1]
- **iii** Draw an equivalent logic circuit to represent the logic gate in **ii** using only NAND gates. [2]

15 a

- **i** Draw the symbol and truth tables for an OR gate and an EXOR gate. [6]
- **ii** Draw a circuit diagram to show how an OR gate can be constructed from a number of NAND gates. [3]
- **iii** Redesign the circuit below so that it may be constructed using ONLY NAND gates. Reduce the circuit to the minimum gate count. [6]

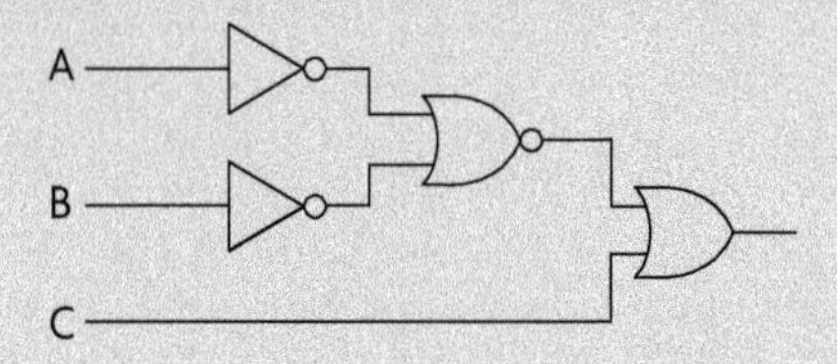

b

- **i** Draw a circuit diagram of a half-adder. Explain with the aid of a truth table, its function and how it performs that function. [4]
- **ii** Explain, with the aid of a diagram, how two half-adders can be used to build a full-adder. [2]

16 The figure below shows three T flip-flops connected together. The output changes when the input detects a negative edged signal (i.e. the input changes from logic 1 to logic 0). The outputs are Q_0, Q_1 and Q_2.

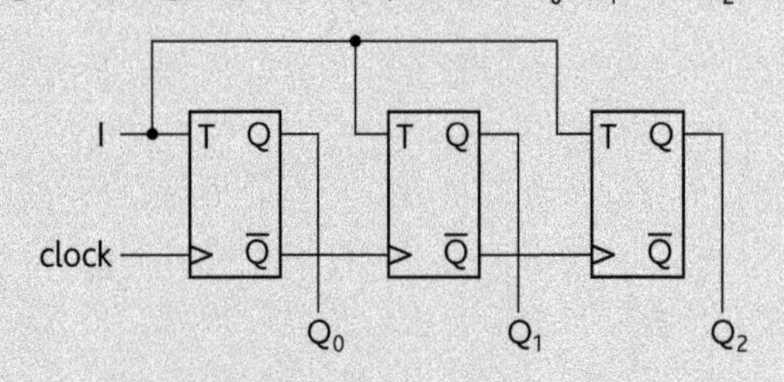

a Copy and complete the truth table to show the outputs Q_0, Q_1 and Q_2 [4]

Clock pulse	Q_0	Q_1	Q_2
0	0	0	0
1	1	0	0
2			
3			
4			
5			

b The outputs from **a** are connected to the inputs of the circuit below. When an input receives a logic 1, a potential of +5 V is applied. When an input receives a logic 0, a potential of 0 V is applied. Q_0, Q_1 and Q_2 apply voltages V_0, V_1 and V_2 respectively to the circuit below.

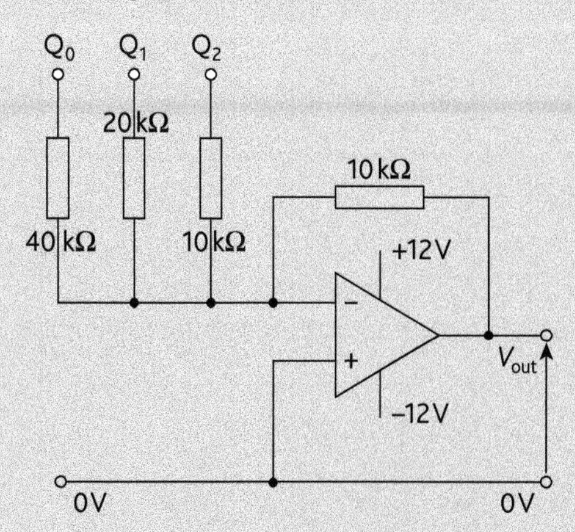

i State the type of circuit shown above. [1]

ii State an equation relating V_{out} to the three input voltages V_0, V_1 and V_2. [2]

iii Copy and complete the following table. [2]

Q_2	Q_1	Q_0	V_{out}/V
0	1	0	−2.50
0	1	1	
1	0	0	
1	0	1	−6.25

iv Explain how the circuit functions as a digital to analogue converter (DAC). [3]

11 The particulate nature of electromagnetic radiation

11.1 The photoelectric effect

Learning outcomes

On completion of this section, you should be able to:

- describe the photoelectric effect
- state the observations associated with the photoelectric effect
- explain the terms *photon*, *work function*, *threshold frequency*, *cut-off wavelength*
- state Einstein's equation to explain the photoelectric effect
- define the electronvolt.

The photoelectric effect

Consider the experiment shown in Figure 11.1.1. A zinc plate is negatively charged and connected to a gold leaf electroscope. When the plate is exposed to weak ultraviolet radiation the plate slowly loses its charge.

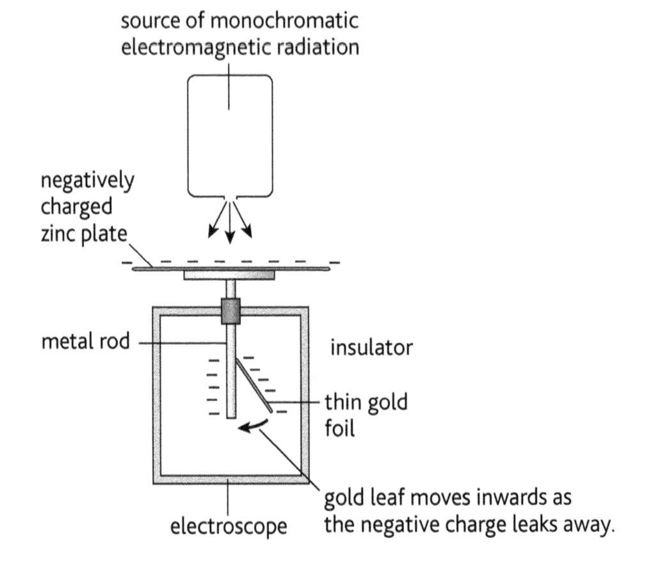

***Figure 11.1.1** Demonstrating the photoelectric effect*

The zinc plate has a negative charge. When it is placed on the gold leaf electroscope it repels the electrons away from the horizontal plate of the gold leaf electroscope. The gold leaf rises, indicating that a charged object is touching the horizontal plate. When ultraviolet (UV) radiation is incident on the zinc plate, the gold leaf begins to fall. This phenomenon is known as the **photoelectric effect**.

Definition

Photoelectric effect

When a metal surface is exposed to electromagnetic radiation, electrons are emitted from the surface.

Experiments performed to investigate the photoelectric effect have resulted in the following observations:

- Electrons are emitted immediately upon being exposed to electromagnetic radiation, even if the intensity of the radiation is weak.
- If the intensity of the radiation is increased, the number of electrons emitted per second also increases.
- Electrons are only emitted when the frequency of the incident radiation is above a minimum frequency (threshold frequency f_0). If the frequency is below this minimum frequency no electrons are emitted, even if the intensity of the radiation is high.
- If the frequency of the incident radiation is greater than the minimum frequency required for electron emission, the maximum kinetic energy of the emitted electrons increases.
- Electrons are emitted with a range of kinetic energies. The kinetic energies range from zero up to a maximum value.

Classical wave theory cannot explain the photoelectric effect. It cannot explain why there is a threshold frequency or how low intensity electromagnetic radiation is able to cause emission of electrons with high kinetic energies. According to the wave theory, only very intense electromagnetic radiation will allow instantaneous emission of electrons from the metal surface. Low intensity electromagnetic radiation should not be able to produce electron emission. Also, according to the wave theory, increasing the intensity of the electromagnetic radiation should increase the kinetic energy of the emitted electrons but not frequency. These questions remained unanswered until Einstein attempted to explain the photoelectric effect using a particle model instead of a wave model.

Einstein's photoelectric equation

In 1905 Albert Einstein provided an explanation for the photoelectric effect. He proposed that the electromagnetic radiation can be thought of as being made up of a stream of tiny bundles or packets of energy. Each packet is discrete and is therefore referred to as a **quantum**. These packets are called **photons**.

Definition

A photon is a quantum of electromagnetic radiation.

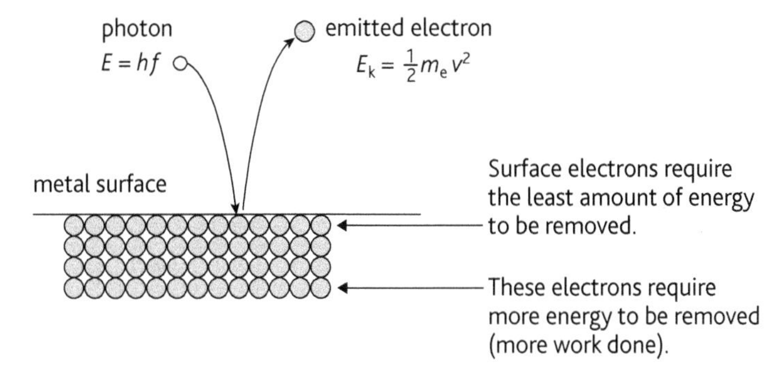

Figure 11.1.2

Equation

The energy of a photon E is given by the equation

$E = hf$

E – energy of a photon/J
h – the Planck constant (6.63×10^{-34} J s)
f – frequency of the incident radiation/Hz

When a single photon reaches the surface of the metal, it interacts with a single electron on the surface (Figure 11.1.2). The photon delivers the amount of energy equivalent to hf to the surface electron. Energy is conserved in the interaction between the photon and the electron. This means that all the energy supplied to the electron will be used to remove it from the metal surface and any remaining amount will be the kinetic energy of the emitted electron.

Definition

The work function ϕ is the minimum energy needed to free an electron from the surface of the metal.

The minimum energy needed to free the electron from the surface is called the **work function** (ϕ). If the energy of the photon is greater than the work function, the electron will escape from the surface and the remaining energy is then transferred to the electron as kinetic energy, $E_k = \frac{1}{2}mv_e^2$.

Einstein's equation to explain the photoelectric effect is given below.

Equation

$hf = \phi + E_{kmax}$

hf – energy of incident photon/J
ϕ – energy needed to free electron from surface (work function)/J
E_{kmax} – kinetic energy of emitted electron/J

Definition

The threshold frequency (f_0) is the minimum frequency of the incident electromagnetic radiation required for electrons to be emitted.

An electron will only be emitted if the energy of the photon is greater than the work function. The energy of the photon is directly proportional to its frequency ($E = hf$). Therefore, the higher the frequency of the incident radiation, the greater is the energy of the photons arriving at the metal surface. Thus a minimum frequency is required in order for electrons to be emitted. This minimum frequency is called the **threshold frequency** (f_0).

The **intensity** of the incident electromagnetic radiation is defined as the power incident per unit area on the surface. ($I = P/A$). Power is defined as the energy per unit time ($P = E/t$). Therefore, the intensity of the incident electromagnetic radiation represents the number of photons incident on the surface per unit time. Increasing the intensity of the incident radiation increases the number of photons striking the surface per unit time. Therefore, increasing the intensity of the incident radiation causes more photons to strike the metal surface. This means that there will be more photon–electron interactions. As a result, more electrons will be emitted from the metal surface.

Equation

$$I = \frac{P}{A}$$

I – intensity of incident electromagnetic radiation/W m^{-2}
P – incident power/W or J s^{-1} (gives a measure of the number of photons incident on the surface per unit time)
A – surface area/m^2

If the frequency of the incident radiation is below the threshold frequency, no electrons will be emitted. This is because the photons will not have sufficient energy to allow the electrons to overcome the work function. Increasing the intensity of the incident radiation will have no effect because it only means that more photons are striking the metal surface per unit time and they still will not have enough energy to allow the electrons to overcome the work function and leave the surface.

The electrons emitted from a metal surface (when exposed to electromagnetic radiation) have a range of kinetic energies. The electrons having the maximum kinetic energy are the ones closest to the surface of the metal, because not much energy is needed to remove them. Electrons further away from the surface of the metal require more energy to remove them and therefore have lower kinetic energies.

Definition

The cut-off wavelength (λ_0) is the maximum wavelength of the incident electromagnetic radiation required for electrons to be emitted.

For a given metal, there is a threshold frequency (f_0). The metal has a corresponding **cut-off wavelength** (λ_0). In order for electrons to be emitted, the wavelength of the incident electromagnetic radiation must be lower than the cut-off wavelength.

The work function of a metal is related to both the threshold frequency and the cut-off wavelength as shown in the equation.

Equation

$$\phi = hf_0 = h\frac{c}{\lambda_0}$$

ϕ – work function /J
h – the Planck constant/J s
f_0 – threshold frequency/Hz
c – speed of light/m s^{-1}
λ_0 – cut-off wavelength/m

Definition

1 eV is the energy transformed by an electron as it moves through a potential difference of 1 V.

1 eV = 1.6×10^{-19} J

The energy associated with photons in electromagnetic radiation is extremely small. The SI unit for energy is the joule (J), but since the energies of photons are much smaller than the joule, another suitable unit is defined. This unit is called the **electronvolt** (eV). When an electron moves through a potential difference, energy is transferred. Since an electron has a charge of 1.6×10^{-19} C, when an electron is moved through a potential difference of 1 V, the work done is:

$$W = QV = 1.6 \times 10^{-19} \times 1 = 1.6 \times 10^{-19}\,\text{J}$$

Example

Ultraviolet radiation of frequency 4.83×10^{15} Hz is incident on a metal surface. The work function of the metal surface is 1.92×10^{-18} J.

Calculate:

a the energy of a photon from this radiation

b the maximum kinetic energy of the photoelectrons emitted from the metal surface.

---

a Energy of a photon $= hf = 6.63 \times 10^{-34} \times 4.83 \times 10^{15}$
$= 3.20 \times 10^{-18}$ J

b $hf = f + E_{kmax}$

$E_{kmax} = hf - \phi$

$E_{kmax} = 3.20 \times 10^{-18} - 1.92 \times 10^{-18} = 1.28 \times 10^{-18}$ J

Example

The work function of a metal is 2.30×10^{-19} J. The metal surface is illuminated with electromagnetic radiation of wavelength 200 nm. Electrons are emitted from the surface.

Calculate:

a the energy of a photon having a wavelength of 200 nm

b the threshold frequency of the metal

c the maximum kinetic energy of the electrons emitted from the metal surface.

---

a Energy of a photon $\frac{hc}{\lambda} = \frac{6.63 \times 10^{-34} \times 3.0 \times 10^{8}}{200 \times 10^{-9}} = 9.95 \times 10^{-19}$ J

b Threshold frequency $\frac{\phi}{h} = \frac{2.30 \times 10^{-19}}{6.63 \times 10^{-34}} = 3.47 \times 10^{14}$ Hz

c $hf = \phi + E_{kmax}$

$E_{kmax} = hf - \phi$

$E_{kmax} = 9.95 \times 10^{-19} - 2.30 \times 10^{-19} = 7.65 \times 10^{-19}$ J

Key points

- When a metal surface is exposed to electromagnetic radiation, electrons are emitted from the surface (photoelectric effect).
- Classical wave theory cannot explain the photoelectric effect.
- Einstein proposed a particle model of electromagnetic radiation to explain the photoelectric effect.
- 1 electronvolt (eV) is the energy transformed by an electron as it moves through a p.d. of 1V.

Learning outcomes

On completion of this section, you should be able to:

- describe an experiment to investigate the photoelectric effect
- provide an explanation of the experiment
- explain the terms *photoelectric current* and *stopping potential*
- explain how the results can be used to measure the Planck constant.

Investigating the photoelectric effect

Figure 11.2.1 shows how the photoelectric effect can be investigated. A zinc plate is exposed to ultraviolet (UV) light. Electrons are emitted from the zinc plate. The emitted electrons are attracted to the gauze because of its positive electric potential. This is because the gauze is connected to the positive terminal of the variable d.c. supply.

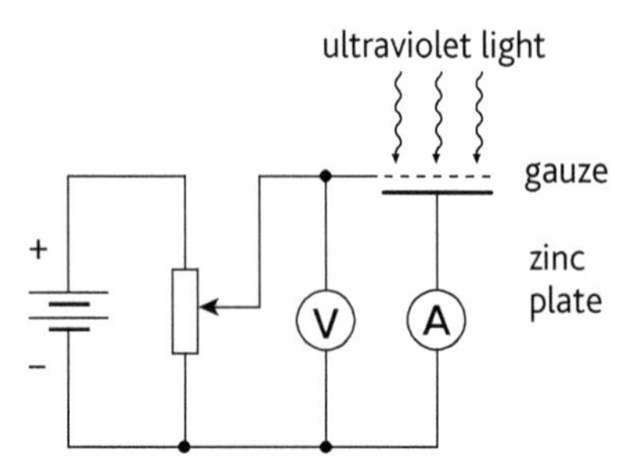

Figure 11.2.1 *Apparatus to investigate the photoelectric effect*

The ammeter measures the **photoelectric current**. The voltmeter measures the potential difference between the zinc plate and the gauze. When the gauze is at a positive potential with respect to the zinc plate a very small current is detected. No matter how much the potential difference between the zinc plate and the gauze is increased, the photoelectric current remains constant. This is because the photoelectric current is dependent on the number of electrons leaving the surface of the zinc. As long as the intensity of the UV radiation remains unchanged, the photoelectric current will remain constant.

Reversing the polarity of the d.c. supply causes the photoelectric current to decrease. In this instance, the gauze has a negative potential with respect to the zinc plate. The gauze repels the electrons emitted from the zinc surface. If the potential of the gauze is made even more negative by adjusting the potentiometer, the photoelectric current eventually reduces to zero. A photoelectric current of zero implies that no electrons are reaching the gauze. The potential difference that causes the photoelectric current to reduce to zero is called the **stopping potential** (V_s).

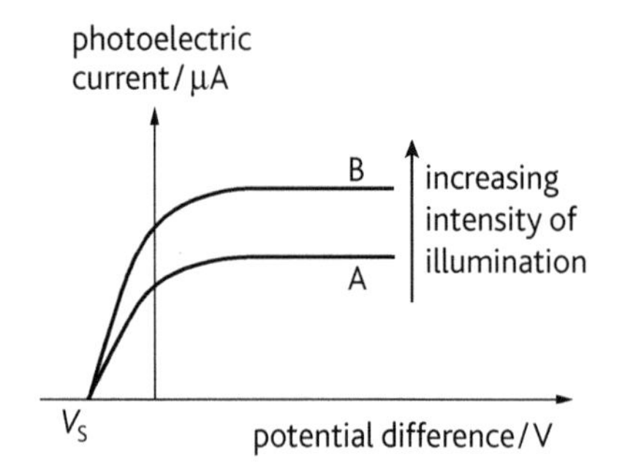

Figure 11.2.2

The graph labelled A in Figure 11.2.2 shows the result of the experiment. Increasing the intensity of the UV radiation produces the graph labelled B. This graph has the same characteristic shape as A. However, the photoelectric current is greater. The stopping potential is the same because the energy of the photon and the work function remains fixed, so the maximum kinetic energy will also remain constant.

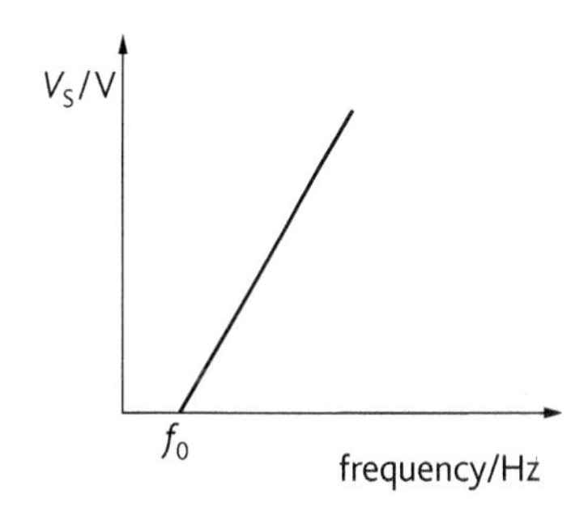

Figure 11.2.3 *Graph of stopping potential against frequency*

In the experiment above, the frequency of the incident UV radiation is known. The same apparatus can be used to determine the work function of the zinc metal and the Planck constant. The frequency of the incident radiation is varied and the stopping potential in each case is measured.

A graph of stopping potential against frequency is plotted as shown in Figure 11.2.3.

According to Einstein's equation:

$$hf = \phi + \tfrac{1}{2}mv^2 \qquad (1)$$

The work function of the zinc metal is given by:

$$\phi = hf_0 \qquad (2)$$

The maximum kinetic energy of the emitted electrons is given by:

$$\tfrac{1}{2}mv^2 = eV_s \qquad (3)$$

Substituting Equations (2) and (3) into Equation (1) and rearranging it,

$$eV_s = hf - hf_0$$

$$\therefore \quad V_s = \frac{h}{e}f - \frac{h}{e}f_0 \qquad (4)$$

Comparing Equation (4) with $y = mx + c$,

$$\text{the gradient of the straight line} = m = \frac{h}{e}$$

$$\text{the } y\text{-intercept} = c = -\frac{hf_0}{e}$$

So according to Equation (4), if a graph of V_s is plotted against f, the gradient will be h/e, the x-intercept will be f_0 and the y-intercept will be $-hf_0/e$.

Applications of the photoelectric effect

One of the best applications of the photoelectric effect is in the photocell (Figure 11.2.4). As electromagnetic radiation (e.g. light) strikes the metal cathode, electrons are emitted from the surface. The electrons are attracted towards the collector and a small current is produced. This device has numerous applications. It can be found in light meters, photocopiers and even a digital camera. Another important application is in the manufacture of photovoltaic cells. These devices harness solar energy and produces electrical energy. Photovoltaic cells can be found in wrist watches, calculators and satellites orbiting the Earth.

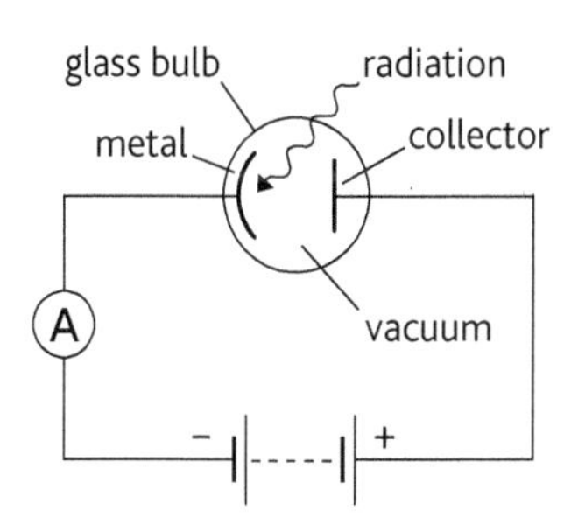

Figure 11.2.4 *A photocell*

Key points

- The photoelectric effect can be investigated by measuring the photoelectric current when the frequency of the incident electromagnetic radiation is varied.
- The stopping potential can be used to determine the work function of the metal.
- The Planck constant can be determined from the data obtained from photoelectric effect experiments.

11.3 Examples on the photoelectric effect

Learning outcomes

On completion of this section, you should be able to:

- solve problems relating to the photoelectric effect.

Example

A parallel beam of ultraviolet radiation of wavelength 240 nm and intensity of 800 W m^{-2} is incident normally on a zinc surface of area 1.2×10^{-4} m^{-2}. The work function of zinc is 4.3 eV.

Calculate:

a the energy of a photon of ultraviolet radiation

b the power of the radiation incident on the surface

c the number of photons incident per second on the surface

d the threshold frequency of zinc

e the maximum kinetic energy of the emitted electrons.

---

a Energy of a photon $E = \dfrac{hc}{\lambda} = \dfrac{6.63 \times 10^{-34} \times 3.0 \times 10^{8}}{240 \times 10^{-9}}$

$= 8.29 \times 10^{-19}$ J

b Power $P = IA = 800 \times 1.2 \times 10^{-4} = 0.096$ W

c Energy incident per second on surface = 0.096 J

One photon has an energy of = 8.29×10^{-19} J

Number of photons incident per second on surface $= \dfrac{0.96}{8.29 \times 10^{-19}}$

$= 1.16 \times 10^{17}$

d Work function of zinc $\phi = 4.3$ eV

Work function in joules $= 4.3 \times 1.6 \times 10^{-19} = 6.88 \times 10^{-19}$ J

Threshold frequency of zinc $f_0 = \dfrac{\phi}{h} = \dfrac{6.88 \times 10^{-19}}{6.63 \times 10^{-34}} = 1.04 \times 10^{15}$ Hz

e Maximum kinetic energy of the photoelectrons $\frac{1}{2}mv^2$

$= hf - \phi$

$= 8.29 \times 10^{-19} - 6.88 \times 10^{-19}$

$= 1.41 \times 10^{-19}$ J

Example

In the photoelectric effect, electrons are emitted from a metal surface when it is illuminated with electromagnetic radiation of a particular frequency. The variation with frequency f of the maximum kinetic energy E_{kmax} of the emitted electrons is shown in Figure 11.3.1.

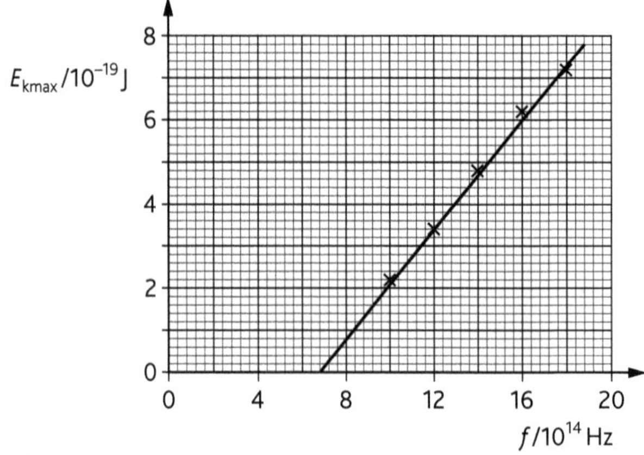

Figure 11.3.1

Use the graph to find:

a the threshold frequency

b the Planck constant

c the work function of the metal.

The photoelectric equation is $hf = \phi + E_{kmax}$

Rearranging the equation, $E_{kmax} = hf - \phi$

The graph in Figure 11.3.1 shows a plot of E_{kmax} against f. Therefore the gradient of the line is equal to the Planck constant and the y-intercept is ϕ, the work function of the metal.

a The threshold frequency corresponds to the point on the graph where the maximum kinetic energy of the photoelectrons is zero.

Therefore, the threshold frequency $= 6.8 \times 10^{14}$ Hz

b Planck constant $= \dfrac{0 - 6.0 \times 10^{-19}}{6.8 \times 10^{14} - 16 \times 10^{14}} = 6.52 \times 10^{-34}$ J s

c Using the point $(16 \times 10^{14}, 6.0 \times 10^{-19})$, the work function is given by

$$\phi = hf - E_{kmax} = (6.52 \times 10^{-34} \times 16 \times 10^{14}) - 6.0 \times 10^{-19}$$
$$= 4.43 \times 10^{-19}\text{ J}$$

Example

In an experiment to investigate the photoelectric effect, the wavelength λ of the electromagnetic radiation incident on a metal surface and the maximum kinetic energy E_{kmax} is measured. The variation with E_{kmax} of $1/\lambda$ is shown in Figure 11.3.2.

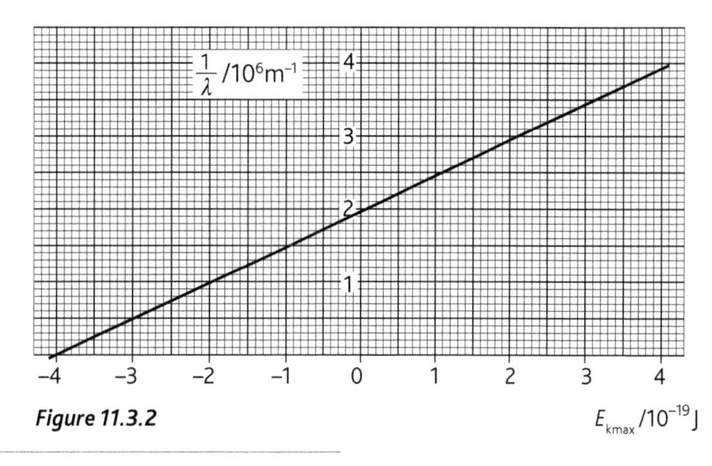

Figure 11.3.2

Determine:

a the work function of the metal surface

b the value of the Planck constant.

Starting with Einstein's equation $\dfrac{hc}{\lambda} = \phi + E_{kmax}$ $\quad \left(E = hf = \dfrac{hc}{\lambda}\right)$

Rewriting the equation we get $\dfrac{1}{\lambda} = \dfrac{\phi}{hc} + \dfrac{E_{kmax}}{hc}$

Therefore, plotting a graph of $\dfrac{1}{\lambda}$ against E_{kmax} will obtain a straight line with gradient $\dfrac{1}{hc}$ and y-intercept $\dfrac{\phi}{hc}$.

a When $\dfrac{1}{\lambda} = 0$, $\phi = -E_{kmax}$. Therefore, $\phi = 4.0 \times 10^{-19}$ J

b Using the points $(2 \times 10^{-19}, 3 \times 10^{6})$ and $(-4 \times 10^{-19}, 0)$,

the gradient of the straight line $= \dfrac{3 \times 10^{6} - 0}{2 \times 10^{-19} - (-4 \times 10^{-19})}$

$= 5 \times 10^{24}$

Therefore, $\dfrac{1}{hc} = 5 \times 10^{24}$

$h = \dfrac{1}{5 \times 10^{24} \times 3 \times 10^{8}} = 6.67 \times 10^{-34}$ J s

Revision questions 6

Answers to questions that require calculation can be found on the accompanying CD.

1 A 1.2 mW laser produces blue light of wavelength 4.75×10^{-7} m.

Calculate:

a the frequency of blue light [2]

b the energy of a photon of blue light [2]

c the number of photons emitted per second from the laser. [2]

2 A photocell is shown in the diagram below.

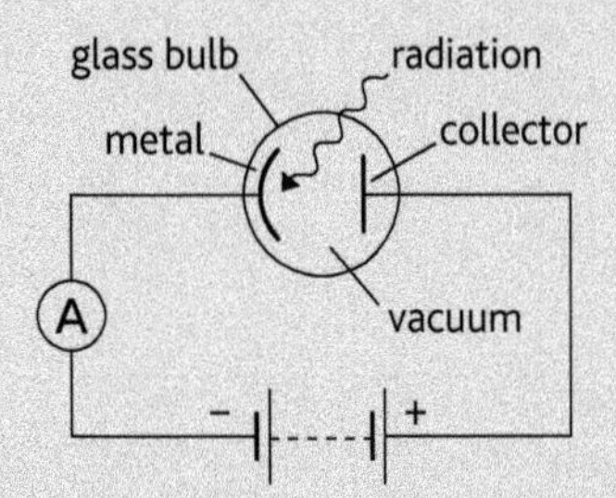

As the metal inside the glass bulb is exposed to electromagnetic radiation, electrons are emitted and then attracted to the collector plate. The ammeter detects a small current of 12 μA. Calculate:

a the quantity of charge reaching the collector in 6 seconds [2]

b the number of photoelectrons reaching the collector during this time. [2]

3 a Explain what is meant by the photoelectric effect. [3]

b Explain what is meant by the terms:

i threshold frequency [2]

ii work function energy. [2]

c Write down an equation for the photoelectric effect. [2]

4 A parallel beam of ultraviolet radiation of wavelength 255 nm and intensity of 600 W m^{-2} is incident normally on a zinc surface of area 8.4×10^{-4} m^{-2}. The work function of zinc is 4.3 eV.

Calculate:

a the energy of a photon of ultraviolet radiation [2]

b the power of the radiation incident on the surface [2]

c the number of photons incident per second on the surface [2]

d the threshold frequency of zinc [2]

e the maximum kinetic energy of the emitted electrons. [2]

5 a State the effect that produces electrons when a clean metal surface is illuminated with electromagnetic radiation. [1]

b The variation with frequency f of the maximum kinetic energy E_{kmax} of the emitted electrons from a metal surface is shown below.

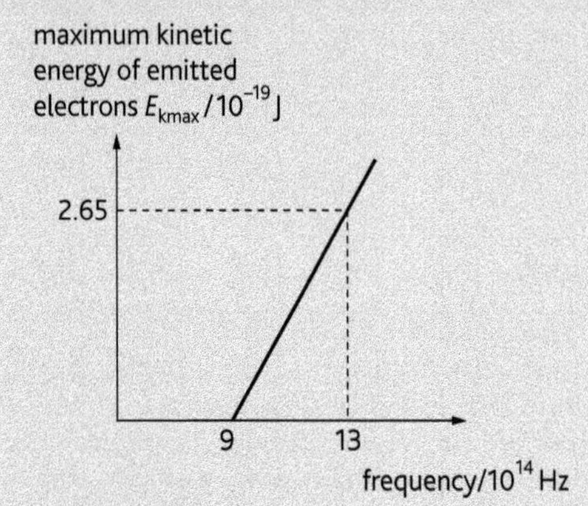

i State the threshold frequency. [1]

ii Use the graph to estimate the Planck constant. [3]

iii Calculate the work function energy for the metal. [2]

c Copy the diagram above. Sketch a new graph showing the effect of using a metal that has a higher work function energy. [2]

d Explain why for a frequency of 13×10^{14} Hz electrons are emitted from the metal surface with a range of kinetic energies from zero to 2.65×10^{-19} J. [2]

6 a Explain what is meant by a photon. [1]

b Calculate the energy of a photon of red light having a wavelength of 650 nm. [2]

7 Describe an experiment to show:

a the particulate nature of electromagnetic radiation [4]

b the wave nature of electromagnetic radiation. [6]

8 a What evidence does the photoelectric effect provide about the nature of electromagnetic radiation? [2]

b State three observations concerning the photoelectric effect. [3]

c Explain the photoelectric effect using Einstein's photoelectric equation. [4]

d Explain the following:

i Doubling the intensity of incident electromagnetic radiation doubles the number of electrons emitted from a metal surface. [3]

ii Increasing the frequency of the incident radiation increases the maximum kinetic energy of the emitted electrons. [3]

9 a Describe a laboratory demonstration which cannot be explained by the wave theory of light but which requires an explanation in which light has a particular nature. [4]

b For light having a wavelength of 640 nm, calculate:

i its frequency [2]

ii the energy of a photon [2]

iii the rate of emission of photons for a light power of 15 W. [2]

10 a State what is meant by the photoelectric effect. [2]

b State three experimental observations associated with the photoelectric effect. [3]

c The radius of an atom is approximately 2.0×10^{-10} m. A lamp is placed above a metal surface and provides energy at a rate of 0.80 W m^{-2}. An electron requires a minimum energy of 5.76×10^{-19} J to be emitted from the metal surface. Assume that the electron can collect energy from a circular area which has a radius equal to that of the atom. On the basis of the wave theory, estimate the time taken for an electron to be emitted from the metal surface. Comment on your answer. [5]

Learning outcomes

On completion of this section, you should be able to:

- describe Millikan's oil drop experiment
- discuss the evidence in Millikan's oil drop experiment for the quantisation of charge.

Millikan's oil drop experiment

In 1909 Robert Millikan performed experiments to measure the elementary electric charge e. Figure 11.4.1 shows a diagram of Millikan's apparatus.

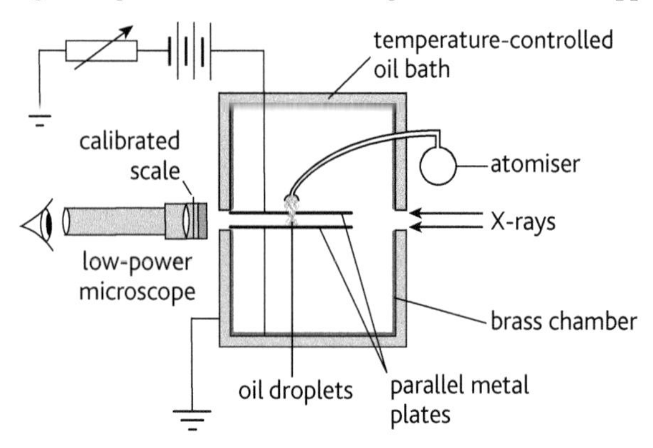

Figure 11.4.1 *Millikan's apparatus*

Two horizontal metal plates approximately 20 cm in diameter were set up so that they were 1.5 cm apart. The upper plate had a small hole at its centre. An atomiser was used to create a fine mist of oil droplets above the hole. Some of the oil drops got charged by friction as they came out of the atomiser. Eventually some of the oil droplets passed through the hole and entered the region between the plates.

The oil droplets were viewed using a low-power microscope. The region between the plates was illuminated and the droplets were seen in the microscope as specks. The eyepiece of the microscope included a calibrated scale that was used to measure the distance travelled by an oil drop. The microscope was focused on one particular oil drop and its terminal velocity v_1 was determined by measuring the time t_1 for it to fall through a known distance x, when there was no electric field applied between the plates.

The forces acting on a falling oil drop are its weight W, the upthrust U due to air and drag force D due to air (Figure 11.4.2).

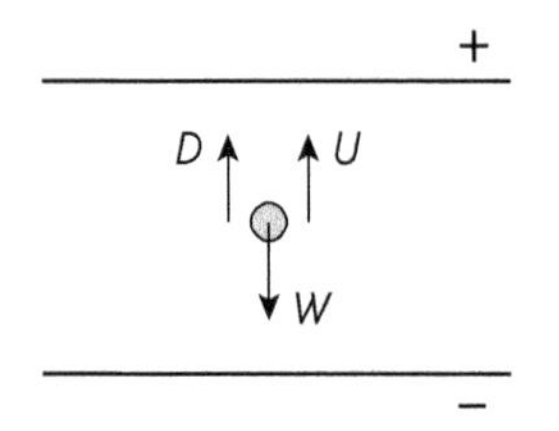

Figure 11.4.2 *Forces acting on a falling oil drop*

r – radius of oil drop

ρ_0 – density of the oil

ρ_a – density of air

η – viscosity of air

g – acceleration due to gravity

v_1 – terminal velocity

$$\text{Weight of oil drop} = \text{volume of drop} \times \text{density of oil} \times g = \tfrac{4}{3}\pi r^3 \rho_0 g$$

$$\text{Upthrust} = \text{weight of air displaced}$$

$$= \text{volume of drop} \times \text{density of air} \times g = \tfrac{4}{3}\pi r^3 \rho_a g$$

$$\text{Drag force} = 6\pi r \eta v_1 \text{ (Stokes' law)}$$

$$\text{Weight of oil drop} = \text{upthrust} + \text{drag force}$$

$$\tfrac{4}{3}\pi r^3 \rho_0 g = \tfrac{4}{3}\pi r^3 \rho_a g + 6\pi r \eta v_1 \qquad (1)$$

An electric field E was then applied between the plates so that the drop moved upward with a terminal velocity v_2. The terminal velocity v_2 was determined by measuring the time t_2 for the oil drop to move through a known distance y.

The electric force F_E acting on an oil drop is Eq, where q is the charge on the oil drop and $E = V/d$, where V is the potential difference across the plates and d is the separation of the plates. Figure 11.4.3 shows the forces acting on an oil drop when it is moving upwards with a terminal velocity.

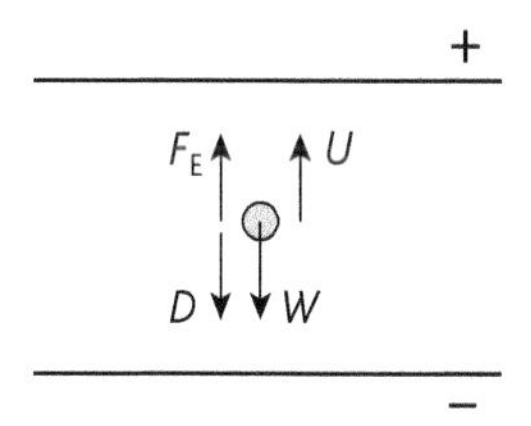

Figure 11.4.3 *Forces acting on an oil drop moving upward with terminal velocity*

Weight of oil drop + drag force = upthrust + electric force

$$\tfrac{4}{3}\pi r^3\rho_0 g + 6\pi r\eta v_2 = \tfrac{4}{3}\pi r^3\rho_a g + Eq \qquad (2)$$

Subtracting Equation (1) from Equation (2) gives:

$$6\pi r\eta v_2 = Eq - 6\pi r\eta v_1$$

$$Eq = 6\pi r\eta(v_2 - v_1)$$

$$q = \frac{6\pi r\eta(v_2 - v_1)}{E}$$

The electric field strength E was measured using $E = V/d$.

The terminal velocity v_1 was measured using $v_1 = x/t_1$.

The terminal velocity v_2 was measured using $v_2 = y/t_2$.

The radius r of an oil drop was determined by calculation using Equation (1).

$$r = \sqrt{\frac{9\eta v_1}{2(\rho_0 - \rho_a)g}}$$

How Millikan improved the accuracy of his experiments

- X-rays were used to ionise the air inside the chamber to increase the charge on the oil drops.
- A constant-temperature enclosure surrounded the apparatus in order to eliminate convection currents.
- A low vapour pressure oil was used in the experiment to reduce evaporation.

Conclusions

Figure 11.4.4 illustrates the type of results obtained when performing Millikan's experiment.

After measuring the charges on hundreds of oil drops Millikan concluded the following:

- Electric charge is quantised. All electric charges are integral multiples of a unique elementary charge e.
- The magnitude of the fundamental charge $e = 1.6 \times 10^{-19}$ C.

Therefore, since charge is quantised, charge can only exist as e, $2e$, $3e$, $4e$, etc.

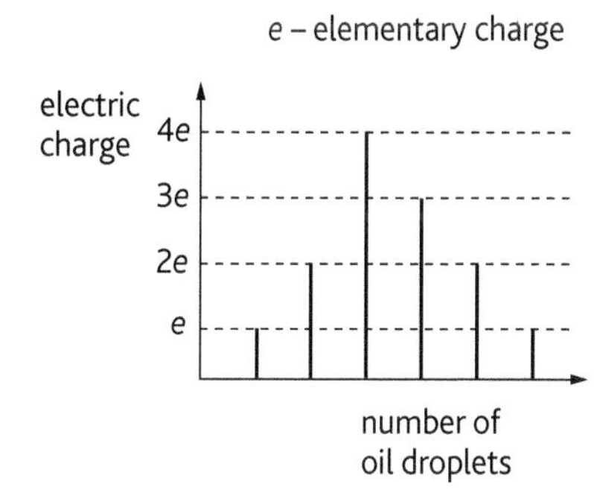

Figure 11.4.4 *Results of the experiment*

Key points

- Millikan's experiments showed that electric charge is quantised.
- All electric charges are integral multiples of a unique elementary charge e.
- The magnitude of the fundamental charge $e = 1.6 \times 10^{-19}$ C.

11.5 Emission and absorption spectra

Learning outcomes

On completion of this section, you should be able to:

- distinguish between absorption and emission spectra
- discuss how line spectra provide evidence for discrete energy levels in isolated atoms.

Line spectra

If white light is passed through a diffraction grating, a spectrum of colours is seen. White light is made up of different colours, each having its respective wavelength (red, orange, yellow, blue, green, indigo, violet). The wavelengths range from 400 nm (violet) to 750 nm (red).

If the light from a lamp that contains sodium vapour is allowed to pass through a diffraction grating, a spectrum is observed. However, unlike the spectrum obtained using white light, this spectrum contains only specific colours. This spectrum is called a **line spectrum** (Figure 11.5.1). More specifically, it is called a **line emission spectrum**.

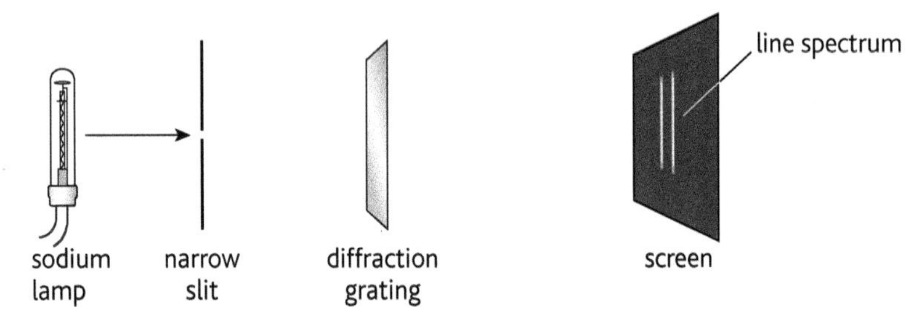

Figure 11.5.1 A line emission spectrum

Suppose white light is passed through a cool gas and the spectrum produced is observed. A series of dark lines against a coloured background is observed. The dark lines correspond to missing colours of the spectrum of white light. This type of spectrum is called a **line absorption spectrum** (Figure 11.5.2).

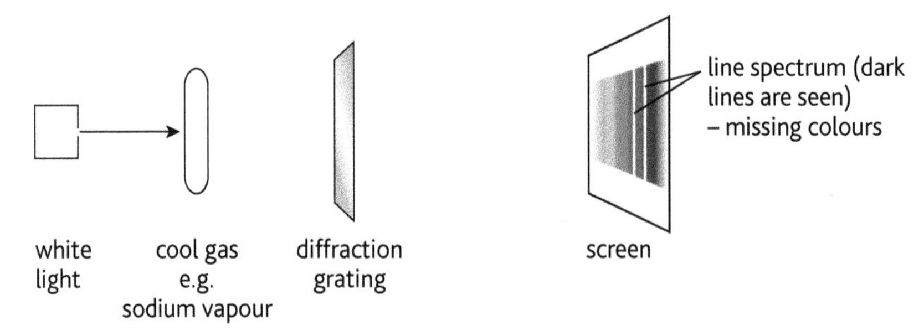

Figure 11.5.2 A line absorption spectrum

(a)

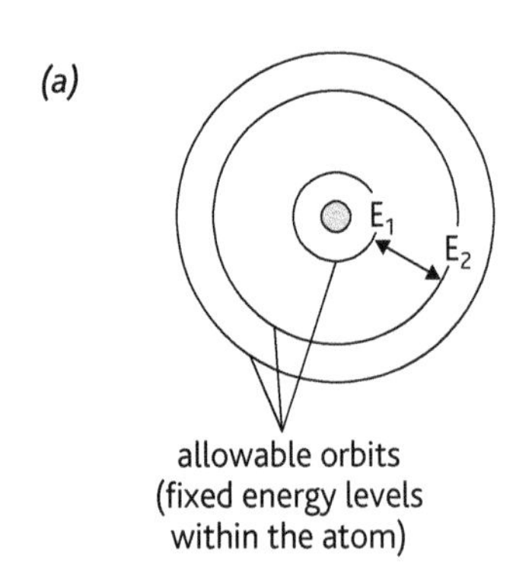

(b) Beyond this point ionisation occurs. The electron is removed from the atom.

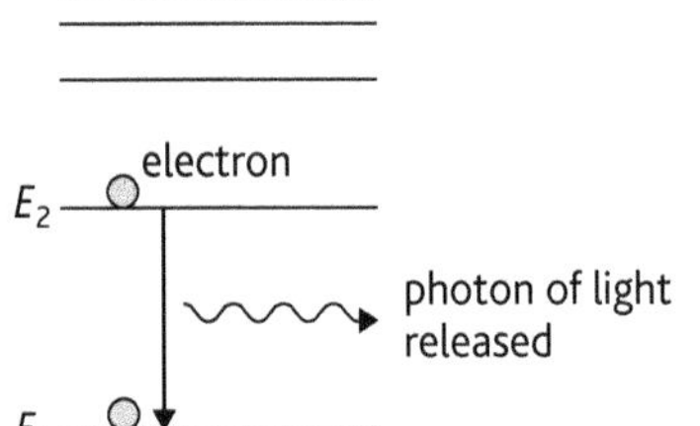

(c) increasing wavelength

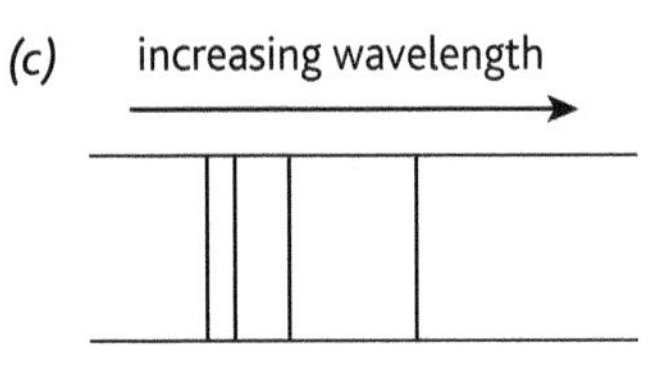

Figure 11.5.3 Discrete energy levels and the line spectrum of an isolated atom

Explaining line spectra

In order to explain line spectra, light needs to be thought of as being made up photons, as described in the photoelectric effect. It is also known that orbital electrons of the atoms can absorb light energy that is incident on them. From experiments, it was observed that atoms of a particular element can only absorb and emit light of certain wavelengths. So, for example, a sodium vapour lamp produces a spectrum that is different from that produced by a neon lamp. Each element produces a unique line spectrum. Scientists can observe a spectrum and identify which elements are present. This is the principle used to determine what elements are present in celestial bodies such as the Sun.

Line spectra provide evidence for discrete energy levels in isolated atoms. Since only fixed wavelengths are being produced in a line spectrum, this suggests that the only discrete or fixed energy levels are possible in isolated atoms (Figure 11.5.3(a) and (b)). It can therefore be said that the energy is **quantised**.

When an electron falls from a higher energy level E_2 to a lower energy level E_1, a photon of light is emitted (Figure 11.5.3(b)). The energy of this photon is exactly equal to $E_2 - E_1$. The greater the difference between E_2 and E_1 the greater is the energy of the emitted photon (shorter wavelength). The movement of an electron from one energy level to another is called a transition. Since there are several discrete energy levels, different transitions are possible. This explains why only certain wavelengths are present in the spectrum of a hot gas like sodium.

Observing Figure 11.5.3(a), it can be seen that inside the atom, there are fixed orbits that electrons can occupy. These orbits have fixed energy values. Figure 11.5.3(b) is called an **energy level diagram**. It is made up of a series of horizontal lines with energy values associated with each one.

If an atom is supplied with energy, an electron may jump from energy level E_1 to energy level E_2. If this electron falls back to energy level E_1, a photon of light is emitted. The wavelength of this photon is given by $\lambda = \frac{hc}{E_2 - E_1}$.

An atom needs to be supplied with the exact amount of energy $E_2 - E_1$ in order for an electron to undergo a transition from E_1 to E_2. The transition will not take place if too little or too much energy is supplied. If too little energy is supplied, the electron may not be able to make it to a higher energy level. If too much energy is supplied, the electron will not occupy one of the allowable energy levels within the atom. The equations shown are used to determine the energy, frequency or wavelength of a photon absorbed or emitted in a line spectrum.

Figure 11.5.3(c) shows a possible line spectrum for this particular atom. Each line indicates a transition on the energy level diagram.

The concept of discrete energy levels inside an atom can therefore be used to explain how line emission and line absorption spectra occur (Figure 11.5.4).

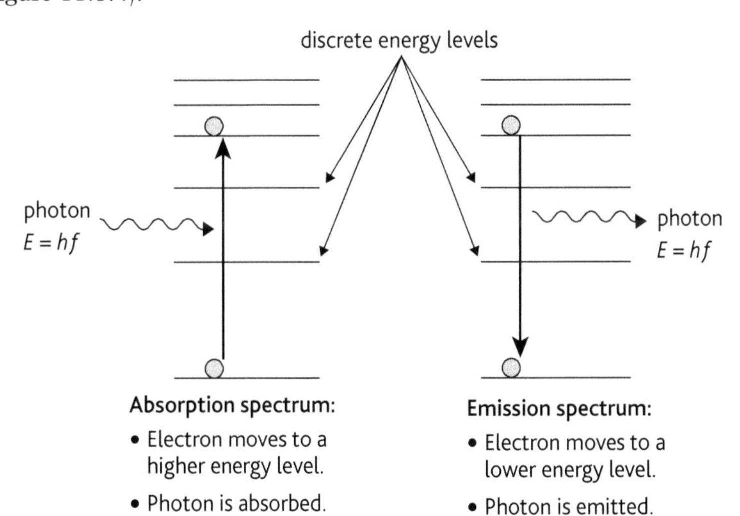

Figure 11.5.4 *Explaining line spectra*

Definition

A line emission spectrum is a series of discrete bright lines on a dark background.

Definition

A line absorption spectrum is a continuous bright spectrum crossed by dark lines.

Equation

$E_2 - E_1 = hf$ or $E_2 - E_1 = \frac{hc}{\lambda}$

E_1 – lower energy level/J
E_2 – higher energy level/J
h – the Planck constant (6.63×10^{-34} J s)
c – speed of light (3×10^8 m s^{-1})
λ – wavelength/m

Key points

- A line emission spectrum is a series of discrete bright lines on a dark background.
- A line absorption spectrum is a continuous bright spectrum crossed by dark lines.
- Line spectra provide evidence for discrete energy levels in isolated atoms.
- Each line represents a photon of a specific energy. The photon is emitted or absorbed as a result of an energy change of an electron. Specific energies (lines) are observed, therefore there are discrete energy levels.
- $E_2 - E_1 = hf$ is used to calculate the frequency of an emitted photon when an electron undergoes an energy change from E_2 to E_1.

11.6 Examples of line spectra

Learning outcomes

On completion of this section, you should be able to:

- use the relationship $hf = E_2 - E_1$ to solve problems.

Example

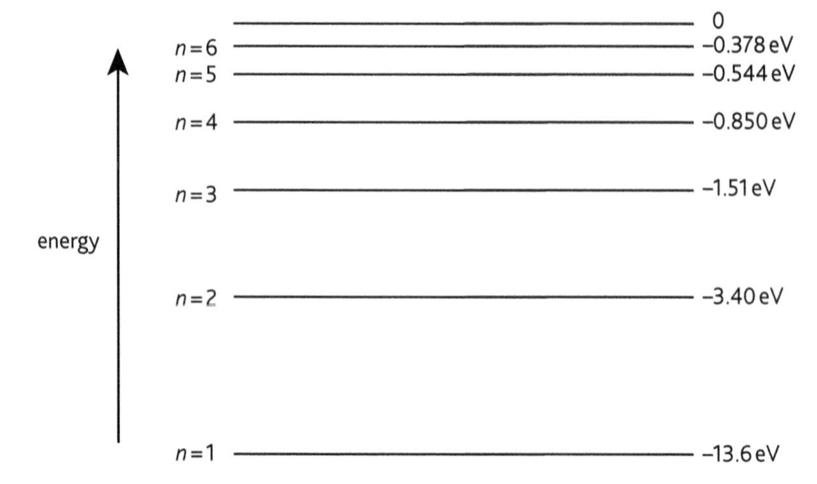

Figure 11.6.1 *Energy level diagram of a hydrogen atom*

Figure 11.6.1 shows part of the energy level diagram of a hydrogen atom.

a Show that when an electron makes a transition from $n = 2$ to $n = 1$, the wavelength of the emitted photon does not lie in the visible region of the electromagnetic spectrum.

b Calculate the minimum wavelength of a photon that could be emitted from the hydrogen atom.

c Sketch the pattern of the visible line emission spectrum. (This occurs when electrons fall to $n = 2$.)

The visible spectrum lies between 400 nm (violet) to 750 nm (red).

a Energy of emitted photon when an electron makes an $n = 2$ to $n = 1$ transition:

$$\begin{aligned} E &= E_2 - E_1 \\ &= -13.6 - (-3.40) \\ &= -10.2\,\text{eV} \end{aligned}$$

Energy is a scalar quantity, so only the magnitude of the energy difference is needed:

$$E = 10.2 \times 1.6 \times 10^{-19}\,\text{J} = 1.632 \times 10^{-18}\,\text{J}$$

Wavelength of photon emitted when an electron makes an $n = 2$ to $n = 1$ transition:

$$\begin{aligned} \lambda &= \frac{hc}{E_2 - E_1} \\ &= \frac{6.63 \times 10^{-34} \times 3.0 \times 10^{8}}{1.632 \times 10^{-18}} \\ &= 1.22 \times 10^{-7}\,\text{m} \\ &= 122\,\text{nm} \end{aligned}$$

This wavelength does not lie in the range 400 nm to 750 nm.

It lies in the ultraviolet region of the electromagnetic spectrum.

b Minimum wavelength of a photon that can be emitted from the hydrogen atom corresponds to the largest transition possible on the energy level diagram. The largest possible transition is from $E = 0$ to $E = -13.6\,\text{eV}$.

$$E_2 - E_1 - 13.6 \times 1.6 \times 10^{-19}\,\text{J} - 2.176 \times 10^{-18}\,\text{J}$$

Wavelength of photon emitted:

$$\lambda = \frac{hc}{E_2 - E_1}$$

$$= \frac{6.63 \times 10^{-34} \times 3.0 \times 10^{8}}{2.176 \times 10^{-18}}$$

$$= 9.14 \times 10^{-8}\,\text{m}$$

c The wavelength of a photon emitted from transition:

$n = 3$ to $n = 2$, $\quad \lambda = \dfrac{6.63 \times 10^{-34} \times 3.0 \times 10^{8}}{(3.40 - 1.51) \times 1.6 \times 10^{-19}} = 658\,\text{nm}$

$n = 4$ to $n = 2$, $\quad \lambda = \dfrac{6.63 \times 10^{-34} \times 3.0 \times 10^{8}}{(3.40 - 0.850) \times 1.6 \times 10^{-19}} = 488\,\text{nm}$

$n = 5$ to $n = 2$, $\quad \lambda = \dfrac{6.63 \times 10^{-34} \times 3.0 \times 10^{8}}{(3.40 - 0.544) \times 1.6 \times 10^{-19}} = 435\,\text{nm}$

$n = 6$ to $n = 2$, $\quad \lambda = \dfrac{6.63 \times 10^{-34} \times 3.0 \times 10^{8}}{(3.40 - 0.378) \times 1.6 \times 10^{-19}} = 411\,\text{nm}$

Now that all the wavelengths have been calculated, the line emission spectrum can be sketched as shown in Figure 11.6.2.

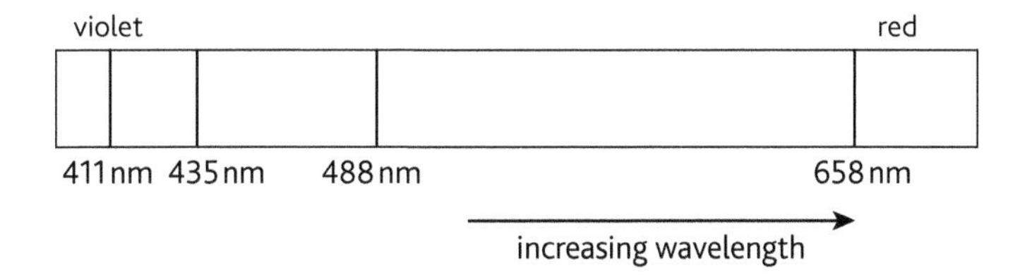

Figure 11.6.2 *The line spectrum of an isolated hydrogen atom*

Example

The diagram below shows some of the outer energy levels of a mercury atom.

0 ionisation
–1.6
–3.7
energy/eV
–5.5
–10.4

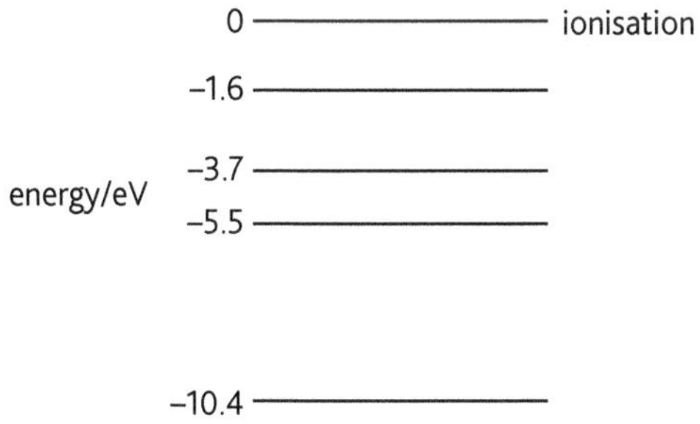

Figure 11.6.3 *Outer energy levels of a mercury atom*

Calculate the ionisation energy in joules for an electron situated in the –10.4 eV energy level.

The ionisation energy for an electron is the energy required to remove it completely from an atom. For the electron at energy level –10.4 eV, the energy required to completely remove it from the atom is $E = -10.4 - 0 = -10.4\,\text{eV}$.

Therefore, energy in joules $= 10.4 \times 1.6 \times 10^{-19} = 1.66 \times 10^{-18}\,\text{J}$

11.7 Wave–particle duality

Learning outcomes

On completion of this section, you should be able to:

- explain the wave–particle duality of matter
- describe and interpret the evidence provided by electron diffraction for the wave nature of particles
- discuss interference and diffraction as evidence of the wave nature of electromagnetic radiation
- use the relation for the de Broglie wavelength $\lambda = \frac{h}{p}$.

Equation

$$\lambda = \frac{h}{p}$$

λ – wavelength of matter/m
h – the Planck constant (6.63×10^{-34} J s)
p – momentum/kg m s^{-1}

The wave–particle duality

The photoelectric effect cannot be explained using classical wave theory. It can only be explained by assuming that the electromagnetic radiation is made up of particles (photons). Similarly line spectra can be explained using a photon model.

The photoelectric effect provides evidence for the particulate nature of electromagnetic radiation. This means that electromagnetic radiation has characteristics that make it seem like particles.

However, it is also known that electromagnetic radiation (or electromagnetic waves) can be diffracted and produce interference patterns. These are properties that are associated with waves.

Light is a form of electromagnetic radiation. Experimental evidence shows that light can be diffracted and can produce interference patterns (Young's double slit experiment). These experiments show that light is a wave. In the case of the photoelectric effect, the idea of light being a wave cannot be used to explain the effect. In order to explain the experimental observations of the photoelectric effect, light has to be thought of as being made of tiny particles called photons. Light can therefore behave as a wave in some instances and as a particle in others. Light is said to have a dual nature, which is referred to as the **wave–particle duality**. The same concept can be applied to all electromagnetic radiation.

Exam tip

1. The photoelectric effect provides evidence for the particulate nature of electromagnetic radiation.
2. Interference and diffraction provide evidence for the wave nature of electromagnetic radiation.
3. Which one is correct? They both are.
4. Under certain conditions, electromagnetic radiation behaves as a particle and under other conditions it behaves as a wave. This phenomenon is called the wave–particle duality.

de Broglie

In 1922 Louis de Broglie suggested that if a wave is able to behave as a particle under certain conditions, particles might behave like a wave under certain conditions.

He proposed the equation shown that relates wavelength and momentum. The equation suggested that a particle is able to have a wavelength.

Electron diffraction

Experimental evidence shows that electrons behave like particles. They have mass and charge and can be deflected by electric and magnetic fields. Newtonian mechanics can also be applied to the movement of electrons. All these features show that an electron behaves as if it were a particle.

However, Figure 11.7.1 shows an experiment that demonstrates that electrons are capable of producing a diffraction pattern.

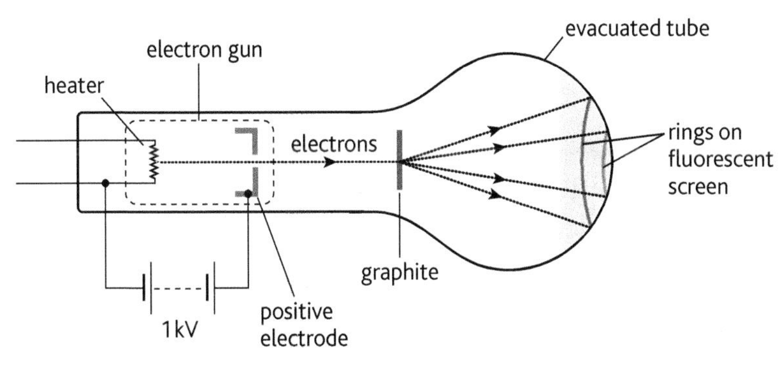

Figure 11.7.1 *Demonstrating electron diffraction*

An electron gun produces a beam of electrons, which are then projected towards a thin layer of graphite. A diffraction pattern of rings is produced on a phosphor-coated screen. Diffraction is a phenomenon associated with waves, and since electrons are able to produce a diffraction pattern they must be behaving like waves. This experiment provides evidence to support de Broglie's theory.

Electrons therefore exhibit a wave–particle duality similar to that of light.

Graphite is a polycrystalline material. Each crystal consists of a large number of carbon atoms arranged in uniform planes. Since there are many layers, the graphite behaves as a diffraction grating. The electrons travel through space like a wave. When the waves pass through the layers of graphite, diffraction occurs. The wavelength of the electrons is comparable to the spacing between the layers of graphite.

If the electrons were behaving like particles, a uniform distribution of dots would be seen on the screen. Increasing the speed of the electrons towards the graphite target increases their momenta. This causes their wavelength to decrease since $\lambda = \frac{h}{p}$. As a result the diameter of the concentric rings decreases.

Exam tip

The electron diffraction experiment provides evidence for the wave nature of small particles such as electrons.

Example

An electron is travelling at a speed of $5.5 \times 10^6\,\mathrm{m\,s^{-1}}$. Calculate the wavelength of this electron. (Mass of an electron = 9.11×10^{-31} kg, the Planck constant = 6.63×10^{-34} J s)

$$\lambda = \frac{h}{p} = \frac{h}{mv} = \frac{6.63 \times 10^{-34}}{9.11 \times 10^{-31} \times 5.5 \times 10^{6}} = 1.32 \times 10^{-10}\,\mathrm{m}$$

Key points

- Wave–particle duality refers to the idea that light and matter have both wave and particle properties.
- de Broglie stated that the wavelength of a particle is given by $\lambda = \frac{h}{p}$.
- Electron diffraction provides evidence that particles are able to behave as waves.
- Interference and diffraction provide evidence for the wave nature of electromagnetic radiation.
- The photoelectric effect provides evidence for the particle nature of electromagnetic radiation.

11.8 X-rays

Learning outcomes

On completion of this section, you should be able to:

- explain the process of X-ray production
- explain the origins of line and continuous X-ray spectra
- use the relationship $I = I_0 e^{-\mu x}$ for the attenuation of X-rays in matter
- discuss the use of X-rays in radiotherapy and imaging in medicine.

X-ray production

When someone has a broken arm or leg the most common way to view the injury, without doing an internal examination, is to get an X-ray. X-rays are simply shadow pictures of internal structures inside the body.

X-rays are produced by bombarding a metal surface with electrons that have been accelerated through a large potential difference. When a charged particle is accelerated, electromagnetic radiation is produced (called Bremsstrahlung radiation). The greater the acceleration of the charged particle, the shorter the wavelength of the electromagnetic radiation produced.

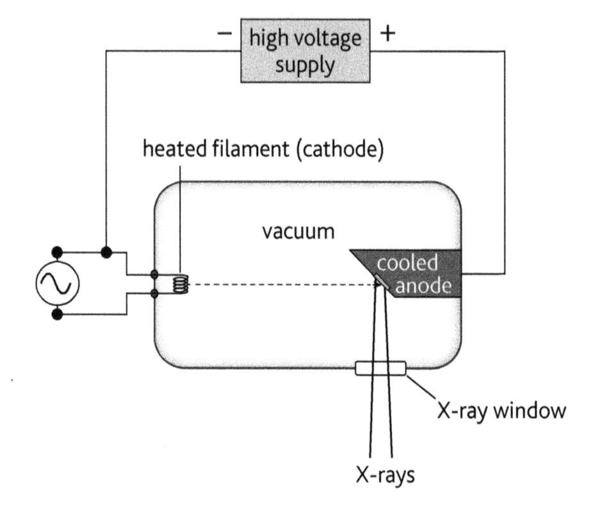

Figure 11.8.1 *A simple diagram showing how X-rays are produced*

Figure 11.8.1 shows a simple diagram of an X-ray tube. Electrons are produced from a heated cathode by a process called **thermionic emission**. In this process, electrons escape from the surface of the cathode as a result of heating.

The electrons produced in this process are accelerated towards the metal anode, using a very large potential difference (20–100 kV). The electrons gain kinetic energy in the process. When the electrons strike the metal surface, they undergo very large decelerations. Most of the kinetic energy of the electrons is converted into thermal energy. Some of the kinetic energy is converted into radiation. The radiation produced lies in the X-ray region of the electromagnetic spectrum. The X-rays then exit through the X-ray window.

X-ray spectrum

Figure 11.8.2 shows a typical spectrum of the X-rays produced in an X-ray tube.

The spectrum produced shows two distinct components. There is a continuous distribution of wavelengths and a series of sharp high-intensity lines. These lines are characteristic of the metal target being used. There is also a cut-off wavelength. This minimum wavelength is the wavelength emitted when all the kinetic energy of the incident electrons is converted into X-rays.

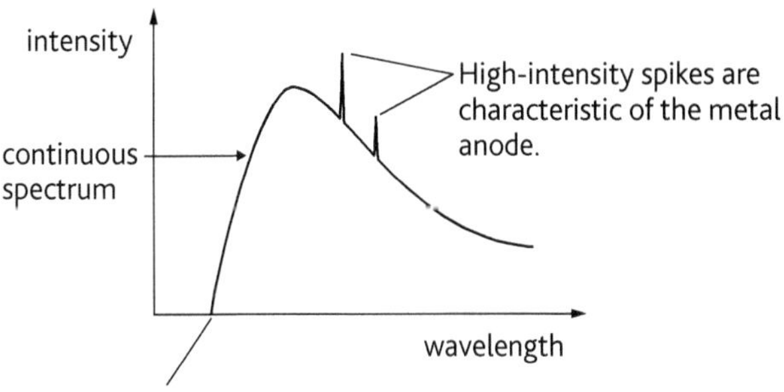

Figure 11.8.2 *Typical spectrum of X-rays during production*

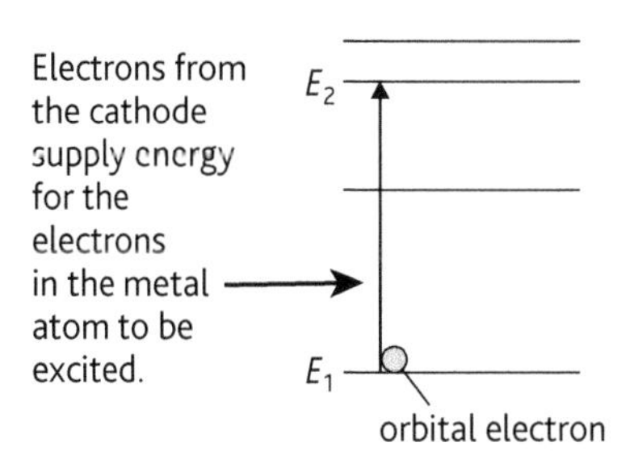

(a) Excitation

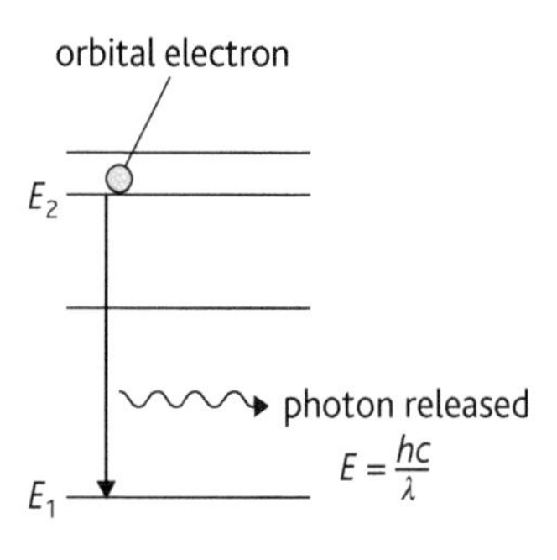

(b) De-excitation

Figure 11.8.3 *Excitation of orbital electrons*

A continuous spectrum is one in which all frequencies are possible within a frequency range. The electrons hitting the metal target have a wide range of decelerations and this is what gives rise to the continuous spectrum.

When electrons strike the metal surface, the electrons that are found in the orbits of the metal atoms become excited and jump to a higher energy level (excitation). When these electrons fall from a high energy level to a lower energy level (de-excitation) photons of energy are emitted (Figure 11.8.3). This gives rise to the line spectrum (spikes). The atoms that make up the metal target have distinct energy levels. This means that orbital electrons can only occupy fixed energy levels so different spikes correspond to photons of different energy being released. If a different metal is used for the anode, the high-intensity spikes occur at different positions because different metal atoms have different energy levels.

Attenuation of X-rays

The intensity of X-rays decreases exponentially as the radiation passes through matter. Figure 11.8.4 shows what happens to the intensity of a parallel beam of X-rays having an initial intensity I_0 as it passes through an absorbing medium. The intensity at the point P is given by $I = I_0 e^{-\mu x}$, where μ is called the linear absorption coefficient of the material.

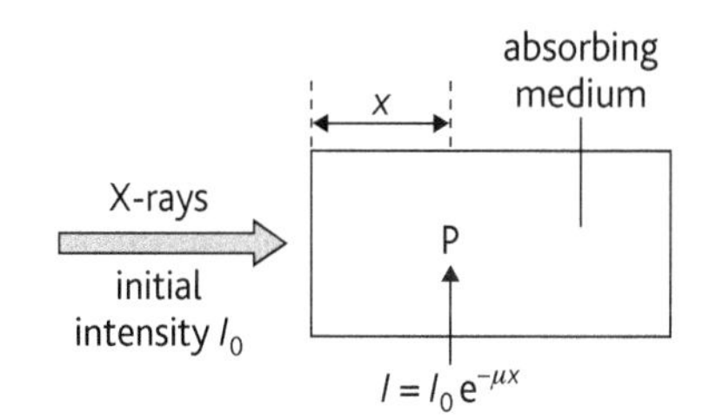

Figure 11.8.4 *Attenuation of X-rays through matter*

Example

An X-ray tube operates at 50 kV and the current through it is 1.1 mA. Calculate:

a the electrical power input

b the speed of the electrons when they hit the target inside the tube

c the cut-off wavelength of the X-rays emitted.

---

a Power input = $IV = 1.1 \times 10^{-3} \times 50 \times 10^3 = 55\,\text{W}$

b Gain in kinetic energy of electrons = loss in electrical potential energy

$$\tfrac{1}{2}mv^2 = QV$$

$$v^2 = \frac{2QV}{m}$$

$$v = \sqrt{\frac{2QV}{m}}$$

$$v = \sqrt{\frac{2 \times 1.6 \times 10^{-19} \times 50 \times 10^3}{9.11 \times 10^{-31}}}$$

$$v = 1.33 \times 10^8\,\text{m s}^{-1}$$

Equation

$I = I_0 e^{-\mu x}$

I – intensity of X-rays at a distance x/W m^{-2}

I_0 – initial intensity of X-rays/W m^{-2}

μ – linear absorption coefficient/m^{-1}

x – distance travelled through medium/m

c The minimum wavelength is the wavelength emitted when all the kinetic energy of the incident electrons is converted into X-rays.

$$E = \frac{hc}{\lambda}$$

$$QV = \frac{hc}{\lambda}$$

$$\text{Cut-off wavelength } \lambda = \frac{hc}{QV} = \frac{6.63 \times 10^{-34} \times 3.0 \times 10^{8}}{1.6 \times 10^{-19} \times 50 \times 10^{3}}$$

$$= 2.49 \times 10^{-11}\,\text{m}$$

Example

The linear absorption coefficient μ for bone and muscle are $2.9\,\text{cm}^{-1}$ and $0.95\,\text{cm}^{-1}$ respectively. A parallel X-ray beam of intensity I_0 is incident on some muscle tissue of thickness 4.0 cm. Below the muscle tissue is some bone of thickness 1.3 cm. Calculate the intensity of the X-ray beam after passing through the muscle tissue and bone. Give the answer in terms of I_0.

Intensity after passing through the muscle tissue $= I = I_0 e^{-\mu x}$

$$= I_0 e^{-(0.95 \times 4.0)} = 0.0224 I_0$$

Intensity after passing through the bone, $I = I_0 e^{-\mu x}$

$$= 0.0224 I_0 e^{-(2.9 \times 1.3)} = 5.16 \times 10^{-4} I_0$$

Uses of X-rays

X-ray photography

X-rays are used to produce shadow photographs of internal structures of the body. They are very penetrating when compared with visible light and are therefore able to pass easily through the body. Different tissues inside the body have different densities. Bone tissue is denser than soft tissue. This means that bones are more effective at absorbing X-rays than soft tissue. As a result there is a contrast between bone and soft tissue on X-ray photographs.

X-ray computed tomography (CT)

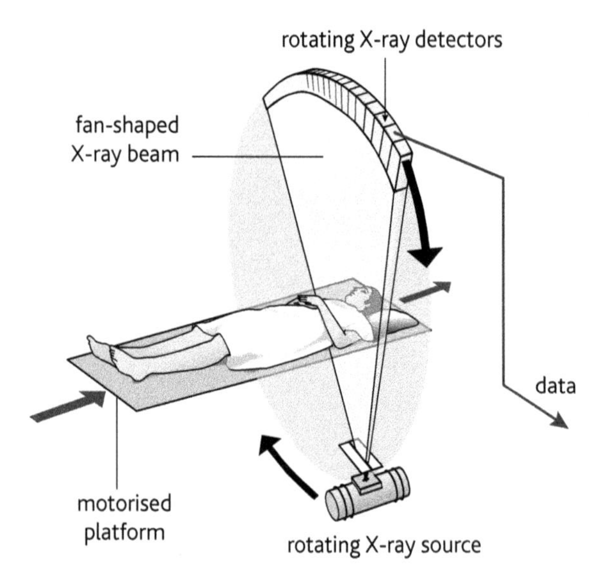

Figure 11.8.5 *A CT scanner*

When X-rays are used to make an image of, for example, a broken leg using photographic film, the photograph is a two-dimensional picture. It gives no perception of depth. Figure 11.8.5 shows a computed tomography (CT) scanner.

CT scanners are able to produce very detailed three-dimensional pictures of the body. The patient lies on a movable table. This allows for the patient to be positioned inside a toroidal (doughnut-shaped) CT scanner. Tomography is a technique by which an image of a slice of the body may be obtained. Inside the scanner there is an X-ray source and multiple X-ray detectors. The X-ray source and detectors are rotated about the patient and multiple images of the slice are taken. The patient is shifted along the axis of the scanner and images of multiple slices are taken. Powerful computers combine the images to produce a three-dimensional picture of the region of the body being investigated.

- Advantages: fast, provides detailed three-dimensional pictures of the body
- Disadvantage: greater exposure to X-rays than standard imaging techniques
- Uses: to detect solid tumours and other problems in the abdomen and chest

Radiotherapy

X-rays are also used in radiotherapy. High-energy X-rays are directed at a person's body to kill cancer cells and keep them from growing and multiplying.

Key points

- X-rays are produced by bombarding a metal surface with electrons that have been accelerated through a large potential difference.
- The typical spectrum of X-rays produced in an X-ray tube consists of a continuous distribution of wavelengths and a series of sharp high-intensity lines.
- The intensity of X-rays decreases exponentially as it passes through matter.
- X-rays are used to obtain shadow pictures of internal structures of the body, three-dimensional images using CT scanners and in radiotherapy in the treatment of cancers.

12 Atomic structure and radioactivity

12.1 Atomic structure

Learning outcomes

On completion of this section, you should be able to:

- give a brief account of the early theories of the atom due to Thomson, Rutherford and Bohr
- describe the Geiger–Marsden experiment
- interpret the Geiger–Marsden experiment
- describe the nuclear model of the atom.

Early theories of the atom

The Ancient Greeks were the earliest to suggest that matter was made up of very small particles called atoms. They thought that if you were to cut something in half and then continue cutting the resulting halves, you would end up with a particle that was so small that it could not be divided any further.

J. J. Thomson was investigating the nature of the particles in cathode rays. He knew that he was dealing with a particle much smaller than the hydrogen atom. Experiments showed that the particle was negatively charged. This particle became known as the **electron**.

In 1903 Thomson suggested that since the atom was electrically neutral, there must be positive charges inside the atom as well. He proposed the 'plum pudding' model of the atom. In this the atom was visualised as being a positively charged sphere with negative electrons distributed throughout it.

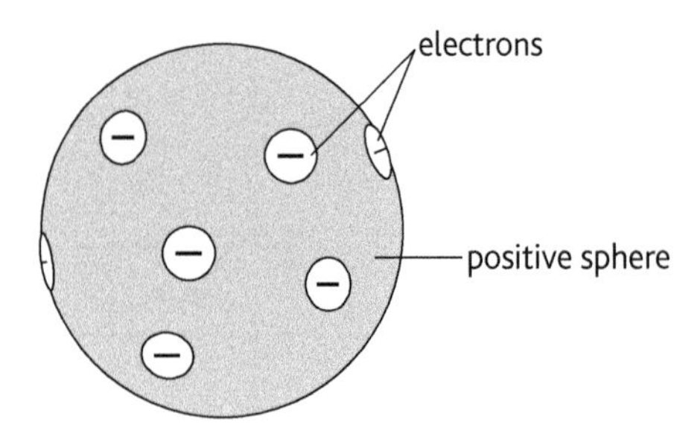

Figure 12.1.1 *The 'plum pudding' model of the atom*

Geiger–Marsden experiment

In 1909 Hans Geiger and Ernest Marsden, under the direction of Ernest Rutherford, performed a now famous experiment in which alpha particles (α-particles) were scattered. (An α-particle is a helium nucleus, i.e. a helium atom that has lost two electrons.) In the experiment a fine beam of α-particles was fired at a very thin sheet of gold foil. The experimental set-up is shown in Figure 12.1.2.

The experiment was performed in an evacuated chamber. This was done to allow the α-particles to reach the gold foil. (If the experiment had been performed in air, the α-particles would have lost all their energy and stopped short of the gold foil.) A glass screen, coated with zinc sulphide was fixed to a microscope. Each time an α-particle struck the glass screen, a tiny flash of light was seen.

A very narrow beam of α-particles was used. This ensured that there was a small collision area, reducing the uncertainty in the scattering angle. The foil used was very thin to prevent too many α-particles from being absorbed, and to ensure that the α-particles were scattered only once.

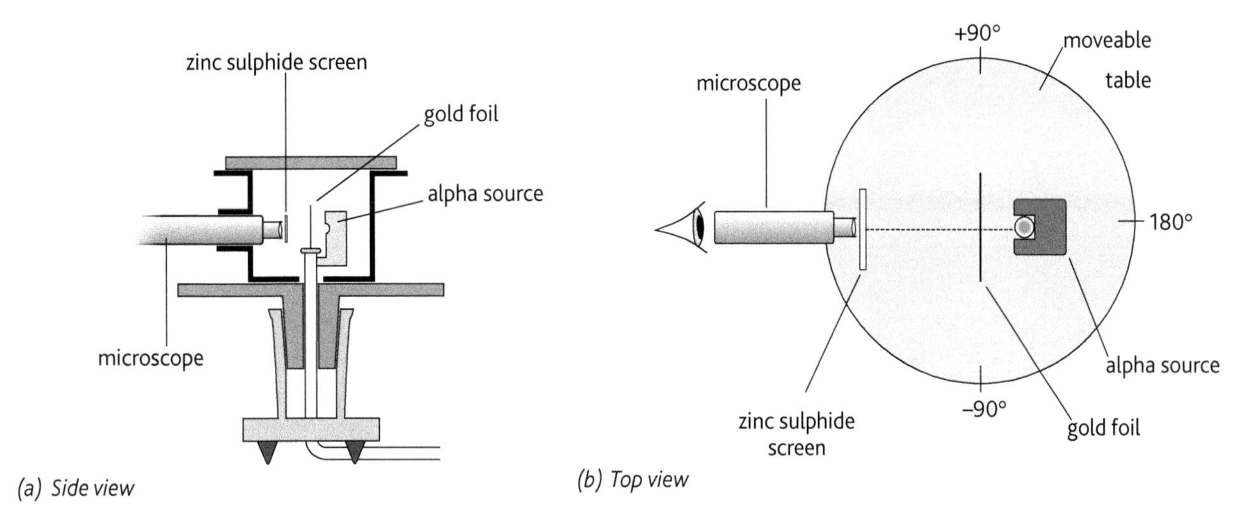

Figure 12.1.2 *The Geiger–Marsden experiment*

The experimental results were as follows (Figure 12.1.3):

- Most of the α-particles went straight through the gold foil.
- Some of the α-particles were deflected slightly.
- A small number of particles were deflected by angles up to almost 180°.

The following is what was actually expected to happen:

- Based on the model of the atom at the time (the 'plum pudding' model), it was expected that most of the particles would pass through the gold foil with little or no deflection.

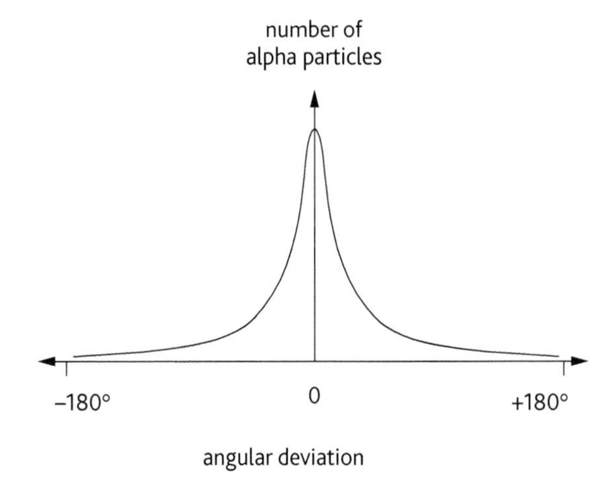

Figure 12.1.3 *The results of the experiment*

Conclusions of the experiment

The experiment suggests that most of the atom is empty space but its positive charge and most of its mass are concentrated in a very small region: the **nucleus**.

Consider the path of four α-particles heading towards a gold nucleus as shown in Figure 12.1.4.

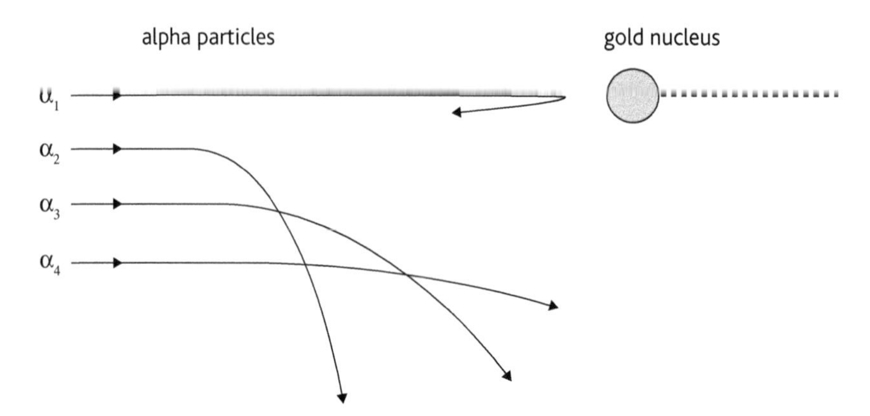

Figure 12.1.4 *The path of α-particles approaching the nucleus*

The α_4-particle is furthest away from the gold nucleus. It therefore experiences the smallest electrostatic (repulsive) force and consequently undergoes very little deflection.

The α_2- and α_3-particles approach closer to the gold nucleus. They therefore experience larger electrostatic forces and consequently undergo larger deflections.

The α_1-particle approaches the gold nucleus head-on. The electrostatic force acts in a direction opposite to that in which the α-particle is travelling, and the α-particle is therefore repelled in the opposite direction.

Most of the α-particles went straight through the gold foil. This suggests that the atom consists mainly of empty space. The size of the nucleus is very small compared with the size of the atom.

The fact that only a small number of α-particles were deflected by angles close to 180° suggests that most of the mass of the atom is contained in the nucleus. The nucleus contains a positive charge and the charge is concentrated in the nucleus.

The conclusions of this experiment led Rutherford to develop a planetary model of the atom (Figure 12.1.5). The atom consists of a central positively charged nucleus with negatively charged electrons orbiting the nucleus. He calculated the diameter of the atom ($\sim 10^{-9}$ m) and the diameter of the nucleus ($\sim 10^{-14}$ m).

There were two problems with this model. Firstly, an atom emits electromagnetic radiation with only specific wavelengths. Secondly, electrons in this model of the atom undergo centripetal acceleration. According to Maxwell's theory of electromagnetism, charged particles accelerating will emit electromagnetic radiation. The effect of this would be that the electrons would eventually fall towards the nucleus, but this does not happen.

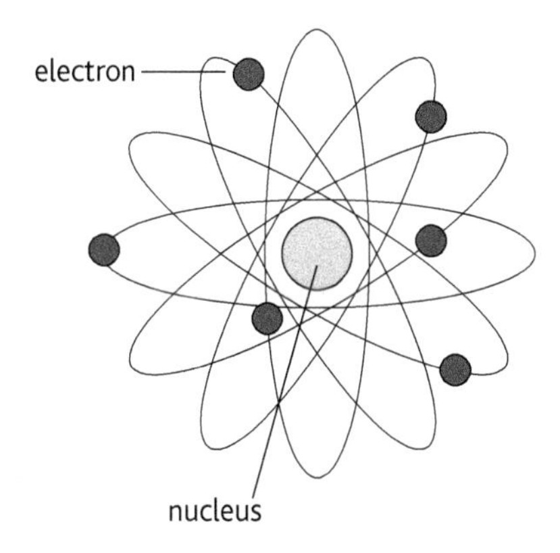

Figure 12.1.5 *The nuclear model of the atom (Rutherford's model)*

Later Niels Bohr was able to provide a model to explain why line spectra are observed and why the electrons do not fall into the nucleus. He suggested that the electrons occupy discrete energy levels or 'shells' within the atom (Figure 12.1.6).

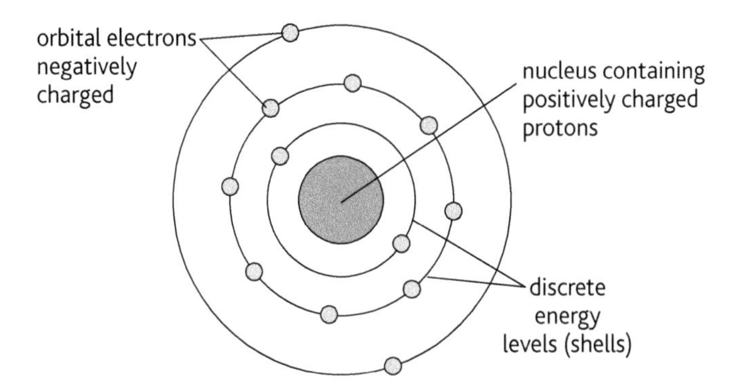

Figure 12.1.6 *Bohr's model of the atom*

Table 12.1.1 *The relative masses and charges of subatomic particles*

	Relative mass	Relative charge
Electron	$\frac{1}{1840}$	–1
Proton	1	+1
Neutron	1	0

Key points

- In J. J. Thomson's model the atom is a positive sphere with negative electrons distributed throughout it.
- The α-scattering experiments provided evidence against this 'plum pudding' model.
- In Rutherford's model the nucleus is very small and positively charged with electrons orbiting around it.
- In Niels Bohr's model of the atom electrons orbit the small positively charged nucleus but occupy discrete energy levels.

Learning outcomes

On completion of this section, you should be able to:

- represent nuclear reactions using nuclear equations
- appreciate that nucleon number, proton number and energy are conserved in nuclear reactions
- describe the processes of nuclear fission and nuclear fusion.

Notation used to represent atoms

An atom consists of a central nucleus surrounded by orbital electrons. The nucleus consists of protons and neutrons. The protons inside the nucleus are positively charged. The neutrons inside the nucleus are uncharged.

Most of the mass of the atom is concentrated in the nucleus. All of the positive charge is located in the nucleus. The electrons orbiting the nucleus contribute very little mass to the atom. The electrons are negatively charged. In a neutral atom, the negative charge of the electrons is equal to the positive charge in the nucleus.

A nuclide is a type of atom whose nuclei have specific numbers of protons and neutrons. The following notation is used to represent different nuclides:

$$^{A}_{Z}\text{X}$$

X represents the symbol for the nuclide.

A represents the nucleon number or mass number of the atom.

Z represents the proton number or atomic number of the atom.

$A = Z + N$, where N is the number of neutrons.

Example

A sodium atom is represented by $^{23}_{11}\text{Na}$.

Nucleon number (see 12.3) or mass number = 23

Proton number or atomic number = 11

Number of electrons = 11

Number of protons = 11

Number of neutrons = 23 − 11 = 12

Nuclear fission

Heavy nuclei such as uranium ($Z = 92$) and plutonium ($Z = 94$) are unstable. They can decay spontaneously. However, if a neutron is projected towards a uranium nucleus it can induce a type of decay called **nuclear fission**. When the neutron collides with the uranium nucleus it makes it even more unstable. The uranium nucleus then splits into two or more stable fragments, accompanied by the emission of several neutrons. An example of a fission reaction is shown in the equation below. In this the two fragments are a krypton nucleus and a barium nucleus.

$$^{1}_{0}\text{n} + {}^{235}_{92}\text{U} \longrightarrow {}^{92}_{36}\text{Kr} + {}^{141}_{56}\text{Ba} + 3^{1}_{0}\text{n} + \text{energy}$$

Energy is released in the fission process. The energy is in the form of kinetic energy of the emitted particles as well as electromagnetic radiation. The neutrons emitted in the reaction can collide with more uranium nuclei and produce even more fission reactions. The result is an uncontrolled chain reaction. In a nuclear reactor, the same reaction takes place, but it is carefully controlled. A material is used to absorb some of the neutrons being released in the reaction.

Definition

Nuclear fission is an induced process whereby an unstable nucleus is bombarded by a neutron. The nucleus splits into two or more stable fragments as well as several neutrons.

In nuclear reactions, mass and charge are always conserved. This means that the total mass number on one side of the equation must equal the total mass number on other side of the equation.

Consider the mass (nucleon) numbers on either side of the fission equation.

$$1 + 235 = 92 + 141 + (3 \times 1) \qquad A = 236$$

The sum of the mass numbers on left-hand side of the equation is equal to the sum of the mass numbers on the right-hand side of the equation.

Consider the atomic (proton) numbers on either side of the fission equation.

$$0 + 92 = 36 + 56 + (3 \times 0) \qquad Z = 92$$

The sum of the atomic numbers on left-hand side of the equation is equal to the sum of the atomic numbers on the right-hand side of the equation.

Nuclear fusion

Light nuclei can become more stable by combining with other light nuclei in a process called **nuclear fusion**. An **isotope** of hydrogen is $^{2}_{1}\mathrm{H}$ (deuterium). Two of these nuclei can combine to form a more stable nucleus ($^{4}_{2}\mathrm{He}$). In this process energy is released.

Fusion is the process by which stars such as the Sun produce energy. An example of a fusion reaction is shown in the equation below.

$$^{2}_{1}\mathrm{H} + {}^{2}_{1}\mathrm{H} \longrightarrow {}^{4}_{2}\mathrm{He} + \text{energy}$$

Consider the mass (nucleon) numbers on either side of the fusion equation.

$$2 + 2 = 4 \qquad A = 4$$

The sum of the mass numbers on left-hand side of the equation is equal to the sum of the mass numbers on the right-hand side of the equation.

Consider the atomic (proton) numbers on either side of the fusion equation.

$$1 + 1 = 2 \qquad Z = 2$$

The sum of the atomic numbers on left-hand side of the equation is equal to the sum of the atomic numbers on the right-hand side of the equation.

Definition

Isotopes of an element have the same atomic number but different mass numbers.

Definition

Nuclear fusion is the process whereby light nuclei become more stable by combining with other light nuclei to form a heavier stable nucleus, accompanied by the release of energy.

Key points

- Nuclear fission is an induced process whereby an unstable nucleus is bombarded by a neutron. The nucleus splits into two or more stable fragments as well as several neutrons.
- Nuclear fusion is the process whereby light nuclei become more stable by combining with other light nuclei to form a heavier stable nucleus, accompanied by the release of energy.
- Nucleon number, proton number and energy are conserved in nuclear reactions.

Table 12.2.1 *Comparison of nuclear fission and nuclear fusion*

	Fission	Fusion
Similarities	Both processes release energy. Charge and mass are conserved in the process. The total rest mass of the resulting nuclide(s) is less than the total rest mass of the original nuclide(s). The binding energy (see 12.3) of the resulting fragments is greater than that of the original nuclide(s).	
Differences	Most of the energy is carried away by massive fragments.	Most of the energy is carried away by light fragments.
	Fission can be easily initiated by neutron bombardment.	Fusion is difficult to achieve.
	There are numerous possible combinations of energy and masses that can be produced.	The energy and masses produced in the reaction are always the same.
Conditions required for process to occur in a sustained manner	The fission rate can be controlled by slowing down the release of neutrons in the nuclear reactor.	Extremely high temperatures and a high density of plasma are required to allow for random fusion collisions to occur.

12.3 Binding energy

Learning outcomes

On completion of this section, you should be able to:

- understand the concept of mass defect
- appreciate the association between energy and mass represented by $E = mc^2$
- illustrate graphically the variation of binding energy per nucleon with nucleon number
- describe the relevance of binding energy per nucleon to nuclear fusion and nuclear fission
- use the atomic mass unit (u) as a unit of energy.

Definition

The mass defect of a nucleus is the difference between the mass of the nucleus and the total mass of its constituent nucleons.

Equation

$E = mc^2$

E – energy/J

m – mass defect/kg

c – speed of light/m s^{-1}

Mass defect

In order to understand why energy is released in the processes of fission and fusion we need to look at the nucleus of an atom more closely. Consider a stable carbon atom, $^{12}_{6}C$. There are six protons and six neutrons inside the nucleus. The protons and neutrons inside the nucleus are referred to as **nucleons**. Therefore, the carbon nucleus contains twelve nucleons. The mass number is often referred to as the nucleon number.

Consider the following hypothetical experiment. The mass of a carbon-12 nucleus is measured with a high degree of accuracy. Inside the nucleus, there are six protons and six neutrons. These nucleons are held together by very strong short-range **nuclear forces**. Suppose you separate the nucleons so that they are an infinite distance apart. The total mass of all the individual nucleons (protons and neutrons) is now measured with a high degree of accuracy. You would expect that the mass of six neutrons and six protons would be exactly equal to the mass of the carbon-12 nucleus. However, this is not the case. Measurements show that the total mass of the nucleons is greater than the mass of the carbon-12 nucleus. This seems to contradict the law of conservation of mass. The missing mass is called the **mass defect**.

In order to separate the nucleons in the nucleus completely, energy has to be supplied. The energy supplied accounts for an apparent increase in mass. In Einstein's famous equation $E = mc^2$, he states that mass and energy are interchangeable.

Binding energy

The energy required to completely separate the nucleons of a nucleus is called the **binding energy** of the nucleus.

The **binding energy per nucleon** is equal to the binding energy of the nucleus divided by the total number of nucleons.

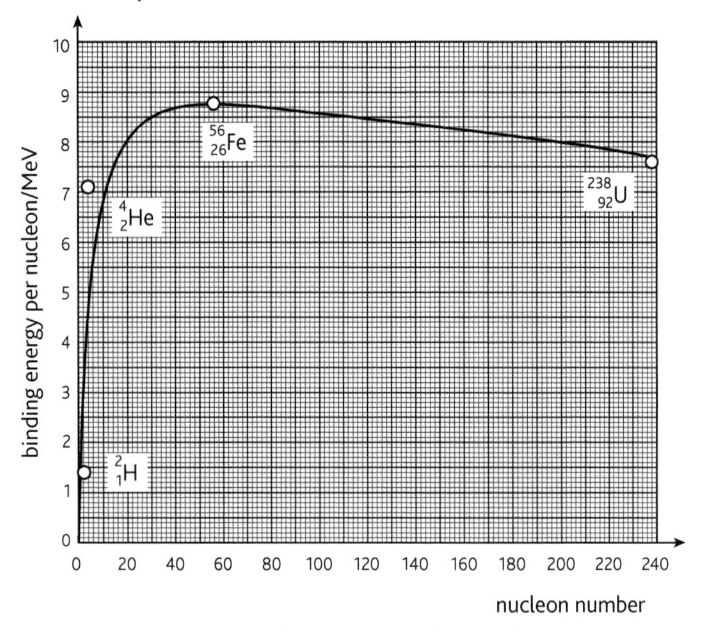

Figure 12.3.1 *Binding energy per nucleon against nucleon number*

Figure 12.3.1 shows the variation of binding energy per nucleon against nucleon number for common nuclei. The larger the value of binding energy per nucleon, the greater is the stability of the nucleus. An isotope of iron $^{56}_{26}\text{Fe}$ has the highest binding energy per nucleon. This means that it is the most stable nucleus on the graph. It requires the most energy to separate the nucleons completely.

The binding energy curve can be used to determine whether fusion or fission is likely to occur.

Consider the following fission reaction.

$$^{1}_{0}\text{n} + ^{235}_{92}\text{U} \longrightarrow ^{92}_{36}\text{Kr} + ^{141}_{56}\text{Ba} + 3^{1}_{0}\text{n}$$

The uranium nucleus (A = 235) splits into two fragments (A = 141 and A = 92). From the binding energy curve, it can be seen that the binding energy of the two fragments is greater than that of the original uranium nucleus. Therefore energy is released in the process.

Consider the following fusion reaction.

$$^{2}_{1}\text{H} + ^{2}_{1}\text{H} \longrightarrow ^{4}_{2}\text{He}$$

Two light nuclei of $^{2}_{1}\text{H}$ (A = 2) fuse together to form $^{4}_{2}\text{He}$ (A = 4). From the binding energy curve, it can be seen that the binding energy of the resulting product is greater than the binding energy of the two $^{2}_{1}\text{H}$ nuclei. Therefore energy is released in the process.

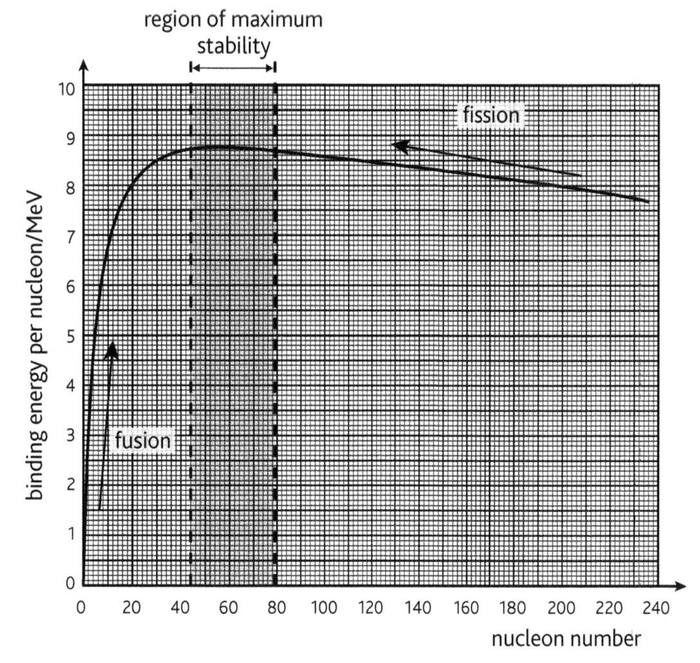

***Figure 12.3.2** Binding energy, fission and fusion*

Unified atomic mass unit (u)

Mass and energy are interchangeable. Therefore, the unit of mass can be used as another unit of energy.

The relative atomic mass A_r of an atom is defined by

$$A_r = \frac{\text{mass of the atom}}{\text{one-twelfth of the mass of a } ^{12}_{6}\text{C atom}}$$

The relative atomic mass of $^{12}_{6}\text{C}$ is exactly 12.

The **unified atomic mass unit** (u) is defined as 1/12 the mass of the carbon atom $^{12}_{6}\text{C}$.

$$\text{Mass of one carbon atom} = \frac{12}{6.02 \times 10^{23}}\text{g} = \frac{12}{6.02 \times 10^{26}}\text{kg} = 12\text{u}$$

$$\text{Therefore,} \quad 1\text{u} = \frac{12}{12 \times 6.02 \times 10^{26}} = 1.66 \times 10^{-27}\text{kg}$$

$$1\text{eV} = 1.6 \times 10^{-19}\text{J}$$

$$1\text{MeV} = 1.6 \times 10^{-19} \times 10^{6} = 1.6 \times 10^{-13}\text{J}$$

A change in mass of 1.0 kg has an energy equivalence of:

$$E = mc^2 = 1.0 \times (3.0 \times 10^{8})^2 = 9.0 \times 10^{16}\text{J}$$

$$\text{Therefore,} \quad 1\text{u} = \frac{1.6 \times 10^{-27} \times 9 \times 10^{16}}{1.6 \times 10^{-13}}\text{MeV}$$

$$1\text{u} = 931\text{MeV}$$

Key points

- The energy required to completely separate the nucleons of a nucleus is called the binding energy of the nucleus.
- The binding energy per nucleon is equal to the binding energy of the nucleus divided by the total number of nucleons.
- Energy is released when heavy nuclei undergo fission.
- Energy is released when light nuclei undergo fusion.

12.4 Calculating energy changes

Learning outcomes

On completion of this section, you should be able to:

- appreciate the association between energy and mass represented by $E = mc^2$ and use this equation to solve problems.

Example

Calculate the binding energy per nucleon of the nucleus $^{238}_{92}U$ using the following data.

Mass of a proton = 1.00728 u

Mass of a neutron = 1.00867 u

Mass of $^{238}_{92}U$ nucleus = 238.05076 u

1 u = 931 MeV

Total mass of nucleons = (92 × 1.00728) + (146 × 1.00867)

= 239.93558 u

Mass defect of the nucleus = 239.93558 – 238.05076 = 1.88482 u

Binding energy of the nucleus $^{238}_{92}U$ in MeV = 1.88482 × 931

= 1755 MeV

$$\text{Binding energy per nucleon} = \frac{1755}{238} = 7.373\,\text{MeV}$$

Example

A possible fission reaction is:

$$^{238}_{92}U + ^{1}_{0}n \longrightarrow ^{144}_{56}Ba + ^{90}_{36}Kr + 2^{1}_{0}n$$

The binding energy per nucleon of the particles is as follows:

$^{235}_{92}U$ 7.62 MeV

$^{144}_{56}Ba$ 8.38 MeV

$^{90}_{36}Kr$ 8.67 MeV

a Estimate the energy released in this fission reaction in joules.

b Calculate the mass equivalent of this energy.

c State two forms of energy the answer in **a** is transformed into.

a Firstly, the binding energy of each nucleus is determined by multiplying the binding energy per nucleon by the nucleon number.

Binding energy of U-235 = $7.62 \times 10^6 \times 235 = 1.7907 \times 10^9$ eV

Binding energy of Ba-144 = $8.38 \times 10^6 \times 144 = 1.2067 \times 10^9$ eV

Binding energy of Kr-90 = $8.67 \times 10^6 \times 90 = 7.803 \times 10^8$ eV

Energy released = $(7.803 \times 10^8 + 1.2067 \times 10^9) - (1.7907 \times 10^9)$

= 1.963×10^8 eV

1 eV = 1.6×10^{-19} J

∴ Energy released = $1.963 \times 10^8 \times 1.6 \times 10^{-19} = 3.141 \times 10^{-11}$ J

b Using $E = mc^2$,

$$m = \frac{E}{c^2}$$

$$m = \frac{3.141 \times 10^{-11}}{(3.0 \times 10^8)^2}$$

$$= 3.49 \times 10^{-28}\,\text{kg}$$

Mass equivalence = 3.49×10^{-28} kg

c The energy is in the form of kinetic energy of the fragments (Kr and Ba) and electromagnetic radiation.

Example

Consider the following reaction:

$$^{4}_{2}\mathrm{He} + ^{9}_{4}\mathrm{Be} \longrightarrow ^{1}_{0}\mathrm{n} + ^{12}_{6}\mathrm{C}$$

The mass of each particle in the reaction is as follows:

$^{4}_{2}\mathrm{He}$	4.00260 u
$^{9}_{4}\mathrm{Be}$	9.01212 u
$^{1}_{0}\mathrm{n}$	1.00867 u
$^{12}_{6}\mathrm{C}$	12.00000 u

Calculate:

a the mass defect

b the energy equivalence of this mass.

---

a Mass defect $= (4.00260 + 9.01212) - (1.00867 + 12.00000)$

$= 0.00605\,\mathrm{u}$

$= 0.00605 \times 1.66 \times 10^{-27}\,\mathrm{kg}$

$= 1.0043 \times 10^{-29}\,\mathrm{kg}$

b $E = mc^2$

$= 1.0043 \times 10^{-29} \times (3.0 \times 10^{8})^2$

$= 9.04 \times 10^{-13}\,\mathrm{J}$

Revision questions 7

Answers to questions that require calculation can be found on the accompanying CD.

1 a Explain how lines in the emission spectrum of gases at low pressure provide evidence for discrete electron energy levels in atoms. [3]

Several lines of the emission spectrum of hydrogen are shown below.

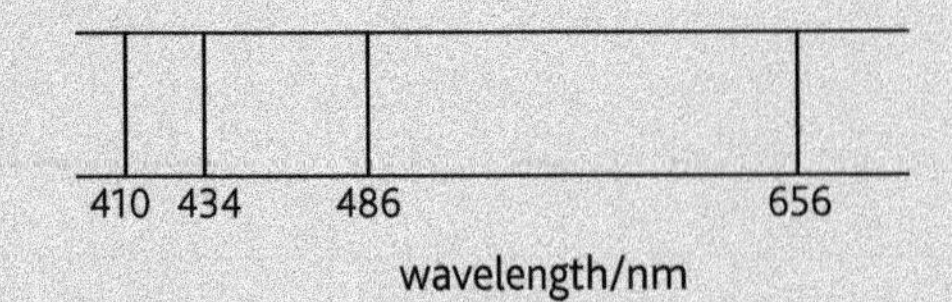

b Calculate the energy of the photons associated with each line in the spectrum. [4]

c The energy levels of an electron in a hydrogen atom are shown below. Copy the diagram and illustrate the transitions that produce the four spectral lines in **b**. [4]

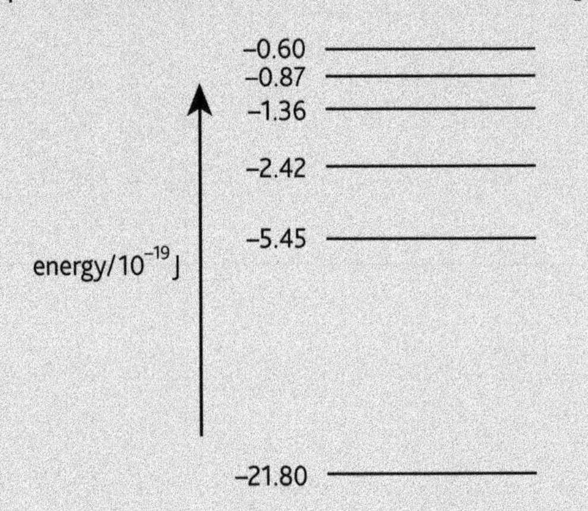

2 Distinguish between a line emission spectrum and a line absorption spectrum. [4]

3 Explain how X-rays are produced in an X-ray machine. [4]

4 In an X-ray tube, the potential difference between the cathode and the metal target is 42 kV. Calculate:

a the speed of the electrons when they hit the metal target [3]

b the cut-off wavelength of the X-rays emitted. [3]

5 The diagram below shows a typical spectrum of the X-rays produced in an X-ray tube.

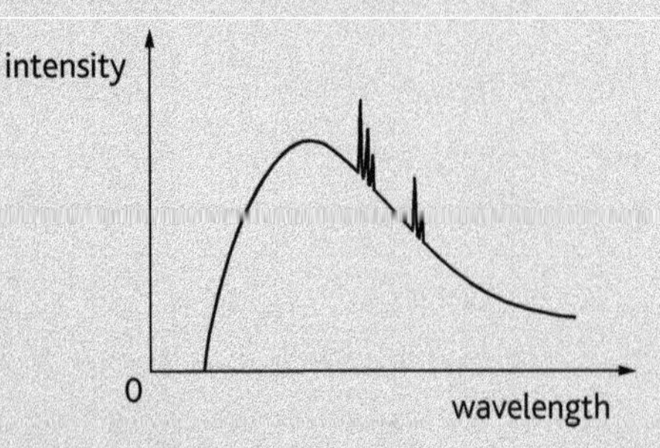

Provide explanations for the following:

a A continuous spectrum of wavelengths is produced. [3]

b The spectrum has a sharp cut-off. [1]

c There are spikes in the spectrum. [2]

6 An electron is accelerated from rest through a potential difference of 5.0 kV. Calculate the de Broglie wavelength of the electron. [4]

7 a State what is meant by the de Broglie wavelength. [1]

b An electron is accelerated from rest through a potential difference of 920 V.

i Calculate the final momentum of the electron. [2]

ii Calculate the de Broglie wavelength of the electron. [2]

c Describe an experiment to demonstrate the wave nature of electrons. [5]

d Explain what is meant by the wave–particle duality. [3]

8 a Explain what is meant by the nucleon number and the nuclear binding energy of the nucleus. [3]

b Sketch a labelled graph to show the variation with nucleon number of the binding energy per nucleon. [3]

c Hence explain why:

i fusion of nuclei having small nucleon numbers releases energy [3]

ii fission of nuclei having large nucleon numbers releases energy. [3]

9 **a** Outline the model of the atom proposed by J. J. Thompson, Rutherford and Bohr. [6]

b In the α-scattering experiments performed under the supervision of Rutherford, α-particles were projected towards a thin sheet of gold foil.

Sketch diagrams to illustrate the path of an α-particle in the following instances:

i The α-particle is moving directly towards the nucleus of a gold atom. [1]

ii The α-particle passes close to the nucleus of a gold atom. [1]

iii The α-particle passes some distance away from the nucleus of a gold atom. [1]

c Explain how the α-particle scattering experiment provides evidence for the existence and small size of the nucleus. [4]

d Give an estimate for the radius of the nucleus of an atom and the radius of an atom. [2]

10 State what is meant by each of the following:

a Proton number [1]

b Nucleon number [1]

c An isotope [1]

d Binding energy of a nucleus [2]

e Binding energy per nucleon [1]

11 Consider the fusion reaction:

$${}^{2}_{1}H + {}^{2}_{1}H \longrightarrow {}^{3}_{2}He + {}^{1}_{0}n$$

The masses of each particle are as follows:

${}^{2}_{1}H$ 2.01355 u

${}^{3}_{2}He$ 3.01493 u

${}^{1}_{0}n$ 1.00867 u

Calculate:

a the mass defect [2]

b the energy released in the reaction in joules. [3]

12 Calculate the binding energy of the nucleus ${}^{238}_{92}U$ in MeV. [6]

Use the following data:

Mass of a proton = 1.00728 u

Mass of a neutron = 1.00867 u

Mass of uranium nucleus = 238.05076 u

1 u = 931.3 MeV

13 Calculate the energy released in the following reaction. [6]

$${}^{241}_{95}Am \longrightarrow {}^{237}_{93}Np + {}^{4}_{2}He$$

Use the following mass data:

${}^{241}Am$ 241.06687 u

${}^{237}Np$ 237.05843 u

${}^{4}He$ 4.00241 u

14 Calculate the energy released in the following reaction. [6]

$${}^{220}_{86}Rn \longrightarrow {}^{216}_{84}Po + {}^{4}_{2}He$$

Use the following mass data:

${}^{220}Rn$ 219.9642 u

${}^{216}Po$ 215.9558 u

${}^{4}He$ 4.00241 u

12.5 Radioactivity

Learning outcomes

On completion of this section, you should be able to:

- relate radioactivity to nuclear instability
- discuss the spontaneous and random nature of nuclear decay
- identify the origins and environmental hazards of background radiation
- describe the operation of simple radiation detectors (G–M tube, cloud chamber and spark counter).

Definition

Radioactive decay is the spontaneous and random process whereby an unstable nucleus attempts to become stable by disintegrating into another nucleus and emitting any one or more of the following: alpha particles, beta particles, gamma rays.

Exam tip

Random nature: the decay is unpredictable. It is impossible to predict which nucleus will decay next or when.

Spontaneous nature: the decay process is not affected by conditions external to the nucleus (e.g. temperature and pressure).

Radioactive decay

In 1896 Henri Becquerel observed that a photographic plate was affected by a uranium compound. He called this phenomenon 'radioactivity'. Whenever a nucleus of an atom is unstable, it will disintegrate to produce a more stable nucleus. In this process ionising radiation is produced. There are three types of ionising radiation. They are alpha particles (α), beta particles (β) and gamma ray photons (γ). **Radioactive decay** is the spontaneous and random process whereby an unstable nucleus attempts to become stable by disintegrating into another nucleus and one or more of the three ionising radiations.

If you were to observe the atoms in a sample of radium you would make some interesting observations. If you were to observe a single atom, you would not be able to predict when it would decay. If it did decay you could still not predict which atom would decay next. This is what is meant when we say that radioactive decay is a random process. Also, if you were to take the same sample of radium and heat it strongly, you would realise that you cannot speed up the decay process. Radioactive decay is a spontaneous process. It is unaffected by conditions external to the nucleus.

Detectors of radiation

The Geiger–Muller tube

Figure 12.5.1 shows a Geiger–Muller (G–M) tube. This particular G–M tube consists of a sealed cylindrical metal tube which acts as the cathode. The thin wire which lies along the axis of the tube acts as the anode. The cylindrical tube is filled with argon gas and bromine vapour at a low pressure. There is a thin mica window at one end of the cylindrical tube. A potential difference about 400 V is applied across the anode and cathode.

When ionising radiation enters through the thin mica window, argon atoms become ionised. The free electrons produced accelerate towards the anode and the positive ions accelerate towards the cathode. The accelerating electrons collide with more argon atoms and produce even more charged particles. There is then an 'avalanche' of electrons moving towards the anode. The result is a large current pulse.

The positive ions are larger than the electrons and travel more slowly. The large number of positive ions around the anode causes the electric field around it to reduce to zero and prevents further ionisation. If the positive ions were to collide with the cathode, electrons would be released. This would create multiple current pulses. In order to prevent this, the bromine vapour acts as a quenching agent. The positive argon ions are neutralised when they collide with bromine molecules. This prevents them from reaching the cathode.

The output of a G–M tube is connected to a scaler or ratemeter. A scaler measures the number of pulses it receives. A ratemeter measures the rate at which it receives pulses.

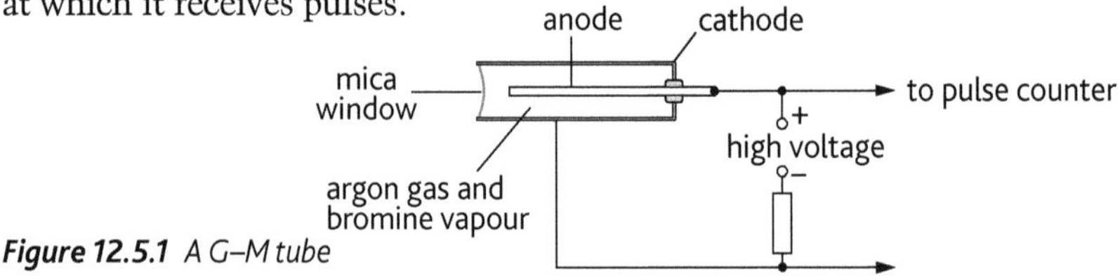

Figure 12.5.1 *A G–M tube*

The cloud chamber

Figure 12.5.2 shows a cloud chamber. In this type of detector, ionising radiation is seen as tracks. The felt ring at the top of the chamber is soaked with alcohol. The alcohol evaporates readily and the chamber saturates with vapour. The temperature of the chamber is reduced by placing dry ice in its base. The alcohol vapour diffuses continuously downwards into the cooler regions of the chamber, and the air there becomes supersaturated with it. When ionising radiation travels through the supersaturated air, ions are produced. The ions act as nucleation sites, and alcohol vapour condenses on these sites. The path travelled by the ionising radiation shows up as tiny droplets of condensation.

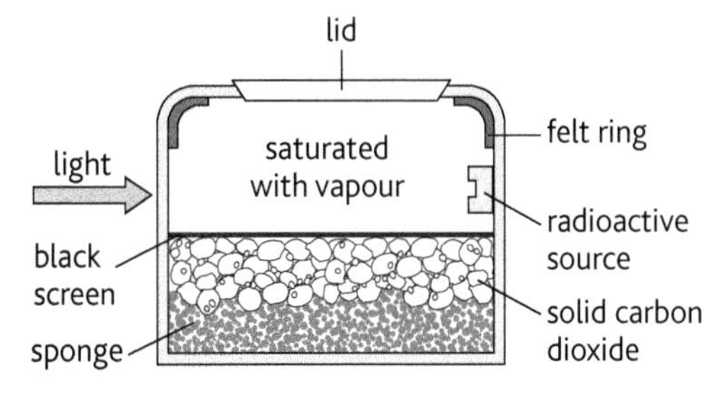

Figure 12.5.2 A cloud chamber

The spark counter

A spark counter (Figure 12.5.3) is a visible and audible way of demonstrating the ionisation effect of different types of radiation. It consists of a metal gauze (cathode) which is situated a few millimetres above a thin wire (anode). The gauze is earthed and a high voltage is applied at the anode. When a radioactive substance is brought close to the gauze, the air in that region becomes ionised. This causes a spark to jump between the wire and the gauze.

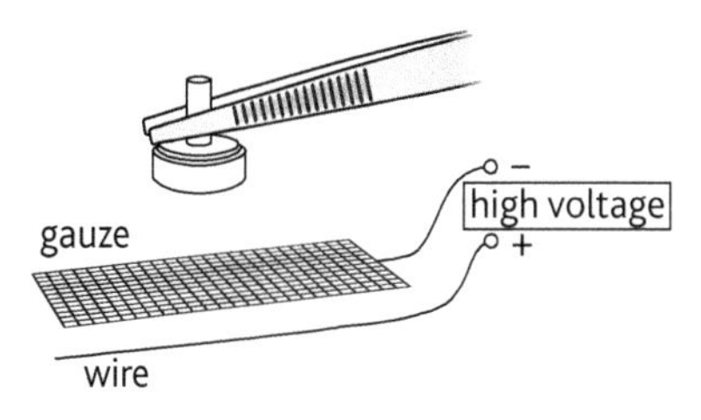

Figure 12.5.3 A spark counter

Background radiation

If a G–M tube is placed in front of a radioactive source, clicks are heard, indicating that the source is emitting ionising radiation. However, if the source is removed from in front of the G–M tube, clicks are still heard. The G–M tube is detecting a type of radiation that exists around us all the time. It is called **background radiation**.

Background radiation is random radioactivity detected from the surroundings (cosmic radiation from space, minerals in the earth, etc.).

The sources of background radiation are:

- cosmic radiation from outer space
- radioactive materials that are naturally found in rocks
- radon gas that accumulates in still air
- nuclear power stations and weapons testing
- medical facilities.

Background radiation can cause cell damage. Prolonged exposure can lead to the formation of cancers, mutations or even death. Background radiation causes us to detect a reading on a G–M tube even though there is no immediate radioactive source present. This means that in order to get the real count rate of a source, the background radiation must first be measured. The actual count rate of a radioactive source is determined by subtracting the background reading from the reading measured on the G–M tube.

Key points

- Radioactive decay is the spontaneous and random process whereby an unstable nucleus attempts to become stable by disintegrating into another nucleus and emitting any one or more of the following: alpha particles, beta particles, gamma rays.
- Random nature: the decay is unpredictable. It is impossible to predict which nucleus will decay next or when.
- Spontaneous nature: the decay process is not affected by conditions external to the nucleus (e.g. temperature and pressure).
- A G–M tube, cloud chamber or spark counter can be used to detect ionising radiation.
- Background radiation is random radioactivity detected from the surroundings.

12.6 Types of radiation

Learning outcomes

On completion of this section, you should be able to:

- describe the nature of α-particles, β-particles and γ-rays as different types of ionising radiation
- distinguish between α-particles, β-particles and γ-rays with reference to charge, mass, speed, effect of electric and magnetic fields, and penetrating properties.

Types of ionising radiation

There are three types of ionising radiation:

- Alpha particles (α)
- Beta particles (β)
- Gamma ray photons (γ)

Whenever radiation passes through matter, it causes electrons to be 'knocked out' of atoms. This results in the formation of ions. For example, a beta particle moving through air will have enough energy to 'knock' electrons out of air molecules and leave a trail of ions behind it. As it ionises the air it loses some of its energy and slows down.

Alpha particles (α)

An alpha particle is a helium nucleus ($^{4}_{2}He^{2+}$). It consists of two protons and two neutrons, and has a positive charge. It is the most intense of the three forms of ionising radiation. As a result, α-particles travel only a few centimetres in air because most of the energy is used up in producing ions. Figure 12.6.1 illustrates what is seen when observing the movement of α-particles in a cloud chamber. Short thick tracks of almost the same length are observed. This means that the α-particles produce a high concentration of ions. It also means that most of the α-particles that are emitted have similar energies.

Americium-241 (^{241}Pa) produces α-particles.

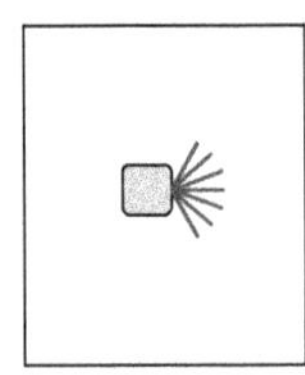

α-particles:
- short thick tracks
- all approximately the same length

Figure 12.6.1 *Alpha particles moving in a cloud chamber*

Beta particles (β)

A beta particle is a fast-moving electron. It has a mass of m_e and charge of e (1.6×10^{-19} C). β-particles are less ionising than α-particles. Figure 12.6.2 illustrates what is seen when observing the movement of β-particles in a cloud chamber. The tracks are much longer and thinner than the ones produced by the α-particles.

Strontium (^{90}Sr) produces β-particles.

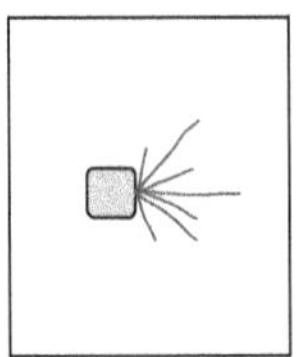

β-particles:
- longer, thinner tracks
- tracks of varying lengths

Figure 12.6.2 *Beta particles moving in a cloud chamber*

Gamma ray photons (γ)

Gamma radiation consists of high-energy electromagnetic waves. It has no mass and no charge. Figure 12.6.3 illustrates what is seen when observing the movement of γ-radiation in a cloud chamber. The tracks are very thin and not well defined. The reason for this is that γ-radiation is weakly ionising compared with alpha or beta particles.

Cobalt (^{60}Co) produces γ-rays.

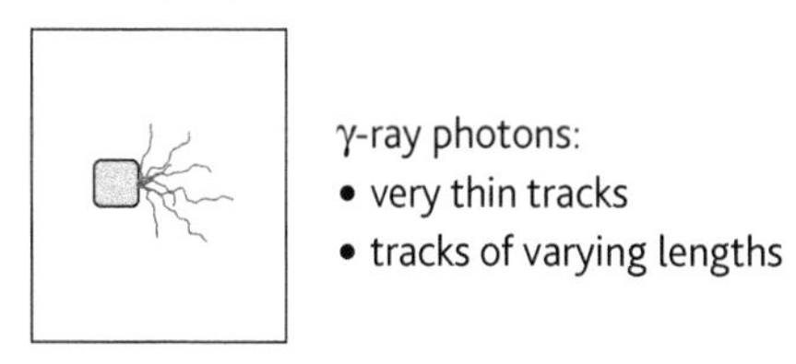

Figure 12.6.3 *Gamma radiation in a cloud chamber*

Penetrating properties

α-particles, β-particles and γ-rays penetrate matter differently (Figure 12.6.4). One way to investigate the penetrating properties is to observe the effect on the count-rate measured on a G–M tube when different materials are placed between the source of the radiation and the G–M tube. It can be shown that:

- α-particles are stopped by a thin sheet of paper (2 mm)
- β-particles are stopped by a thin sheet of aluminium (1–10 mm)
- γ-rays are the most penetrating but most of the radiation is stopped by a sheet of lead.

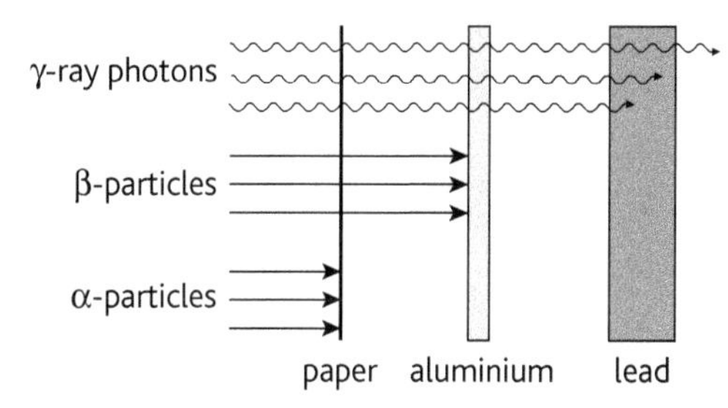

Figure 12.6.4 *The penetrating properties of ionising radiation*

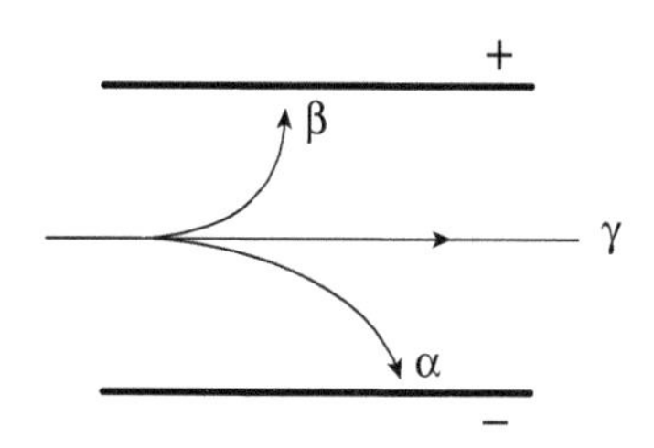

Figure 12.6.5 *Behaviour of α-particles, β-particles and γ-rays in an electric field*

Effect of electric fields

Figure 12.6.5 shows the paths taken by α-particles, β-particles and γ-rays as they pass through an electric field. An α-particle has a positive charge and moves toward the negative plate. A β-particle has a negative charge and moves toward the positive plate. The β-particle has a smaller mass than the α-particle and is deflected more in the electric field. The γ-rays have no charge and therefore pass through the electric field undeviated.

Effect of magnetic fields

Figure 12.6.6 shows the paths taken by α-particles, β-particles and γ-rays as they pass through a magnetic field. The magnetic field is perpendicular to the paths and points into the plane of the paper. A magnetic force is experienced by the α-particles and the β-particles. The direction of the force is determined by Fleming's left-hand rule. The γ-rays have no charge and therefore pass through the magnetic field undeviated.

Figure 12.6.6 is a composite diagram. It illustrates the results obtained separately with each type of radiation. The α-particles show very little deflection compared to β-particles in the same magnetic field.

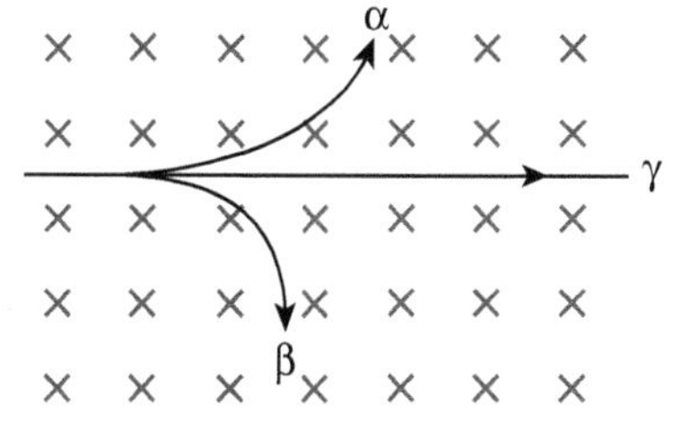

The magnetic field is perpendicular and points into the paper

Figure 12.6.6 *Behaviour of α-particles, β-particles and γ-rays in a magnetic field*

The inverse square law for γ-rays

A point source of γ-rays emits in all directions. The intensity I varies as the inverse square of the distance d from the source.

$$I \propto \frac{1}{d^2}$$

The intensity of the γ-rays thus decreases with distance from the source.

Table 12.6.1 compares the properties of the various ionising radiations.

Table 12.6.1 *Comparison of the properties of α-particles, β-particles and γ-rays*

		α-particles	β-particles	γ-rays
1	Nature	Helium nuclei	Fast-moving electron	Photon of electromagnetic radiation
2	Charge	+2e	−e	None
3	Mass	$4m_p$	m_e or $\left(\frac{1}{1836}m_p\right)$	None
4	Energy	3–7 MeV	1–2 MeV	1–2 MeV
		Monoenergetic from given nuclide	Range of emission of energies from given nuclide from 0 to a maximum	Monoenergetic from given nuclide
5	Speed	Up to 0.01c	(0.01–0.9)c	c
6	Range in air	3–10 cm	1–2 m	Order of kilometres
7	Ionising property	Strongly ionising	Weakly ionising	Very little ionisation
8	Penetrating ability	Stopped by a thin sheet of paper (2 mm)	Stopped by a thin sheet of aluminium (1–10 mm)	Most of it is stopped by a sheet of lead (1–10 cm)
9	Affected by an electric field	Yes	Yes	No
10	Affected by a magnetic field	Yes	Yes	No

Dangers of ionising radiation

Ionising radiation can damage DNA in the nucleus of cells. Damaged DNA can affect the functioning of the cell. It can also affect the ability of the cell to divide. Ionising radiation can lead to mutations, increased risk of developing cancer, radiation sickness (nausea, fatigue and loss of hair) and even sterility.

Although α- and β-particles cannot travel far into living tissue, they can cause severe damage to the skin. As α-radiation is strongly ionising it can cause severe surface burns. Direct damage to internal organs of the body can be caused by γ-rays, because, although less ionising, they are more penetrating.

In 2011 an earthquake of magnitude 9.0 caused a tsunami that devastated parts of Japan. Nuclear power plants were damaged, leading to radiation leaks. It may take some years to determine the impact the radiation leaks have had on the surrounding countryside. Consequently, developed nations have had to rethink their use of nuclear energy. Caribbean leaders have protested against the use of the Caribbean Sea as a transportation route for radioactive waste from the United Kingdom to Japan. An accident or terrorist act could seriously impact on the welfare and livelihood of the Caribbean region. The Caribbean community does not have the facilities to deal with any type of nuclear disaster.

Safety precautions when handling radioactive material

- Forceps should be used when handling radioactive material.
- Always turn a radioactive source away from your body and never point it at anybody.
- Radioactive material should be stored in lead containers when not in use.
- Avoid eating or drinking where radioactivity experiments are being performed.
- A radiation detector should be present in order to monitor radiation levels.
- Radioactive materials should never be disposed of in an ordinary waste bin.

Key points

- Three types of ionising radiation are alpha particles, beta particles and gamma rays.
- An α-particle is a helium nucleus.
- A β-particle is a fast-moving electron.
- γ-rays consist of high-energy electromagnetic waves.
- α-particles are the strongest ioniser and γ-rays are the weakest ioniser.
- α-particles are stopped by a thin sheet of paper (2 mm).
- β-particles are stopped by a thin sheet of aluminium (1–10 mm).
- γ-rays are the most penetrating but most of the radiation is stopped by a sheet of lead.
- α-particles and β-particles are deflected by electric and magnetic fields.
- γ-rays are not deflected by electric or magnetic fields.
- The intensity of γ-rays from a point source is inversely proportional to the square of the distance from the source.

Learning outcomes

On completion of this section, you should be able to:

- represent α-, β- and γ-decay using simple nuclear equations
- define the terms *activity* and *decay constant*, and recall $A = -\lambda N$
- recognise, use and represent graphically solutions of the decay law based on $x = x_0 e^{-\lambda t}$ for activity, number of undecayed particles and received count rate
- define half-life
- use the relation $\lambda = \frac{0.693}{T_{1/2}}$.

Decay equations

Alpha-decay

The following equation illustrates what happens during α-decay.

$$^{A}_{Z}\mathrm{X} \longrightarrow {}^{A-4}_{Z-2}\mathrm{Y} + {}^{4}_{2}\mathrm{He}$$

The mass number of the parent nuclide decreases by four and the atomic number decreases by two. For example:

$$^{234}_{90}\mathrm{Th} \longrightarrow {}^{230}_{88}\mathrm{Ra} + {}^{4}_{2}\mathrm{He}$$

Beta-decay

The following equation illustrates what happens during β-decay.

$$^{A}_{Z}\mathrm{X} \longrightarrow {}_{Z+1}^{\;\;A}\mathrm{Y} + {}^{\;\;0}_{-1}\mathrm{e}$$

The mass number of the parent nuclide remains the same and the atomic number increases by one. For example:

$$^{14}_{6}\mathrm{C} \longrightarrow {}^{14}_{7}\mathrm{N} + {}^{\;\;0}_{-1}\mathrm{e}$$

During β-decay a neutron spontaneously changes into a proton and an electron (β-particle). The β-particle is emitted from the nucleus. The β-particle is not produced from electrons orbiting the nucleus of an atom.

$$^{1}_{0}\mathrm{n} \longrightarrow {}^{1}_{1}\mathrm{p} + {}^{\;\;0}_{-1}\mathrm{e}$$

Gamma-decay

The following equation illustrates what happens during γ-decay.

$$^{A}_{Z}\mathrm{X} \longrightarrow {}^{A}_{Z}\mathrm{X} + {}^{0}_{0}\gamma$$

The mass number and atomic number of the parent nuclide remain the same, since the γ-ray photon has no mass and no charge. Since a γ-ray photon is emitted, the nucleus becomes more stable and less excited.

Activity, decay constant and half-life

The **activity** A of a radioactive sample is the number of nuclei decaying per second. The SI unit is the becquerel (Bq). The activity of a radioactive sample is proportional to the number of nuclei N present in the sample.

$$A \propto N \qquad \text{therefore} \qquad A = -\lambda N$$

λ is the proportionality constant in the equation and is called the **decay constant.**

The decay constant λ is the probability of decay of a nucleus per unit time.

Suppose the number of radioactive nuclei present in a sample is recorded over a period of time as shown in Figure 12.7.1.

Equation

$A = -\lambda N$

A – activity/Bq or s^{-1}

λ – decay constant/s^{-1} or min^{-1} or h^{-1}, etc.

N – number of undecayed nuclei

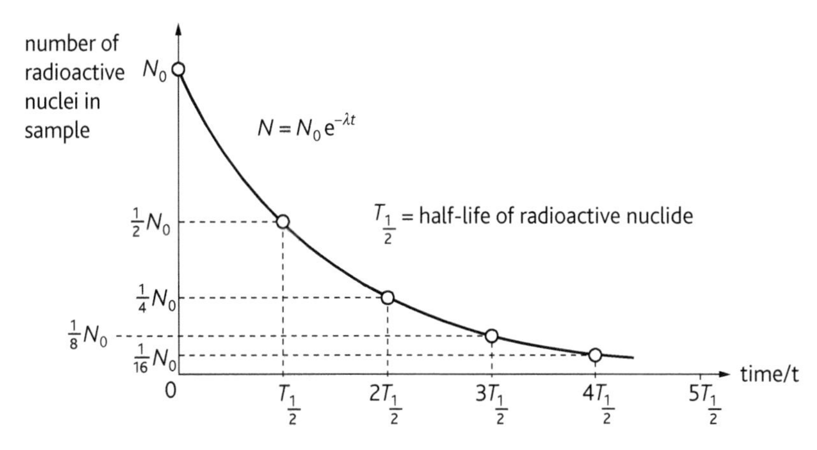

Figure 12.7.1

It can be seen that the graph is exponential in nature. The number of nuclei present in the sample at time $t = 0$ is N_0. As nuclei decay, the number remaining decreases over time. The **half-life** $T_{1/2}$ of a radioactive substance is the average time taken for the number of undecayed nuclei to decrease to half of its initial value.

Half-life can also be defined in terms of activity. The half-life of a radioactive substance is then the time taken for the activity of the sample to decrease to half of its initial value.

- After one half-life ($T_{1/2}$), the number of undecayed nuclei present is $N_0/2$.
- After two half-lives ($2T_{1/2}$), the number of undecayed nuclei present is $N_0/4$.
- After three half-lives ($3T_{1/2}$), the number of undecayed nuclei present is $N_0/8$.

$$N_0 \xrightarrow{T_{1/2}} \frac{N_0}{2} \xrightarrow{T_{1/2}} \frac{N_0}{4} \xrightarrow{T_{1/2}} \frac{N_0}{8}$$

Deriving the decay equation

Let the number of undecayed nuclei present in a radioactive sample at time t be N.

The activity of the sample is proportional to the number of undecayed nuclei present.

Therefore,
$$-\frac{dN}{dt} \propto N$$
$$-\frac{dN}{dt} = \lambda N$$

Separating the variables,

$$dN = -\lambda N \, dt$$
$$\frac{1}{N} dN = -\lambda \, dt$$

Integrating both sides of the equation,

$$\int \frac{1}{N} dN = \int -\lambda \, dt$$
$$\ln N = -\lambda t + c$$
$$N = e^{-\lambda t + c}$$
$$N = e^{-\lambda t} \times e^{c}$$

Let $e^c = A$

Then $N = Ae^{-\lambda t}$

At time $t = 0$, let $N = N_0$

Therefore $N = N_0 e^{-\lambda t}$

This equation is known as the decay equation. It can be used to determine the number of undecayed nuclei present at any time, once the decay constant is known. The equation can also be written in terms of activity: $A = A_0 e^{-\lambda t}$.

Equation

$N = N_0 e^{-\lambda t}$

N – number of undecayed nuclei at time t
λ – decay constant/s^{-1}
N_0 – initial number of undecayed nuclei
t – time/s

The relationship between half-life and the decay constant

The decay equation is given by $N = N_0 e^{-\lambda t}$.

At time $t = 0$, $N = N_0$

At time $t = T_{1/2}$, $N = \frac{N_0}{2}$

Therefore
$$\frac{N_0}{2} = e^{-\lambda T_{1/2}}$$
$$\frac{1}{2} = e^{-\lambda T_{1/2}}$$

Taking natural logarithms on both sides of the equation:

$$\ln\frac{1}{2} = \ln\left(e^{-\lambda T_{1/2}}\right)$$
$$-\ln 2 = -\lambda T_{1/2} \ln e$$
$$\ln 2 = \lambda T_{1/2}$$
$$T_{1/2} = \frac{\ln 2}{\lambda} = \frac{0.693}{\lambda}$$

Equation

$$T_{1/2} = \frac{\ln 2}{\lambda} = \frac{0.693}{\lambda}$$

$T_{1/2}$ – half-life
λ – decay constant

Example

Sodium-24 (^{24}Na) is a radioactive isotope with a half-life of 15.0 hours. A pure sample of sodium-24 has a mass of 5.0 g.

Calculate:

a the number of sodium-24 atoms present in the sample

b the decay constant λ

c the initial activity A_0 of the sample

d the activity of the sample after 45 hours

e the activity of the sample after 30 hours.

(Avogadro constant $N_A = 6.02 \times 10^{23}$ per mole)

---

a 24 g contains 6.02×10^{23} sodium-24 atoms.

5.0 g contains $\dfrac{5.0 \times 6.02 \times 10^{23}}{200} = 1.25 \times 10^{23}$ sodium-24 atoms

b Half-life $= 15 \times 60 \times 60 = 54\,000\,\text{s}$

$$\lambda = \frac{0.693}{T_{1/2}} = \frac{0.693}{54\,000} = 1.28 \times 10^{-5}\,\text{s}$$

c $A = -\lambda N$

$$A_0 = -1.28 \times 10^{-5} \times 1.25 \times 10^{23}$$
$$= 1.60 \times 10^{18}\,\text{Bq}$$

d Number of half-lives $= \dfrac{45}{15} = 3$

After 3 half-lives, the activity of the sample

$$= A_0\left(\frac{1}{2}\right)^n, \text{ where } n \text{ is the number of half-lives}$$
$$= 1.60 \times 10^{18} \times \left(\frac{1}{2}\right)^3$$
$$= 2.0 \times 10^{17}\,\text{Bq}$$

e Activity of the sample after 30 hours is given by:

$$A = A_0 e^{-\lambda t}$$
$$= 1.60 \times 10^{18} \times e^{-(1.28 \times 10^{-5} \times 30 \times 3600)}$$
$$= 4.02 \times 10^{17}\,\text{Bq}$$

Key points

- Mass number and atomic number are conserved in nuclear reactions.
- The activity A of a radioactive sample is the number of nuclei decaying per second.
- The activity of a radioactive sample is proportional to the number of nuclei present in the sample.
- The decay constant λ is the probability of decay of a nucleus per unit time.
- The half-life of a radioactive substance is the average time taken for the number of undecayed nuclei to decrease to half of its initial value.
- Radioactive decay can be expressed mathematically by $x = x_0 e^{-\lambda t}$.
- Half-life and decay constant are related by the equation $T_{1/2} = \dfrac{0.693}{\lambda}$.

Learning outcomes

On completion of this section, you should be able to:

- describe an experiment to measure the half-life of a radioactive substance with a short half-life.

Experiment to measure the half-life of radon-220

Radon-220 ($^{220}_{86}$Rn) is a gas with a half-life of 55 seconds. Its half-life can be measured in a laboratory. When it decays, it emits α-particles. A G–M tube with a very thin mica window can be used to measure the activity of the gas.

$^{220}_{86}$Rn is one of the decay products of an isotope of thorium $^{232}_{90}$Th. All the other nuclides in the decay series have half-lives that are either much longer or much shorter than radon-220, so they do not contribute to the activity of the sample of the gas. The $^{232}_{90}$Th is present in a solid, in the form of thorium hydroxide and $^{220}_{86}$Rn is a gas, which means that the two can be easily separated.

Figure 12.8.1 illustrates the apparatus that can be used to measure the half-life of radon-220. The experiment is performed in a fume cupboard.

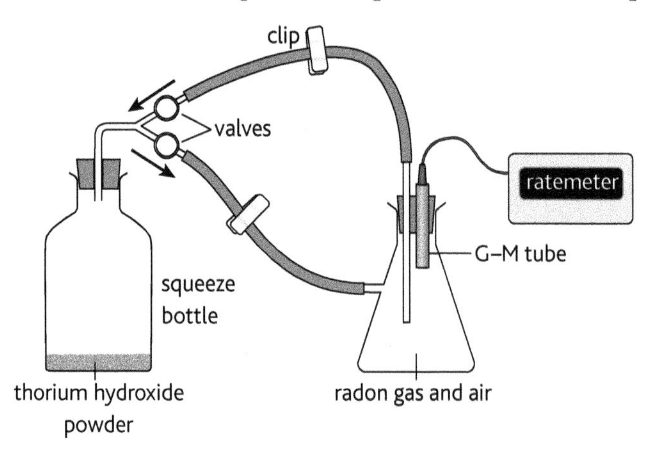

Figure 12.8.1 *Experiment to measure the half-life of radon-220*

1. A G–M tube is used to measure the background count rate.
2. Some thorium hydroxide (which is a solid) is placed in a squeeze bottle.
3. The screw clips are opened.
4. The bottle is squeezed so that it forces radon-220 into the collecting vessel.
5. The screw clips are then closed. This prevents any freshly produced radon-220 gas produced by the thorium hydroxide from entering the collecting vessel.
6. Immediately after closing the screw clips, the scalar and stop watch are started.
7. The count rate R is recorded every 10 seconds for a period of 5 minutes.
8. The background count rate is subtracted from the measured count rate to find the actual count rate A.
9. A graph of A against time is plotted.

Sample data

A student performed an experiment to measure the half-life of radon-220 and obtained the following results.

Background count rate = 2 counts per second

Table 12.8.1 *Sample data*

Time/s	Measured count rate R/s^{-1}	Corrected count rate A/s^{-1}	$\ln A/s^{-1}$
10	53	51	3.93
20	47	45	3.81
30	35	33	3.50
40	36	34	3.53
50	32	30	3.40
60	29	27	3.30
70	21	19	2.94
80	23	21	3.04
90	19	17	2.83
100	18	16	2.77
110	15	13	2.56
120	13	11	2.40
130	13	11	2.40
140	10	8	2.08
150	9	7	1.95
160	9	7	1.95
170	8	6	1.79
180	9	7	1.95
190	8	6	1.79
200	6	4	1.39
210	6	4	1.39
220	5	3	1.10
230	4	2	0.69
240	5	3	1.10
250	5	3	1.10
260	5	3	1.10
270	4	2	0.69
280	4	2	0.69
290	3	1	0
300	3	1	0

Measuring half-life directly from the curve

The half-life can be measured directly from a decay curve. After plotting all the data points on the graph, a curve of best-fit is drawn (Figure 12.8.2).

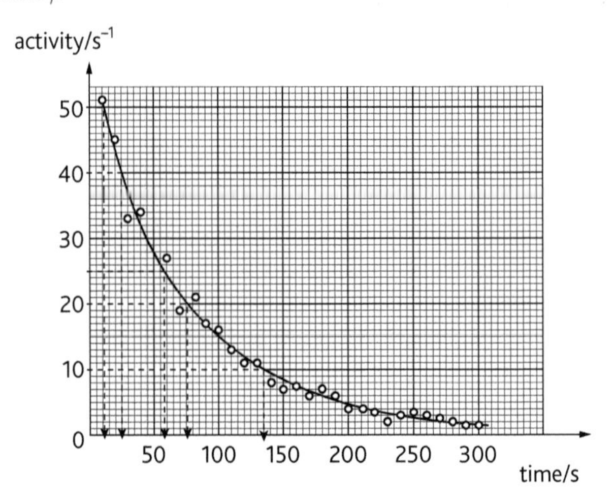

Figure 12.8.2 *Plotting a decay curve to measure half-life*

Several points of interest are worth noting:

- The activity is decreasing with time. This indicates that the number of radon-220 nuclei is decaying.
- The curve does not pass through all the points. The fluctuations about the curve indicate that radioactive decay is a random process.
- The half-life from a decay curve is constant. It is not necessary to start with the activity at time $t = 0$ in order to measure half-life.

Any point on the curve is chosen and the time taken for its value to decrease by half its value is measured. This will be the half-life of radon-220.

Initial activity $= 50\,s^{-1}$:

time taken for activity to reduce to $25\,s^{-1} = 60 - 10 = 50\,s$

Initial activity $= 40\,s^{-1}$:

time taken for activity to reduce to $20\,s^{-1} = 75 - 25 = 50\,s$

Initial activity $= 20\,s^{-1}$:

time taken for activity to reduce to $10\,s^{-1} = 130 - 75 = 55\,s$

Average half-life $= \dfrac{50 + 50 + 55}{3} = 51.7\,s$

Measuring half-life by plotting a straight line

Consider the decay law equation:

$$A = A_0 e^{-\lambda t}$$

Taking $\log_e$ (ln e) on both sides (natural logarithms),

$$\ln A = \ln A_0 e^{-\lambda t}$$

$$\ln A = \ln A_0 - \ln e^{-\lambda t}$$

$$\ln A = \ln A_0 - \lambda t \ln e \quad \text{(but } \ln e = 1\text{)}$$

$$\ln A = \ln A_0 - \lambda t$$

If a graph of $\ln A$ against t is drawn, a straight line with a negative gradient will be obtained, where:

$$\text{gradient} = \lambda \text{ and } y\text{-intercept} = \ln A_0$$

Figure 12.8.3 shows a plot of $\ln A$ against t.

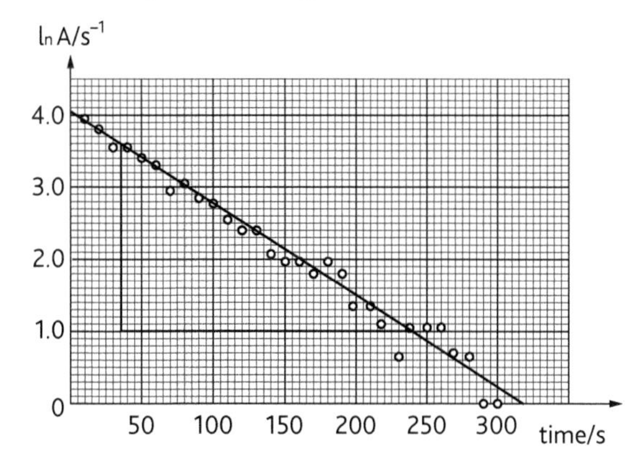

Figure 12.8.3 *Plotting a straight line to measure half-life*

The data points are plotted and a line of best fit is drawn. The gradient is found by drawing a large triangle such that its hypotenuse is greater than half the length of the line of best fit.

From the graph:

$$\text{Gradient of line} = \frac{3.60 - 1.0}{35 - 240} = -0.0127$$

Therefore, $\lambda = 0.0127$

$$\text{Half-life} = \frac{0.693}{\lambda} = \frac{0.693}{0.0127} = 54.6 \text{ seconds}$$

Notice that the half-life measured by both methods is not the same. The half-life measured by the straight line method is more accurate because it is easier to plot a line of best fit rather than a curve of best fit.

Key points

- The half-life of radon-220 can be determined experimentally. The count rate is measured over a period of time.
- The half-life can be measured by measuring the average time taken for the activity to decrease by half its initial value.
- The half-life can also be calculated from the gradient of a suitably plotted straight line.

12.9 Uses of radioisotopes

Learning outcomes

On completion of this section, you should be able to:

- discuss the uses of radioisotopes as tracers, for carbon dating and in radiotherapy.

Uses of radioisotopes

Isotopes are atoms of the same element that have the same atomic number but a different mass number. For example, $^{12}_{6}C$, $^{13}_{6}C$ and $^{14}_{6}C$ are isotopes of carbon. Isotopes have the same number of protons and so the same number of electrons. This means they are chemically similar. Because of this, a physical method is usually used to separate them.

Some isotopes of elements are radioactive and have many useful applications.

Radioactive dating

Radioisotopes can be used to date rocks and artefacts found during archaeological expeditions. Uranium-238 has a half-life of 4500 million years. When uranium-238 decays it eventually forms a stable isotope of lead. The age of a rock sample can be determined by comparing the ratio of the uranium-238 to lead present in the sample.

A very small but constant proportion of carbon dioxide in the atmosphere contains carbon-14 atoms. Carbon-14 is a β-emitter and has a half-life of 5730 years. Carbon-12 is stable and is found in all living organisms. When plants are alive, they absorb carbon dioxide through photosynthesis. Animals feed on these plants. In this way, plants and animals absorb some of the radioactive carbon-14. When they die, the carbon-14 is not replaced but decays. The age of a discovered artefact is determined by comparing the ratio of the number of carbon-14 atoms to the number of carbon-12 atoms present.

Tracers

Radioisotopes can be used as tracers. A small amount of a radioisotope is added to the material to be studied. The radioisotope emits radiation and can be detected. It can be traced and located.

Iodine-131 is used as a tracer. The thyroid gland located in the neck is responsible for regulating growth and the body's metabolism. The thyroid gland requires iodine for proper functioning. The patient is injected with the radioactive iodine and the uptake by the thyroid gland is measured by using a radiation detector.

Tracers can also be used to locate blood clots and tumours in the body.

Iron-59 is a β-emitter with a half-life of 45 days. It is used to measure the amount of wear in engine components. During the manufacturing process, the engine component is made using non-radioactive iron (iron-56), together with a uniform distribution of iron-59. The initial mass and activity of the component is determined before the test. The engine is run over an extended period of time (e.g. 30 days). During the test period, any metal worn off the component is deposited in the engine oil. Immediately after the test, the activity of the engine oil is measured. The amount of material worn off from the engine component can then be calculated.

Radioactive tracers are also used to detect leaks in underground pipes. A radioactive isotope with a short half-life is added to the water. When the water leaks from the underground pipe, the water emits radiation. The radiation can be detected from the surface of the ground and the leak can be isolated.

Thickness control

β-particles with a long half-life are used to monitor the thickness of sheets of materials (e.g. the thickness of paper in paper mills or the thickness of sheets of aluminium).

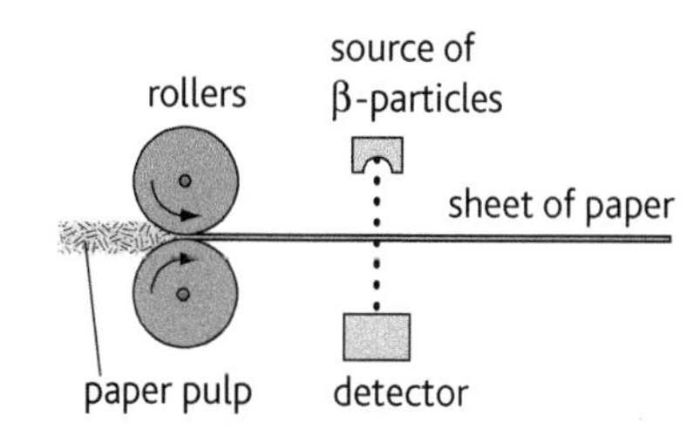

Figure 12.9.1 *Monitoring the thickness of paper sheets*

Figure 12.9.1 shows how this is achieved. A source of β-particles is placed on one side of a sheet of paper and a detector is placed on the other side. The detector controls the spacing between the rollers. When the thickness of the paper sheet increases, the detector measures a reduction in the number of β-particles. The detector sends a signal to the rollers to reduce the spacing between them. When the thickness of the paper sheet decreases, the detector measures an increase in the number of the β-particles. The detector sends a signal to the rollers to increase the spacing between them.

An α-source cannot be used for this because α-particles have only a short range in air and would not reach the detector. A γ-source cannot be used because γ-rays are too penetrating, and would not be stopped by the paper sheet. The detector would measure a constant reading and would be unable to detect any changes in the thickness of the paper.

Smoke detectors

Americium-241 is used in smoke detectors. This is an α-emitter that has a very long half-life. The α-particles ionise the air inside the smoke detector, enabling the air to conduct a small current when a potential difference is applied across an air gap. When smoke particles enter the detector, they absorb some of the α-particles. The current decreases and the alarm is triggered.

Imaging

γ-rays are very penetrating and are used in non-invasive imaging techniques. For example, they are used to check the quality of welded pipes. For this a γ-ray source is placed on one side of the pipe and a photographic film is placed on the other side. This technique is able to detect any faults in the welded joint.

Other applications include the detection of hairline fractures in engine blocks in the car manufacturing industry.

γ-rays are used to check baggage at airports without having to search it.

Sterilisation

γ-rays are used to sterilise food. An intense beam of γ-rays can be used to kill microorganisms on vegetables, meats and grains. It prolongs the shelf life of these items and reduces the need for using preservatives or refrigeration. A similar method uses γ-rays to sterilise medical instruments.

Cancer treatment

Cobalt-60 is source of γ-rays. A large dose of γ-rays is focused on the site of cancerous cells in the body. The high dose of energy is able to kill cancerous cells.

Key points

- Isotopes are atoms of the same element that have the same atomic number but different mass number.
- Radioisotopes have many uses such as radioactive dating, tracers, thickness control, imaging, smoke detectors and cancer treatment.

Revision questions 8

Answers to questions that require calculation can be found on the accompanying CD.

1 a Explain why radioactive decay is considered *spontaneous* and *random*. [3]

b Explain what is meant by the terms *activity* and *decay constant*. [3]

c Show that the number of undecayed nuclei N present in a radioactive sample at time t is given by $N = N_0 e^{-\lambda t}$, where N_0 is the initial number of nuclei present in the sample. [5]

2 a Explain what is meant by the half-life of a radioactive sample. [2]

b Derive an equation relating the decay constant λ and half-life of a radioactive sample, starting with the decay equation $N = N_0 e^{-\lambda t}$. [3]

c A radioactive isotope has a half-life of 15.0 hours. For this isotope, calculate:

- **i** the decay constant [2]
- **ii** the activity of a sample containing 6.02×10^{23} nuclei. [3]

3 A radium nuclide $^{226}_{88}Ra$ emits an α-particle to produce an isotope of radon. State the number of protons and neutrons in the isotope of radon. [2]

4 a Explain what is meant by the term *decay constant*. [2]

b A radioactive isotope has a half-life of 18 days. A sample of the isotope contains 3.0×10^{12} atoms.

Calculate:

- **i** the decay constant for the isotope [3]
- **ii** the initial activity of the sample [3]
- **iii** the activity of the sample after 72 days. [2]

5 Distinguish between α-particles, β-particles and γ-ray photons, making reference to charge, mass, speed and penetration. [6]

6 ^{40}K is an isotope of potassium. It has a half-life of 1.26×10^9 years and decays to form an isotope of argon which is stable. In a sample of rock, the ratio of potassium atoms to argon atoms is 1:3. Calculate the age of the rock by assuming that originally there were no argon atoms present in the sample. [6]

7 The half-life of cobalt-60, $^{60}_{27}Co$, is 5.26 years. Work out:

a the number of protons and neutrons in a cobalt-60 nucleus [2]

b the decay constant of cobalt-60 [2]

c the activity of 10.0 grams of cobalt-60. [3]

8 An isotope of bismuth, $^{212}_{83}Bi$, has a decay constant of $1.15 \times 10^{-2}\,min^{-1}$. $^{212}_{83}Bi$ decays by emitting a β-particle to produce polonium (Po). Polonium then emits an α-particle to produce an isotope of lead (Pb).

a Write two nuclear equations to illustrate the two decay processes. [4]

b Calculate the half-life of $^{212}_{83}Bi$. [2]

c Given that a sample of $^{212}_{83}Bi$ contains $2.5\,\mu g$ of the isotope, calculate the activity of the sample. [3]

9 a Describe an experiment to show that radioactive decay is a random process. [5]

b On a single diagram, illustrate the following:

- **i** radioactive decay is an exponential process [1]
- **ii** radioactive decay is a random process [1]
- **iii** background radiation. [1]

c Explain what is meant by the term *background radiation*. State two sources of background radiation. [3]

10 a Draw a diagram to show the path of a stream of α-, β- and γ-radiation as they pass through a uniform electric field. [3]

b Draw a diagram to show the path of a stream of α-, β- and γ-radiation as they pass through a magnetic field. [3]

c State an instrument that can be used to detect β-particles. [1]

11 Radioisotopes can be dangerous to living organisms but they can also be useful.

a Discuss dangers of the ionising radiation emitted by radioisotopes. [5]

b i Discuss two useful applications of radioisotopes. [4]

ii State the properties of these radioisotopes that make them suited for their use. [3]

12 State the changes to the number of protons and number of neutrons that occur within nuclei when they emit:

a α-particles [2]

b β-particles [2]

c γ-radiation. [1]

13 A radioactive source contains 7.0×10^{16} radioactive nuclei and has an activity of 4.5×10^{7} Bq. For this source, calculate:

a the decay constant [2]

b the half-life [2]

c the time taken for the activity to fall to 2.0×10^{5} Bq. [4]

14 Radioactive decay usually results in the emission of one or more of the following types of ionisation radiation: α-particles, β-particles or γ-radiation. State the type of emission that:

a has a range of energies, rather than discrete values [1]

b produces the greatest density of ionisation in a medium [1]

c produces the least density of ionisation in a medium [1]

d is not affected by electric and magnetic fields [1]

e does not directly result in a change in proton number of the nucleus [1]

f has a range of a few cm in air [1]

g produces thick tracks in a cloud chamber. [1]

15 The radioactive gas radon-220 emits α-particles. It is known to have a short half-life and is easily separated from its solid parent nuclide. With the aid of a diagram, explain how you would determine the half-life of radon-220. Explain how the data would be collected and used to determine the half-life. [7]

16 The half-life of manganese-56 is 2.6 hours. A sample of manganese-56 has a mass of 1.0 μg and is known to decay β-particles.

a State the nucleon number and proton number of the daughter nuclide produced in the decay process. [2]

b Sketch a graph to show the decay of the sample over a 15-hour period. [3]

c Calculate the number of manganese-56 atoms present in the sample. [2]

d Calculate the initial activity of the sample. [2]

e Determine the time at which the number of manganese-56 is equal to 1.35×10^{15}. [3]

17 Strontium-90 is a radioactive isotope with a half-life of 28.0 years. A sample of strontium-90 has an activity of 4.5×10^{9} Bq. Calculate:

a the decay constant in s^{-1} [3]

b the mass of strontium-90 in the sample. [4]

18 Radium-224 has a half-life of 3.6 days. A sample of radium-224 has a mass of 1.82 mg. Calculate:

a the decay constant of radium-224 in s^{-1} [2]

b the activity of the sample. [4]

19 Americium-241 is an artificially produced radioactive element. A particular sample of americium-241 of mass 3.6 μg has an activity of 4.2×10^{5} Bq. Calculate:

a the number of americium-241 atoms present in the sample [2]

b the decay constant [2]

c the half-life of americium-241. [2]

Module 3 Practice exam questions

Answers to the multiple-choice questions and to selected structured questions can be found on the accompanying CD.

Multiple-choice questions

1 Which of the following is true?

- **i** The photoelectric effect provides evidence for the wave nature of electromagnetic radiation.
- **ii** The electron diffraction experiment provides evidence for the wave nature of particles.
- **iii** Interference and diffraction provides evidence for the wave nature of electromagnetic radiation.

a i only **b** i and ii only
c i, ii and iii **d** ii and iii only

2 The potential difference across an X-ray tube is maintained at 80 kV. What is the cut-off wavelength of the X-rays emitted?

a 1.55×10^{-11} m **b** 8.00×10^{-11} m
c 2.49×10^{-30} m **d** 1.55×10^{-10} m

3 Which of the following cannot be explained by the wave theory of electromagnetic radiation?

a Interference
b The photoelectric effect
c Diffraction
d Polarisation

4 When ultraviolet radiation is incident on a zinc plate photoelectrons are emitted.

How would the number of photoelectrons emitted per second N and the maximum kinetic energy E of the photoelectrons be affected when the ultraviolet source is replaced with a less intense source of the same wavelength?

	N	E
a	Decreased	Increased
b	Increased	Decreased
c	Decreased	Unchanged
d	Unchanged	Decreased

5 An electron has a mass 9.11×10^{-31} kg and kinetic energy 2.85×10^{-18} J. What is the de Broglie wavelength?

a 1.82×10^{-10} m **b** 2.85×10^{-10} m
c 2.91×10^{-10} m **d** 2.91×10^{-11} m

6 In an laboratory experiment to measure the charge on an oil drop, the following results were obtained:

16.24, 8.12, 24.36, 32.53, 32.45

16.51, 24.35, 8.12, 32.48, 16.24

The unit used for charge was not the SI unit. What value do the results suggest for the magnitude of the elementary charge e on an electron as measured in the units used?

a 21.14 **b** 7.83 **c** 8.12 **d** 4.06

7 The light in a beam has a continuous spectrum of wavelengths from 410 nm to 680 nm. The light is incident on some cool hydrogen gas and is then viewed through a diffraction grating. Which of the following best describes the spectrum that is seen?

a Dark lines on a coloured background
b Coloured lines on a black background
c Dark lines on white background
d Coloured lines on a white background

8 An electron moves from an energy level E_1 to energy level E_2 within an atom. E_1 is at a higher energy level than E_2. What is the wavelength of the photon that is released?

a $\frac{hc}{E_2 - E_1}$ **b** $\frac{E_1 - E_2}{hc}$ **c** $\frac{E_2 - E_1}{E_1 - E_2}$ **d** $\frac{hc}{E_1 - E_2}$

9 Consider the following reaction.

$${}^{1}_{1}\text{p} + {}^{2}_{1}\text{H} \longrightarrow {}^{3}_{2}\text{He}$$

What is the energy released in the reaction?

Use the following data to work out your answer.

Mass of ${}^{2}_{1}$H nucleus = 2.01355 u

Mass of proton = 1.00728 u

Mass of ${}^{3}_{2}$He = 3.01494 u

1 u = 931.3 MeV

a 5.49 MeV **b** 3.02 MeV
c 6.04 MeV **d** 5621 MeV

10 Two radioactive samples P and Q each have an activity of A at time $t = 0$. P has a half-life of 18 days and Q has a half-life of 12 days. P and Q are mixed together. The total activity of the mixture after 36 days is:

a $\frac{1}{4}A$ **b** $\frac{3}{8}A$ **c** $\frac{1}{16}A$ **d** $\frac{5}{8}A$

Structured questions

11 a Briefly describe the models of the atom proposed by:

- **i** J. J. Thomson [2]
- **ii** Rutherford [3]
- **iii** Neils Bohr. [3]

b State the observations from the alpha scattering experiments performed by Geiger and Marsden, under the supervision of Rutherford. [3]

c Explain how these observations led to the conclusion that an atom was composed mainly of empty space, with a very small positive nucleus. [3]

d State the approximate value for:

- **i** the diameter of a gold atom [1]
- **ii** the diameter of a gold nucleus. [1]

12 a Outline Millikan's oil drop experiment and summarise the experimental evidence it provides for the quantisation of charge. [6]

b State two measures taken by Millikan to improve the accuracy of his oil drop experiment. [2]

c Explain how he was able to change the charge on an oil drop. [1]

d Two parallel metal plates are 1.5 cm apart. The potential difference between the plates is 1400 V. A small oil drop of mass of 7.61×10^{-15} kg remains stationary between the plates. The density of the air is negligible in comparison with that of the oil.

- **i** Draw a diagram to show the forces acting on the oil drop. [2]
- **ii** Calculate the electric field strength between the plates. [2]
- **iii** Calculate the charge on the oil drop. [3]
- **iv** Calculate the number of electrons attached to the oil drop. [2]

13 a Outline the experimental evidence that suggests that electromagnetic radiation is:

- **i** a wave [4]
- **ii** a particle. [4]

b Explain why line emission spectra provide evidence for the existence of discrete energy levels inside atoms. [3]

14 The diagram below represents energy levels within a hydrogen atom.

energy level	energy/eV
6	−0.37
5	−0.54
4	−0.85
3	−1.5
2	−3.4
(ground state) 1	−13.6

a State the amount of energy required to ionise an electron from the ground state. [1]

b An electron makes a transition from the $n = 3$ to $n = 2$ state.

- **i** Calculate the frequency of the photon emitted. [2]
- **ii** Calculate the wavelength of the photon emitted. [1]
- **iii** In which region of the electromagnetic spectrum would the radiation be found? [1]

c An electron in the ground state of a hydrogen atom is struck by a photon. State and explain what happens to the electron and what happens to the photon when the energy of the photon is:

- **i** 8 eV [2]
- **ii** 15 eV. [2]

15 a Explain what is meant by the wave–particle duality of electromagnetic radiation. [2]

b Describe an experiment which demonstrates that particles can demonstrate a wave nature. [3]

c Calculate the wavelength of a particle of mass 9.11×10^{-31} kg when travelling with a speed equal to 20% the speed of light. [3]

16 a Explain what is meant by the photoelectric effect. [2]

b State three experimental observations of the photoelectric effect. [3]

c Explain what is meant by the terms *work function energy* and *threshold frequency* when applied to the photoelectric effect. [2]

d Electromagnetic radiation of intensity $8.5 \times 10^3\,\text{W m}^{-2}$ and wavelength 240 nm is incident on an iron surface of area $2.0\,\text{cm}^2$. The iron surface reflects 80% of the electromagnetic radiation. The threshold frequency of iron is 1.1×10^{15} Hz.

Calculate:

i the intensity of the electromagnetic radiation absorbed by the metal surface [2]

ii the power of the radiation incident on the surface [2]

iii the energy of a photon incident on the surface [2]

iv the number of photons incident on the surface per second. [2]

Given that only 0.001% of the photons hitting the surface result in the production of photoelectrons, determine:

v the photoelectric current [2]

vi the work function for iron [2]

vii the stopping potential for this radiation. [4]

17 a Outline the principle of production of X-rays in an X-ray tube. [6]

b Sketch a diagram to show the spectrum of the X-rays emitted. Label:

i the continuous X-ray spectrum [1]

ii the characteristic X-ray spectrum [1]

iii the cut-off wavelength. [1]

iv Explain the origin of each feature of the spectrum. [4]

c An X-ray tube operates at 100 kV and the current through it is 1.0 mA. Calculate:

i the power input [2]

ii the speed of the electrons as they hit the target [3]

iii the cut-off wavelength of the X-rays emitted. [3]

18 a Explain what is meant by the nucleon number and the nuclear binding energy of the nucleus. [3]

b Sketch a labelled diagram to show the variation with nucleon number of the binding energy per nucleon. [4]

c Explain what is meant by nuclear fusion and nuclear fission. [3]

d Using the sketch in **b**, explain why energy is released during nuclear fusion and nuclear fission. [4]

19 Calculate the energy released in the following reaction. [6]

$${}^{235}_{92}\text{U} + {}^{1}_{0}\text{n} \longrightarrow {}^{144}_{56}\text{Ba} + {}^{90}_{36}\text{Kr} + 2\,{}^{1}_{0}\text{n}$$

Use the following data:

Mass of a neutron = 1.00867 u

Mass of uranium-235 nucleus = 235.124 u

Mass of barium-144 nucleus = 143.923 u

Mass of krypton-90 nucleus = 89.920 u

20 Calculate the binding energy of the nucleus ${}^{7}_{4}\text{Be}$ in MeV. [6]

Use the following data:

Mass of a proton = 1.00728 u

Mass of a neutron = 1.00867 u

Mass of beryllium nucleus = 7.01693 u

1 u = 931.3 MeV

21 a An isotope of lead is ${}^{214}_{82}\text{Pb}$.

i Explain what is meant by the term *isotope*. [1]

ii State the number of protons and neutrons present in a nucleus of ${}^{214}_{82}\text{Pb}$. [2]

b ${}^{214}_{82}\text{Pb}$ has a half-life of 27 minutes. A nucleus of ${}^{214}_{82}\text{Pb}$ decays by emitting a β-particle to form element ${}^{214}_{83}X$. Element ${}^{214}_{83}X$ can decay by one of two methods:

$$X \longrightarrow Y + \alpha$$

$$X \longrightarrow Z + \beta$$

Write a nuclear reaction for each of the three decays described above. [5]

c A sample of ${}^{214}_{82}\text{Pb}$ has a mass of 1.8 μg at time $t = 0$. Calculate:

i the number of atoms in the sample [2]

ii the decay constant of ${}^{214}_{82}\text{Pb}$ [2]

iii the activity of the sample at time $t = 0$ [2]

iv the time at which the activity of the sample is 4.5×10^5 Bq. [3]

22 a Explain what is meant by the following:

i The activity of a radioactive source [1]

ii The decay constant of a nuclide [2]

iii The half-life of a nuclide [2]

b A radioactive source contains 7.5×10^{14} radioactive nuclei and has an activity of 4.8×10^5 Bq. For this source, calculate:

i the decay constant [2]

ii the half-life [2]

iii the time taken for the activity to fall to 4.0×10^4 Bq. [3]

23 a Explain what is meant by the term *half-life*. [2]

b Describe an experiment to measure the half-life of radon-220, which is a gas with a short half-life. [7]

c Explain how you would use the data collected to determine the half-life of radon-220. [3]

24 a Explain what is meant by a radioisotope. [1]

b Discuss two uses of radioisotopes. [6]

c Indicate the properties of each radioisotope that make them suited for their use. [2]

25 a Draw a labelled diagram of G–M tube. [5]

b A mixture of argon and bromine is sometimes used inside the G–M tube. State the function of each gas. [3]

c Explain the principle of operation of a G–M tube. [4]

26 a Explain the principle of operation of a cloud chamber. [6]

b Sketch a pattern of what is observed when each the following ionising radiations is present in a cloud chamber:

i α-radiation [2]

ii β-radiation [2]

iii γ-radiation [2]

27 A radium nuclide $^{226}_{88}Ra$ decays by alpha-particle emission to form an isotope of radon Rn. The half-life of radium-226 is 1600 years.

a Write a nuclear equation for the decay. [2]

b Calculate the decay constant of the radium nuclide. [2]

c Calculate the energy released in the reaction, using the following data. [3]

Mass of radium-226 = 226.0254 u

Mass of radon-222 = 222.0175 u

Mass of an alpha particle = 4.0026 u

28 a Explain what is meant by the terms half-life and decay constant. [2]

b Show that the decay constant λ and the half-life $T_{1/2}$ of an isotope are related by the expression

$$\lambda T_{1/2} = 0.693$$ [3]

c On separate diagrams, sketch a graph to show:

i that radioactive decay is exponential [1]

ii that radioactive decay is random. [1]

iii What happens to the graph in **i** when a radioactive isotope is heated strongly? [1]

29 The radioactive nuclide of plutonium $^{239}_{94}Pu$, decays by alpha-particle emission with a half-life of 2.4×10^4 years. The energy of the alpha-particle is 8.0×10^{-13} J.

a Write an equation for the decay. [2]

b State the number of protons and neutrons in the nucleus of the decay product. [2]

c Calculate the decay constant of the plutonium nuclide. [2]

d The plutonium isotope is used as a power source in a particular device. The power required by the device is 1.8 W.

i Calculate the minimum rate of decay of the plutonium for it to produce 1.8 W of power. [3]

ii Calculate the number of plutonium atoms needed to produce the activity in **d i**. [3]

30 In the context of an atomic nucleus:

a State what is meant by binding energy, and explain how it arises. [3]

b State what is meant by mass difference. [1]

c State the relationship between binding energy and mass difference. [1]

'Binding energy of the nucleus is related to the stability of a nucleus.'

d Explain what is meant by this statement. [2]

31 Fusion is the process by which the Sun produces energy. The Sun produces energy by a sequence of three separate fusion reactions. This process effectively combines four protons into one helium nucleus. The fusion process releases an amount of energy that is equivalent to the mass difference between four $^{1}_{1}H$ atoms and one $^{4}_{2}He$ atom. Using the data below, calculate the energy released in the reaction in MeV.

Mass of $^{1}_{1}H$ = 1.00783 u

Mass of $^{4}_{2}He$ = 4.00260 u [4]

13 Analysis and interpretation

13.1 Analysis and interpretation

Learning outcomes

On completion of this section, you should be able to:

- draw straight line graphs using experimental data
- determine gradients and intercepts
- rearrange equations to obtain a straight line form.

Plotting graphs

In experimental work, data is often collected. This data must be analysed. One way of analysing data is by drawing a suitable graph to determine unknown constants.

The following guidelines should be used when drawing graphs.

- Choose a scale that is suitable (e.g. 1 cm to 5 units, 1 cm to 10 units). Do not use scales such as 1 cm to 3 units.
- The scale should be chosen such that the points being plotted occupy at least half the graph paper.
- The x and y axes should be labelled with the quantities being plotted and the units (e.g. d/cm, T/s, height/m, temperature/°C).
- The graph should be given an appropriate title.
- A × or a ⊙ should be used to represent the points being plotted.
- If the data appears to follow a linear relationship, a line of best fit should be drawn so that there are equal numbers of data points on either side of the line.
- If the data appears to follow a non-linear relationship, a smooth curve should be drawn through the data points. Do not use a straight line between adjacent points.

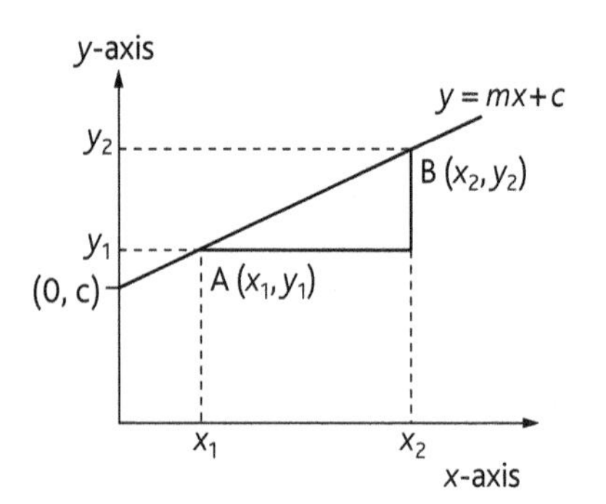

Figure 13.1.1 *A straight line graph*

Two quantities are required when plotting a graph (x-axis and y-axis). In an experiment to investigate the relationship between two quantities, only one quantity can be changed at a time. This quantity is called the **manipulated variable**. It is usually plotting on the x-axis. The second quantity is called the **responding variable**. The responding variable is usually plotted on the y-axis. Since only one quantity is being changed in the experiment, all other quantities are known as **controlled variables**.

The general equation of a straight line is $y = mx + c$. The gradient or slope of the line is m. The intercept on the y-axis is c. Consider Figure 13.1.1.

A and B lie on the straight line. The coordinates of A and B are (x_1, y_1) and (x_2, y_2) respectively. The gradient of the line passing through A and B is given by:

$$\text{Gradient (slope)} = \frac{y_2 - y_1}{x_2 - x_1}$$

Plotting linear graphs from non-linear relationships

In practical work it is often required to establish relationships between two quantities. If two quantities x and y are related such that they have a linear relationship, a straight line graph would be obtained when y is plotted against x.

It is often required that an expression be rewritten so that it resembles the form of the equation of a straight line. The table shows some examples.

Expression	What to plot?	Constants	Gradient	y-intercept
$y = ax^2 + b$	y against x^2	a and b	a	$(0, b)$
$T = kl^n$	$\lg T$ against $\lg l$	k and l	n	$(0, \lg k)$
$y^2 = ax^2 + bx$	$\frac{y^2}{x}$ against x	a and b	a	$(0, b)$
$N = Ae^{-kt}$	$\ln N$ against t	A and k	$-k$	$(0, \ln A)$

Suppose T and l are related by the following equation:

$$T = kl^n$$

Taking log_{10} on both sides of the equation gives:

$$\lg T = \lg(kl^n)$$

$$\lg T = \lg k + \lg(l^n)$$

$$\lg T = \lg k + n \lg l$$

$$\lg T = \lg k + n \lg l \qquad \text{This is the linear form.}$$

A straight line graph will be obtained if $\lg T$ is plotted against $\lg l$. The gradient of the line is n and the y-intercept is $\lg k$.

Suppose N and t are related by the following equation:

$$N = Ae^{-kt}$$

Taking $\log_e$ on both sides of the equation gives:

$$\ln N = \ln(Ae^{-kt})$$

$$\ln N = \ln A + \ln(e^{-kt})$$

$$\ln N = \ln A - kt \ln e$$

$$\ln N = \ln A - kt \qquad \text{This is the linear form.}$$

A straight line graph will be obtained if $\ln N$ is plotted against t. The gradient of the line is $-k$ and the y-intercept is $\ln A$.

Example

Two capacitors are connected in series. One capacitor has a capacitance of 2.2μF and the other is an unknown capacitor of capacitance C. The capacitors are charged using a battery of e.m.f. E. When fully charged, the capacitors are discharged through a 10 MΩ resistor. The potential difference V across the both capacitors is measured over a period time. The results are shown in the table below.

t/s	5.0	10.0	15.0	20.0	25.0	30.0	35.0	40.0
V/V	6.45	4.62	3.31	2.37	1.70	1.22	0.87	0.63

The potential difference V across the capacitors varies according to the equation:

$$V = V_0 e^{-t/CR}$$

where C represents the combined capacitance of the two capacitors.

a Plot a graph of $\ln V$ against t.

b Calculate the gradient of the graph and hence determine the time constant.

c Calculate the value of the unknown capacitance.

d Determine the value of the e.m.f. E of the battery.

Exam tip

Recall the rules for logarithms:

1. $\log_b(A)^n = n \log_b A$
2. $\log_b(A) + \log_b(B) = \log_b(AB)$
3. $\log_b(A) - \log_b(B) = \log_b\left(\frac{A}{B}\right)$

Exam tip

1. $\log_{10}$ is usually written as lg.
2. $\log_e$ is usually written as ln, where e = 2.718

a $V = V_0 e^{-t/CR}$

Taking natural logarithms of both sides of the equation, ($\log_e$ or ln)

$$\ln V = \ln(V_0 e^{-t/CR})$$
$$\ln V = \ln V_0 + \ln(e^{-t/CR})$$
$$\ln V = \ln V_0 - t/CR$$
$$\ln V = -(1/CR)t + \ln V_0$$

When a graph of $\ln V$ against t is plotted, a straight line is obtained.

The gradient of the straight line $= -(1/CR)$

The y-intercept $= \ln V_0$

t/s	5.0	10.0	15.0	20.0	25.0	30.0	35.0	40.0
V/V	6.45	4.62	3.31	2.37	1.70	1.22	0.87	0.63
ln (V/V)	1.86	1.53	1.20	0.86	0.53	0.20	−0.14	−0.46

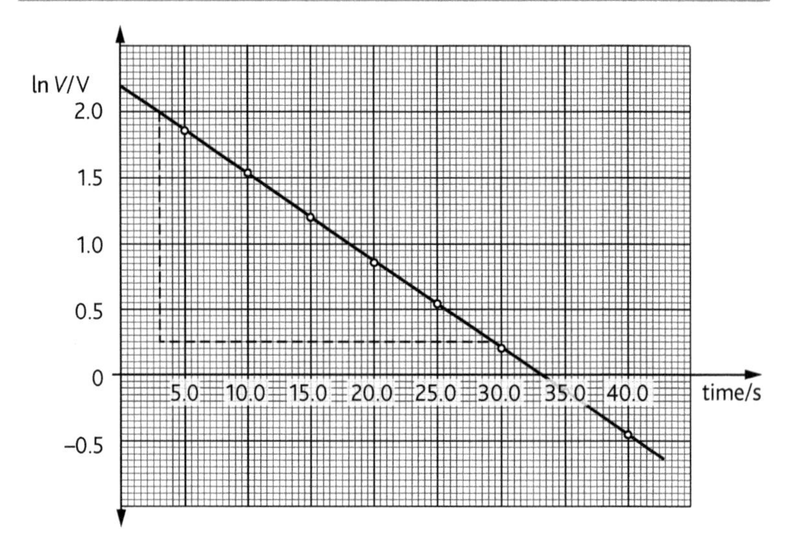

Figure 13.1.2

b Choosing two points on the straight line,

(3.0, 2.0) and (29.0, 0.25)

$$\text{Gradient of straight line} = \frac{0.25 - 2.0}{29.0 - 3.0} = -0.0673$$
$$-(1/CR) = -0.0673$$
$$CR = \frac{1}{0.0673}$$
$$CR = 14.9\,\text{s}$$

The time constant $\tau = CR = 14.9\,\text{s}$

c $CR = 14.9\,\text{s}$

$$C \times 10 \times 10^6 = 14.9$$
$$C = \frac{14.9}{10 \times 10^6}$$
$$C = 1.49\,\mu\text{F}$$

The combined capacitance of both capacitors $= 1.49\,\mu\text{F}$

For capacitors in series, $\frac{1}{C_{\text{total}}} = \frac{1}{C_1} + \frac{1}{C_2}$

$$\frac{1}{4.9 \times 10^{-6}} = \frac{1}{2.2 \times 10^{-6}} + \frac{1}{C_2}$$

$$\frac{1}{C_2} = \frac{1}{4.9 \times 10^{-6}} - \frac{1}{2.2 \times 10^{-6}}$$

$$\frac{1}{C_2} = 2.166 \times 10^{5}$$

$$C_2 = 4.6 \times 10^{-6}$$

Therefore, the capacitance of the unknown capacitor = $4.6\,\mu\text{F}$

d When the capacitors were fully charged, the potential difference across them was equal to the e.m.f. of the battery E.

At time $t = 0$, $V_0 = E$

From the graph $\ln V_0 = 2.2$

$$V_0 = e^{2.2}$$

Therefore $E = 9.0\,\text{V}$

Key points

- Experimental data can be used to determine relationships between two variables.
- In an experiment, the manipulated variable is the variable that is altered in the experiment.
- The responding (dependent) variable is the variable that changes when some action is taken in the experiment.
- Controlled variables are those kept constant throughout an experiment.
- Some non-linear equations can be rearranged to obtain linear equations.

1 A student performed an experiment to investigate how the current through a silicon diode is related to the potential difference across it when the diode is maintained at a constant temperature of 80 °C. He used the circuit in Figure 13.1.3 to obtain the following results.

I/mA	5	10	20	30	40	60
V/V	0.56	0.59	0.63	0.65	0.66	0.68

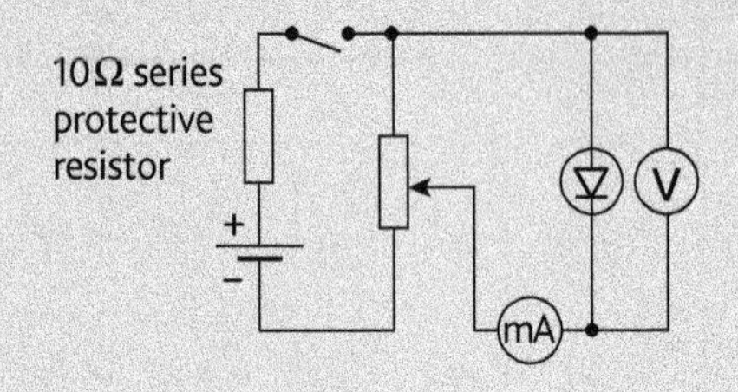

Figure 13.1.3

The current I in the diode is related to the potential difference V by the expression

$$I = I_0 e^{kV/T}$$

Where I_0 and k are constants and T is the temperature measured in kelvin.

a Plot a graph of ln (I/mA) against V and draw a line of best fit. [6]

b Determine the gradient and the y-intercept. [4]

c Determine the value of k and I_0. [4]

2 In an experiment to investigate the variation of current with potential difference for a filament lamp, the circuit in Figure 13.1.4 was used.

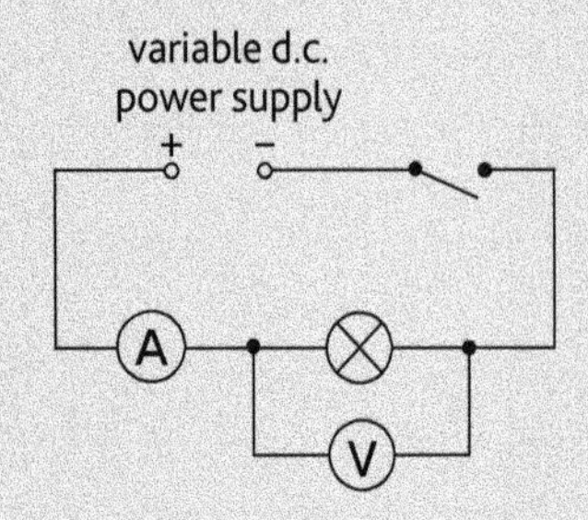

Figure 13.1.4

The following results were obtained.

V/V	1.25	2.40	3.20	4.20	5.50	6.20
I/A	0.85	1.20	1.34	1.56	1.81	1.84

It is suggested that I and V are related by the equation $I = kV^n$.

a State and explain how the resistance of a filament lamp changes with increasing current. [3]

b Plot a graph of lg (I/A) against lg (V/V). [6]

c Comment on whether the relationship is valid. [2]

d Determine the gradient and y-intercept of the line of best fit. [4]

e Determine the values of n and k. [2]

3 An experiment is performed to investigate how $t_{1/2}$, the time taken for a capacitor to discharge to half its initial charge, varies with capacitance C of the capacitor. In an experiment six different capacitors (C) are available. The circuit in Figure 13.1.5 is used to obtain the following data.

C/μF	500	666	1000	1500	2000	3000
$t_{1/2}$/s	15.3	19.8	28.7	45.6	60.4	87.1

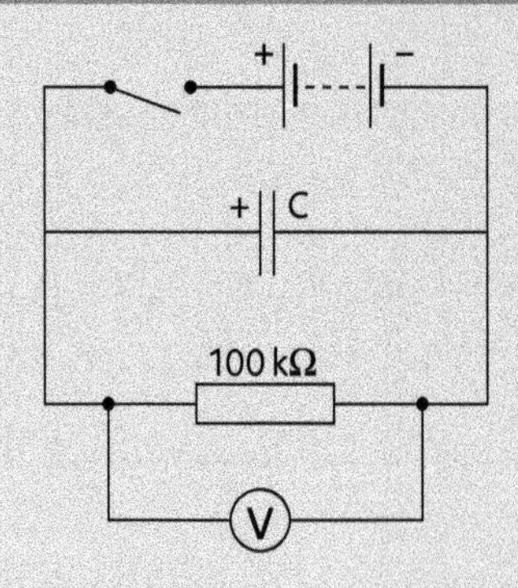

Figure 13.1.5

a Explain how the data in the table could be obtained. [3]

b State how the experimenter could reduce the random errors in the measurement of $t_{1/2}$. [1]

c Plot a graph of $t_{1/2}$/s against C/μF. [6]

d Draw a line of best fit through the data points. [1]

e Determine the gradient of the line. [2]

The variation of potential difference across a capacitor with time t as the capacitor discharges through a resistor is given by the expression $V = V_0 e^{-t/CR}$, where V_0 is the initial potential difference across the capacitor and R is the combined resistance of the 100 kΩ resistor and the voltmeter.

f Show that the time $t_{1/2}$ for the potential difference to fall to half its initial value is given by $t_{1/2} = CR \ln 2$. [3]

g Use your answer to **e** to calculate the value of R. [2]

h Determine the value of the resistance of the voltmeter. [3]

4 a Einstein's equation used to explain the photoelectric effect can be written as $hf = \phi + E_{max}$. Explain what is meant by each of the symbols used in the equation. [3]

b In an experiment to investigate the photoelectric effect, the wavelength λ of the electromagnetic radiation incident on a metal surface is varied and the stopping potential V_s is measured. The following data was obtained from one such experiment.

Wavelength/nm	250	280	310	365	405
V_s/V	2.41	1.72	1.20	0.71	0.40
Frequency/Hz					

i Draw a diagram to show how the data could be obtained. [4]

ii Copy and complete the table to include frequency. [4]

iii Plot a graph of stopping potential V_s against frequency f. [4]

iv Use your graph to determine:

1 the value of the Planck constant [3]

2 the threshold frequency [2]

3 the work function of the metal under investigation. [2]

5 Figure 13.1.6 shows how an operational amplifier is used as a non-inverting amplifier. A student wishes to plot the transfer characteristic (a plot of V_{out} against V_{in}).

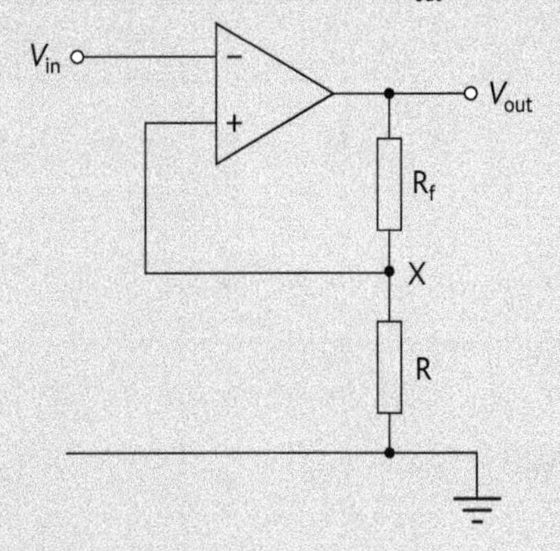

Figure 13.1.6

a Explain how you would use Figure 13.1.6 and any other circuit components to collect data in order to plot the transfer characteristic of the amplifier. [4]

The following data was obtained in one particular investigation.

V_{in}/V	3.00	2.50	2.20	1.80	1.20	1.00	0.75	0.55	0.15
V_{out}/V	14.80	14.80	14.80	14.40	9.60	8.00	6.00	4.40	1.20

V_{in}/V	−0.55	−0.90	−1.20	−1.50	−1.90	−2.50	−3.00
V_{out}/V	−4.40	−7.20	−9.60	−12.00	−14.80	−14.80	−14.80

b Use the data to plot a graph of V_{out} against V_{in}. [5]

c Determine the gain of the amplifier from your graph. [2]

d What is the range of possible input voltages for the output of the amplifier not to be saturated? [2]

6 In an experiment to measure the half-life of a radioactive sample, the following data was obtained.

Time/min	1.0	2.0	3.0	4.0	5.0	6.0
Count rate/min^{-1}	2843	2619	2420	2220	2040	2070

The background count rate was 100 counts/min.

a Use the data to plot a suitable straight line graph. (Remember to allow for the background count rate.) [8]

b Write down an equation for the straight line. [2]

c Determine the decay constant λ of the radioactive sample. [2]

d Determine the half-life of the radioactive sample. [2]

e Calculate the activity of a sample containing 2.0×10^{15} atoms of the radioactive substance. [3]

7 The intensity I of γ-rays through a medium varies according to the equation:

$$I = I_0 e^{-\mu x}$$

where μ is the linear absorption coefficient of the medium and x is the distance travelled inside the medium. An experiment is set up to measure the linear absorption coefficient of lead. The following data is obtained.

Thickness of lead x/cm	0.15	0.25	0.35	0.45	0.55	0.65
Intensity I counts per second	2338	1980	1690	1428	1220	1011

a Plot a graph of ln I against x and draw a line of best fit. [8]

b Determine the gradient and y-intercept. [4]

c State the absorption coefficient μ of lead. [1]

d Write down the equation of the line of best fit. [1]

List of physical constants

Universal gravitational constant	G	=	$6.67 \times 10^{-11}\,\mathrm{N\,m^2\,kg^{-2}}$
Acceleration due to gravity	g	=	$9.80\,\mathrm{m\,s^{-2}}$
Radius of the Earth	R_E	=	$6380\,\mathrm{km}$
Mass of the Earth	M_E	=	$5.98 \times 10^{24}\,\mathrm{kg}$
Mass of the Moon	M_M	=	$7.35 \times 10^{22}\,\mathrm{kg}$
Atmosphere	atm	=	$1.00 \times 10^{5}\,\mathrm{N\,m^{-2}}$
Boltzmann's constant	k	=	$1.38 \times 10^{-23}\,\mathrm{J\,K^{-1}}$
Coulomb constant		=	$9.00 \times 10^{9}\,\mathrm{N\,m^2\,C^{-2}}$
Mass of the electron	m_e	=	$9.11 \times 10^{-31}\,\mathrm{kg}$
Electron charge	e	=	$1.60 \times 10^{-19}\,\mathrm{C}$
Density of water		=	$1.00 \times 10^{3}\,\mathrm{kg\,m^{-3}}$
Resistivity of steel		=	$1.98 \times 10^{-7}\,\Omega\,\mathrm{m}$
Resistivity of copper		=	$1.80 \times 10^{-8}\,\Omega\,\mathrm{m}$
Thermal conductivity of copper		=	$400\,\mathrm{W\,m^{-1}\,K^{-1}}$
Specific heat capacity of aluminium		=	$910\,\mathrm{J\,kg^{-1}\,K^{-1}}$
Specific heat capacity of copper		=	$387\,\mathrm{J\,kg^{-1}\,K^{-1}}$
Specific heat capacity of water		=	$4200\,\mathrm{J\,kg^{-1}\,K^{-1}}$
Specific latent heat of fusion of ice		=	$3.34 \times 10^{5}\,\mathrm{J\,kg^{-1}}$
Specific latent heat of vaporisation of water		=	$2.26 \times 10^{6}\,\mathrm{J\,kg^{-1}}$
Avogadro constant	N_A	=	6.02×10^{23} per mole
Speed of light in free space	c	=	$3.00 \times 10^{8}\,\mathrm{m\,s^{-1}}$
Permeability of free space	μ_0	=	$4\pi \times 10^{-7}\,\mathrm{H\,m^{-1}}$
Permittivity of free space	ε_0	=	$8.85 \times 10^{12}\,\mathrm{F\,m^{-1}}$
The Planck constant	h	=	$6.63 \times 10^{-34}\,\mathrm{J\,s}$
Unified atomic mass constant	u	=	$1.66 \times 10^{-27}\,\mathrm{kg}$
Rest mass of proton	m_p	=	$1.67 \times 10^{-27}\,\mathrm{kg}$
Molar gas constant	R	=	$8.31\,\mathrm{J\,K^{-1}\,mol^{-1}}$
Stefan–Boltzmann constant	σ	=	$5.67 \times 10^{-8}\,\mathrm{W\,m^{-2}\,K^{-4}}$
Mass of neutron	m_n	=	$1.67 \times 10^{-27}\,\mathrm{kg}$

Glossary

A

Activity The number of nuclei decaying per second.

Ammeter An instrument used to measure electric current.

Asynchronous circuit Sequential digital circuits in which the circuit operations are driven by changes in the input signals.

B

Background radiation The random radioactivity detected from the surroundings.

Bandwidth The range of frequencies for which the gain of an amplifier remains constant.

Barrier potential The potential difference across a p-n junction when a p-type material is placed against an n-type material.

Binding energy The energy required to completely separate the nucleons of a nucleus.

Binding energy per nucleon The binding energy of the nucleus divided by the total number of nucleons.

Bistable A device that has two stable states.

C

Capacitance The charge stored per unit potential difference in a capacitor.

Capacitor An electrical component that stores charge.

Cascaded The arrangement of connecting several amplifiers such that the output of one amplifier is connected to the input of another amplifier.

Charge $Q = I \times t$

Combinational circuit A digital circuit in which the output of the circuit is determined by the current inputs.

Conductor A material that allows electrons to flow through it easily, e.g. a metal.

Controlled variable Variables that are kept the same throughout an experiment.

Coulomb 1 coulomb (C) is the quantity of charge that passes through any section of a conductor in 1 second when a current of 1 ampere is flowing.

Coulomb's law The force acting between two point charges is proportional to the product of their charges and inversely proportional to the square of the distance between them.

Counter A register that is able to count the number of clock pulses arriving at its clock input.

Cut-off wavelength The maximum wavelength of the incident electromagnetic radiation required for electrons to be emitted.

D

Decay constant The probability of decay of a nucleus per unit time.

Depletion region (or layer) The region on either side of a p-n junction where there are no net charge carriers.

Dielectric The material placed between the plates of a capacitor.

Diffusion current A current that occurs because of a difference in the concentration of holes and electrons in a p-n junction.

Doping The process by which an impurity (another element) is added to a semiconductor to enhance its conductive properties.

Drift current A current that flows when a potential difference is applied across a p-n junction.

Drift velocity The net velocity of electrons along a metal when a potential difference is applied across it.

E

Eddy currents Currents that flow in a conductor that is situated in a changing magnetic field.

Electrical resistance See *resistance*.

Electric current A flow of charged particles.

Electric field A region around a charged body where a force is experienced.

Electric field strength The force acting per unit positive charge.

Electric potential The potential at a point is the work done in moving unit positive charge from infinity to that point.

Electromagnetic induction The effect of producing an electric current using magnetism.

Electromotive force (e.m.f.) The energy converted from chemical (or mechanical) energy into electrical energy per unit charge flowing through it.

Electron A negatively charged particle that orbits the nucleus of an atom.

Electronvolt 1 electronvolt (eV) is the energy transformed by an electron as it moves through a potential difference of 1 volt.

Energy level diagram A diagram that shows the various energy levels within an atom.

Equipotiential line A line joining points of equal potential.

F

Farad A capacitor has a capacitance of 1 farad (F) if the charge stored is 1 coulomb when a potential difference of 1 volt is applied across it.

Faraday's law The magnitude of the induced e.m.f. is proportional to the rate of change of magnetic flux linkage.

Feedback The process of taking a fraction of the output signal and adding it to the input signal being fed into an amplifier.

Ferromagnetic Substances such as iron, steel and nickel which show strong magnetic properties when subjected to a magnetising force.

Fleming's left hand rule A rule used to predict the direction of a force on a current-carrying conductor placed at right angles to a magnetic field.

Fleming's right hand rule A rule used to predict the direction of the induced current in a conductor moving at right angles to a magnetic field.

Flip-flop The basic memory element in sequential circuits.

Forward-biased When a cell is connected across a p-n junction such that the positive terminal is connected to the p-type material and the negative terminal is connected to the n-type material.

Frequency The number of complete cycles generated per second.

Full-adder A circuit that adds three one-bit numbers.

G

Gain The ratio of the output voltage to the input voltage.

H

Half-adder An arithmetic circuit used to add two one-bit numbers.
Half-life The average time taken for the number of undecayed nuclei to decrease to half of its initial value.
Hall voltage The potential difference set up transversely across a current-carrying conductor when a perpendicular magnetic field is applied.
Hole A covalent bond that is missing one electron.

I

Induced current A current that flows in a conductor as a result of electromagnetic induction.
Insulator A material that does not allow an electric current to flow through it easily, e.g. rubber.
Intensity The power incident per unit area on a surface.
Internal resistance The resistance internal to a cell.
Intrinsic semiconductor Pure elements such as silicon and germanium.
Ion A charged atom or molecule.
Isotopes Atoms of the same element that have the same atomic number but different mass number.

K

Kirchhoff's first law The sum of the currents flowing into any point in a circuit is equal to the sum of the currents flowing out of that point.
Kirchhoff's second law The algebraic sum of the e.m.f.s around any loop in a circuit is equal to the algebraic sum of the p.d.s around the loop.

L

Lenz's law The induced e.m.f. (or current) acts in such a direction to produce effects to oppose the change causing it.
Line absorption spectrum A continuous bright spectrum crossed by dark lines.
Line emission spectrum A series of discrete bright lines on a dark background.
Line spectrum A spectrum that consists of discrete lines.
Logic gates The building blocks of digital circuits.

M

Magnetic field A region around a magnet where a magnetic force is experienced.
Magnetic flux The product of magnetic flux density and the area through which is passes.
Magnetic flux density Numerically equal to the force per unit length on a straight conductor carrying unit current normal to the field.
Magnetic flux linkage The magnetic flux linking or passing through a coil. It is numerically equal to $N\phi$.
Magnetic pole One of the two ends of a magnet.
Manipulated variable A variable that is changed in an experiment.
Mass defect The difference between the mass of the nucleus and the total mass of its constituent nucleons.

N

n-type material A material in which the majority of charge carriers are electrons.
Negative charge An excess of electrons.
Negative feedback The process of taking a negative fraction of the output signal of an amplifier and adding it to the input signal being fed into the amplifier.
Nuclear fission An induced process whereby an unstable nucleus is bombarded by a neutron. The nucleus splits into two or more stable fragments as well as several neutrons.
Nuclear force Very strong short-range force inside the nucleus of an atom.
Nuclear fusion A process whereby light nuclei become more stable by combining with other light nuclei to form a heavier stable nucleus, accompanied by the release of energy.
Nucleons The protons and neutrons inside the nucleus of an atom.
Nucleus The central part of an atom.

O

Ohm 1 ohm (Ω) is the resistance of a conductor through which a current of 1 ampere flows when there is a potential difference of 1 volt across it.
Ohm's law The current flowing through a conductor is directly proportional to the potential difference across it provided that there is no change in the physical conditions of the conductor.
Open-loop gain The gain of an amplifier without feedback.

P

p-type material A material in which the majority of charge carriers are 'holes'.
Peak value The maximum instantaneous value of an alternating current.
Period The time taken for one complete cycle of an alternating current.
Permeability of free space A measure of the ability of a medium to transmit a magnetic field.
Permittivity of free space A measure of how easy it is to transmit an electric field through space.
Photoelectric current The rate at which electrons are emitted from a metal surface when exposed to electromagnetic radiation.
Photoelectric effect The emission of electrons from a metal surface when exposed to electromagnetic radiation.
Photon A quantum of electromagnetic radiation.
Positive charge A deficiency of electrons.
Positive feedback The process of taking a positive fraction of the output signal of an amplifier and adding it to the input signal being fed into the amplifier.
Potential difference The potential difference V between two points in a circuit is the work done (energy converted from electrical energy to other forms of energy) in moving unit positive charge from one point to the other.
Power The rate at which energy is converted.

Q

Quantised Refers to energy levels that can only take discrete values.
Quantum A packet of energy.

R

Radioactive decay The spontaneous and random process whereby an unstable nucleus attempts to become stable by disintegrating into another nucleus and emitting any one or more of the following: alpha particles, beta particles, gamma rays.

Rectification The process by which an alternating current is converted into a direct current.

Register Several flip-flops connected together.

Resistance The opposition to the flow of an electric current. Resistance is the ratio of the potential difference (V) across the conductor to the current (I) flowing through it.

Resistivity $\rho = \frac{RA}{l}$

Responding variable A variable that changes when some action is taken in an experiment.

Reverse-biased When a cell is connected across a p-n junction such that the positive terminal is connected to the n-type material and the negative terminal is connected to the p-type material.

Root mean square (value of an alternating current) That steady direct current which delivers the same average power to a resistive load as the alternating current.

S

Saturated The condition whereby the output of an amplifier is at its maximum possible value (limited by the power supply voltage).

Semiconductor A material that is neither a good conductor nor a good insulator.

Sequential circuits Digital circuits in which the output of the circuit is determined by the current inputs and previous outputs.

Slip ring A device used in a.c. motors and generators.

Split-ring commutator A device used in d.c. motors or generators to allow an electric current to flow in one direction.

Stopping potential The potential difference that causes a photoelectric current to reduce to zero.

Synchronous circuits A sequential digital circuit in which a clock input is used to drive all the circuit operations.

T

Temperature coefficient The relative change of a physical property when the temperature is changed by 1 kelvin.

Terminal potential difference The potential difference across a cell.

Tesla 1 tesla (T) is the magnetic flux density if a force of 1 newton acts on a wire of length 1 metre, carrying a current of 1 ampere placed perpendicular to the magnetic field.

Thermionic emission The process by which electrons are emitted from a metal surface by heating it.

Threshold frequency The minimum frequency of the incident electromagnetic radiation required for electrons to be emitted from a metal surface.

Time constant The time taken for the charge on a capacitor to fall to 1/e (0.368) of its initial value.

Timing diagram A diagram used to illustrate the logic states of various inputs and outputs over a period of time.

Transformer A device used to change the voltage of an alternating power supply.

Truth table A table representing the logic function of a digital circuit.

U

Unified atomic mass unit (u) 1/12 the mass of the carbon atom $^{12}_{6}C$.

V

Virtual earth A concept whereby a terminal of an op-amp is at zero potential, even though it is not physically connected to earth.

Volt 1 volt (V) is the potential difference between two points in a circuit when 1 joule of energy is converted when 1 coulomb of charge flows between the two points.

Voltmeter An instrument used to measure potential difference.

W

Watt 1 watt (W) is a rate of conversion of energy of 1 joule per second.

Wave–particle duality The idea that light and matter have both wave and particle properties.

Weber 1 weber (Wb) is the magnetic flux when a flux density of 1 tesla passes perpendicularly through an area of 1 m^2.

Work function The minimum energy needed to free an electron from a metal surface.

Index

Headings in **bold** indicate glossary terms.

F

G

H

I

K

L

M

N

O

P